Applications of Fuzzy Logic

Towards High Machine Intelligence Quotient Systems®

PRENTICE HALL SERIES ON ENVIRONMENTAL AND INTELLIGENT MANUFACTURING SYSTEMS
M. Jamshidi, Editor

Volume 1: ***ROBOTICS AND REMOTE SYSTEMS FOR HAZARDOUS ENVIRONMENTS***, edited by M. Jamshidi and P. J. Eicker, 1993.

Volume 2: ***FUZZY LOGIC AND CONTROL: Software and Hardware Applications***, edited by M. Jamshidi, N. Vadiee, and T. J. Ross, 1993.

Volume 3: ***ARTIFICIAL INTELLIGENCE IN OPTIMAL DESIGN AND MANUFACTURING***, edited by Z. Dong, 1995.

Volume 4: ***SOFT COMPUTING: Fuzzy Logic, Neural Networks, and Distributed Artificial Intelligence***, edited by F. Aminzadeh and M. Jamshidi, 1994.

Volume 5: ***NONLINEAR CONTROL OF ROBOTIC SYSTEMS FOR ENVIRONMENTAL WASTE AND RESTORATION***, by D. Dawson, M. Bridges, and Z. Qu, 1995.

Volume 6: ***DESIGN AND IMPLEMENTATION OF INTELLIGENT MANUFACTURING SYSTEMS: From Expert Systems, Neural Networks, to Fuzzy Logic***, by H. R. Parsaei and M. Jamshidi, 1995.

Volume 7: ***ADVANCES IN DISTRIBUTED SENSOR TECHNOLOGY***, Edited by S. S. Iyengar, L. Prasad, and H. Min, 1995.

Volume 8: ***LARGE-SCALE SYSTEMS: Modeling, Control, and Fuzzy Logic***, M. Jamshidi, 1995.

Volume 9: ***APPLICATIONS OF FUZZY LOGIC: Towards High Machine Intelligence Quotient Systems***®, Edited by M. Jamshidi, A. Titli, L.A. Zadeh, and S. Boverie, 1997.

Applications of Fuzzy Logic

Towards High Machine Intelligence Quotient Systems®

Editors

Mohammad Jamshidi
At&T Professor and Director
NASA Center for Autonomous Control Engineering (ACE)
University of New Nexico
Albuquerque, New Mexico

Lotfi Zadeh
Professor Emeritus and Director
Berkeley Initiative on Soft Computing (BISC)
University of California
Berkeley, California

André Titli
Professor, Institut National des Sciences Appliquees (INSA) and Laboratoire d'Architecture et d'Analyse des Systèmes (LAAS)
Centre National de la Recherche Scientifique
Toulouse, France

Serge Boverie
Chef de Projet
Siemens Automotive, S.A.
Toulouse, France

To join a Prentice Hall PTR internet mailing list, point to http://www.prenhall.com/register

Prentice Hall PTR
Upper Saddle River, New Jersey 07458
http://www.prenhall.com

Library of Congress Cataloging in Publication Data

Applications of fuzzy logic: towards high machine intelligence quotient systems / edited by Mohammad Jamshidi...[et al.].
p. cm.
Includes bibliographical references and index.
ISBN 0-13-362831-0
1. Intelligence. I. Jamshidi, Mohammad.
TJ217.5.A66 1997
629.8'9--dc21 96-47268
CIP

Editorial/production supervision: *Nicholas Radhuber*
Manufacturing manager: *Alexis Heydt*
Acquisitions editor: *Bernard Goodwin*
Marketing manager: *Miles Williams*
Editorial assistant: *Kate Hargett*

Prentice-Hall, Inc.
A Simon & Schuster Company
Upper Saddle River, New Jersey 07458

The Publisher offers discounts on this book when ordered in bulk quantities.
For more information, contact:

Corporate Sales Department
Prentice Hall PTR
1 Lake St.
Upper Saddle River, NJ 07458
Phone: 800-382-3419 Fax: 201-236-7141
E-mail: dan_rush@prenhall.com

Printed in the United States of America
10 9 8 7 6 5 4 3 2

ISBN 0-13-362831-0

Prentice-Hall International (UK) Limited, *London*
Prentice-Hall of Australia Pty. Limited, *Sydney*
Prentice-Hall Canada Inc., *Toronto*
Prentice-Hall Hispanoamericana, S.A., *Mexico*
Prentice-Hall of India Private Limited, *New Delhi*
Prentice-Hall of Japan, Inc., *Tokyo*
Simon & Schuster Asia Pte. Ltd., *Singapore*
Editora Prentice-Hall do Brasil, Ltda., *Rio de Janeiro*

TABLE OF CONTENTS

ABOUT THE EDITORS

Mohammad Jamshidi is the AT&T Professor and Director, NASA Center for Autonomous Control Engineering (ACE) at the University of New Mexico at Albuquerque, USA. He was Directeur du Recherche Associé at Laboratoire d'Analyse et d'Architecture des Systèmes (LAAS) of Centre National de la Recherche Scientifique, Toulouse, France

André Titli is Professor, Institut National des Sciences Appliquées (INSA) and Laboratoire d'Analyse et d'Architecture des Systèmes (LAAS) of Centre National de la Recherche Scientifique, Toulouse, France. He is currently on leave at NASA ACE Center, the University of New Mexico at Albuquerque, USA.

Lotfi A. Zadeh, father of fuzzy logic, is Professor Emeritus and Director, Berkeley Initiative on Soft Computing (BISC) at the University of California, Berkeley, CA, USA.

Serge Boverie is Chef de Projet at Siemens Automotive, S. A., Toulouse, France.

PREFACE

Fuzzy set theory has rapidly grown to become a major scientific domain collectively called "*fuzzy systems*" or "*fuzzy logic*." In a **broad sense**, fuzzy logic refers to fuzzy sets — a set with unsharp boundaries. Examples of fuzzy sets are "hot," "tall," "medium," etc. In a **narrow sense**, fuzzy logic is a logical system that aims to formalize approximate reasoning. As such, it is multivalued logic, but its agenda is quite different from that of traditional multivalued logical systems such as three-level logic of Lukasiewicz. In fact, fuzzy sets are much broader than fuzzy logic in the narrow sense and contain the latter as one of its branches. Other branches are fuzzy arithmetic, fuzzy mathematical programming, fuzzy topology, fuzzy pattern recognition, fuzzy optimization, fuzzy data analysis, fuzzy control, etc. The year 1990 was considered as the beginning of an era and trend in the design of new products — appliances, consumer electronic devices, and like. The trend refers to the marked increase with MIQ® — Machine Intelligence Quotient. Applications of fuzzy systems technology are so vast that a single book can no longer give a comprehensive treatment of even a majority of relevant applications.

Since the launching of the Sputnik four decades ago, great progress has been achieved in the understanding of how to model, identify, control, and implement digital controllers for complex large-scale systems. However, to be able to design systems having high MIQ,® a profound change in the orientation of control theory may be required. Fuzzy logic—unlike classical logic—is tolerant to imprecision, uncertainty, and partial truth. One of the earliest applications of fuzzy logic was the use of this technology to control a steam engine by Mamdani and his colleagues in 1974. Another major milestone, which actually started in 1977, was the application of fuzzy logic in the control of cement kilns by the Danish manufacturer of these plants, pioneered by Holblad and Ostergaard in 1982. Today, fuzzy technology is a multibillion dollar industry spanning the entire globe. From *automatic transmission* of car engines to *zoom control of hand-held cameras.*

The objective of **this volume** is to bring to the attention of the readers some of the latest and perhaps nontraditional applications of fuzzy logic. This is one of the first volumes in which MIQ® is brought to the attention of the scientific community.

Most of the applications in this volume are hardware oriented, even though software applications are just as important. This volume assumes that the reader has a basic knowledge of fuzzy systems technology. Those who may need more knowledge can find it in Volumes 3 and 7 of this book series.

This volume was organized first while the first author was on sabbatical leave at the Laboratoire d'Analyse d'Architecture des Systèmes (LAAS) of Centre National de la Recherche Scientifique (CNRS), Toulouse, France and at the same time visiting Siemens Automotive S. A., also in Toulouse. The volume was finished while the first author returned to the NASA Center for Autonomous Control Engineering (ACE) at the University of New Mexico The first author wishes to express his sincere appreciation to the French National Science Foundation (CNRS) and LAAS as well to the NASA Headquarters especially the Office of Equal Opportunity for their financial support . The support of NASA OEO headed by its Associate Administrator George Reese and the URC Program Director Bettie White is very much appreciated.

The editors wish to thank all the authors of this volume for their valuable contributions to the field in general and to this volume in particular. We would like to acknowledge the editorial assistance of Milton Lau of *MKK* for his excellent and professional typesetting of the manuscript and patience with all its revisions and alterations. The support and cooperation of Prentice Hall, especially Bernard Goodwin, Senior Editor and Vice President and the production staff of Prentice Hall PTR are appreciated. Last, but by no means least, we would like to thank our families for their support and understanding.

Mo Jamshidi
Albuquerque, NM, USA

André Titli
Toulouse, France

Lotfi Zadeh
Berkeley, CA, USA

Serge Boverie
Toulouse, France

LIST OF CONTRIBUTORS

R. A. Aliev
Azerbaijan Oil Academy
AZERBAIJAN

R. R. Aliev
Azerbaijan Oil Academy
AZERBAIJAN

R. Babuška
Delft University of Technology
THE NETHERLANDS

S. Boverie
SIEMENS Automotive S.A.
FRANCE

A. Fichera
Università de Catania
ITALY

L. Fortuna
Università de Catania
ITALY

A. García-Cerezo
Universidad de Málaga
SPAIN

F. Graf
Siemens AG
GERMANY

S. Graziani
Università de Catania
ITALY

W. Hauptmann
Siemens AG
GERMANY

M. Jamshidi
University of New Mexico - ACl
USA

A. Kandel
University of South Florida
USA

R. E. King
University of Patras
GREECE

F. Klawonn
University of Braunschweig
GERMANY

M. Koch
Technical University of Ilmenau
GERMANY

R. Kruse
University of Braunschweig
GERMANY

T. Kuhn
Technical University of Ilmenau
GERMANY

JM. Le Quellec
SIEMENS Automotive S. A.
FRANCE

A. Mandow
Universidad de Málaga
SPAIN

J. L. Martínez
Universidad de Málaga
SPAIN

A. Ollero
Universidad de Sevilla
SPAIN

G. Oriolo
Università di Roma "La Sapienza"
ITALY

R. Palm
Siemens AG
GERMANY

R. Petersen
VOLKSWAGEN AG
GERMANY

Y. A. Phillis
Technical University of Crete
GREECE

F. G. Pin
Oak Ridge National Laboratory
USA

C. Raymond
SIEMENS Automotive S. A.
FRANCE

M. Schröder
VOLKSWAGEN AG
GERMANY

N. Sepehri
The University of Manitoba
CANADA

K. Storjohann
Siemens AG
GERMANY

S. R. Swanson
NASA
USA

A. Titli
INSA/LAAS CNRS 7
FRANCE

N. C. Tsourveloudis
Technical University of Crete
GREECE

E. W. Tunstel Jr.
University of New Mexico - ACE
USA

E. Uchino
Kyushu Institute of Technology
JAPAN

G. Ulivi
Università di Roma "La Sapienza"
ITALY

M. Vendittelli
Università di Roma "La Sapienza"
ITALY

H. B. Verbruggen
Delft University of Technology
THE NETHERLANDS

N. Vojdani
University of Dortmund
GERMANY

J. Wernstedt
Technical University of Ilmenau
GERMANY

M. G. Xibilia
Università di Catania
Catania, ITALY

T. Yamakawa
Kyushu Institute of Technology
JAPAN

Y.-Q. Zhang
University of South Florida
USA

1

INTELLIGENT CONTROL AND EVOLUTION OF MOBILE ROBOT BEHAVIOR

E. W. Tunstel, Jr.[†] and M. Jamshidi
NASA Center for Autonomous Control Engineering
University of New Mexico
USA

The objective of this chapter is to describe mobile robot control systems that integrate fuzzy reasoning with behavior control for intelligent sensor-based navigation. A hierarchical fuzzy control architecture is introduced that is useful for realizing complex fuzzy controllers and, in particular, complex fuzzy behavior-based systems. Issues of behavior synthesis, and design via the genetic programming approach to computational evolution are discussed. Finally, applications that demonstrate the practicality of this integration are covered which include tethered semi-autonomous control and autonomous embedded control.

1.1 INTRODUCTION

A research aim of intelligent robot control is to develop autonomous mobile robot systems that can dynamically interact with the real world. The notion of intelligence in

[†] NASA Jet Propulsion Laboratory Fellow.

such systems dictates the need for cognitive, or knowledge-based control. System autonomy implies that the mobile robot is to be self-contained. That is, it must be capable of sensing the real world, reasoning about the real world, and physically influencing the real world. In this context, real world is meant to refer to physical environments that may be unstructured, unpredictable, dynamic, noisy, and/or unknown. As such, the real world presents an immense level of uncertainty that complicates the endeavor of developing and controlling autonomous mobile robot systems.

In attempts to formulate approaches that can handle real-world uncertainty, researchers are frequently faced with the necessity of considering tradeoffs between developing complex cognitive systems that are difficult to control, or adopting a host of assumptions that lead to simplified models which are not sufficiently representative of the system *or* the real world. The latter option is a popular one which often enables the formulation of viable control laws. However, these control laws are typically valid only for systems that comply with the imposed assumptions, and furthermore, only in neighborhoods of some nominal state. The option that involves complex systems has been less prevalent due to the lack of analytical methods that can adequately handle uncertainty and concisely represent knowledge in practical control systems. Recent research and applications employing non-analytical methods of soft computing such as fuzzy logic, evolutionary computation, and neural networks have demonstrated the utility and potential of these paradigms for intelligent control of complex systems [1]. In particular, fuzzy logic has proven to be a convenient tool for handling real-world uncertainty and knowledge representation [2].

Fuzzy logic control is one of the more active areas of application of fuzzy logic and the underlying fuzzy set theory introduced by Zadeh [3]. In the context of mobile robot control, a fuzzy logic-based system has the advantage that it allows the intuitive nature of sensor-based navigation to be easily modeled using linguistic terminology. The computational loads of typical fuzzy inference systems are relatively light. As a result, reactive fuzzy control systems permit intelligent decisions to be made in real time, thus allowing for smooth and uninterrupted motion. Some architectural structure for mobile robot control systems can be imposed by forming a hybrid between fuzzy logic control and behavior-based control which is a product of artificial intelligence research in mobile robotics. We will discuss architectural issues of fuzzy behavior controllers and applications that demonstrate the practicality of integrating fuzzy reasoning with behavior control for sensor-based navigation. A novel fuzzy behavior control architecture will be introduced, as well as a fuzzy behavior design methodology based on computational evolution.

1.2 OVERVIEW OF BEHAVIOR CONTROL

In recent years we have witnessed a shift in ideology regarding problem decomposition for mobile robot control. Traditional architectures are modularized according to a functional decomposition of tasks into an overall sense-plan-act cycle. The robot control system executes functional modules sequentially, and will malfunction if any module is missing. Due to this sequential processing computational bottlenecks are common, particularly in the planning modules. Mobile robot research is now leaning towards control approaches that are modeled after natural processes, and that advocate a *behavioral* decomposition of tasks. The approach has come to be known as the behavior control paradigm and has grown out of an amalgamation of ideas from ethology and control theory [4,5,6]. Under this paradigm, the high-level motion control system is decomposed into a set of special-purpose behaviors that achieve distinct tasks when subjected to particular stimuli. Clever coordination of individual behaviors results in the emergence of more intelligent behavior suitable for dealing with complex situations. In principal, behavior control can be applied in situations where classical control is not feasible, usually due to tremendous modeling difficulties.

The behavior control paradigm was initially proposed in the seminal paper by Brooks [5] where it was realized in the *subsumption architecture*. In this architecture, a number of behaviors (implemented as finite state automata) execute in parallel in response to instantaneous sensory data. The behaviors comprise a distributed system that controls a mobile robot through arbitrated competition and collaboration. Since the introduction of this architecture, a number of variants have been proposed for behavior control. This chapter presents one of the most recent variants among those that employ approximate reasoning via fuzzy logic.

1.3 FUZZY BEHAVIOR CONTROL SYSTEMS

Fuzzy logic control systems and behavior control systems share common developmental advantages for intelligent systems. Namely, they both require short development times and are intrinsically flexible in their control structure and design. Individual fuzzy rules (and behaviors) can be formulated independently, and additional rules (and behaviors) can easily be added to the control system if necessary without altering the existing system. Until recently, these common advantages have perhaps been overlooked judging from the fact that the majority of implementations of behavior control are based on crisp (i.e., nonfuzzy) data processing and binary logic-based reasoning.

Advantages are gained in a fuzzy behavior control hybrid with regard to representation and handling of uncertain and imprecise knowledge about the robot's environment, and in behavior coordination and conflict resolution. Nonfuzzy behavior controllers explicitly account for real-world uncertainty by augmenting crisp reasoning with heuristics in a manner akin to production systems of artificial intelligence. Fuzzy control, on the other hand, implicitly accounts for uncertainty by virtue of the approximate reasoning capability of fuzzy logic. When behaviors compete for control of the robot, or when competing behaviors recommend conflicting control actions, an important decision problem arises. This occurs frequently during any sufficiently complex navigation task. In the jargon of behavior control the problem is referred to as behavior arbitration. It consists of coordination of multiple task-achieving behaviors and conflict resolution among them. The typical approach to this problem, taken by nonfuzzy behavior controllers, employs a prioritization scheme wherein the control recommendation of only one behavior among several competing behaviors is taken; recommendations from the remaining behaviors are ignored [5,6,7]. With fuzzy reasoning, coordination and conflict resolution are implicitly handled by the computational aspects of aggregation and defuzzification. As a result, coordination and conflict resolution are enhanced since *all* control recommendations (conflicting or collaborating) contribute to the overall control decision.

1.3.1 Synthesis of Fuzzy Behaviors

In many of the nonfuzzy behavior control implementations, behaviors are synthesized as finite state automata or augmented finite state machines. In contrast to this, fuzzy behaviors are synthesized as fuzzy rule bases, i.e., collections of a finite set of fuzzy *if-then* rules. Each behavior is encoded as a fuzzy rule base with a distinct mobile robot control policy governed by fuzzy inference. Thus, each fuzzy behavior is similar to the conventional fuzzy logic controller in that it performs an inference mapping from some input space to some output space. If X and Y are input and output universes of discourse of a behavior with a rule base of size n, the usual fuzzy if-then rule takes the following form:

$$\text{IF } x \text{ is } \tilde{A}_i \text{ THEN } y \text{ is } \tilde{B}_i \qquad (1\text{-}1)$$

where x and y represent input and output fuzzy linguistic variables, respectively, and $\tilde{A}_i \in X$ and $\tilde{B}_i \in Y$ ($1 \leq i \leq n$) are fuzzy sets representing linguistic values of x and y. Typically, the input x refers to sensory data and y to actuator control signals. In general, the rule antecedent consisting of the proposition "x is $\tilde{A}_i$" could be replaced by a conjunction of similar propositions; the same holds for the rule consequent "y is $\tilde{B}_i$." In its simplest form, we can formally define a fuzzy behavior as a function (B), from sensor space (S) to actuator space (A), i.e. B: $S \rightarrow A$, where the universes of discourse for S and A are such that $S \subset \Re^n$ and $A \subset \Re^n$. Embodied in this function is a fuzzy relation between fuzzy sets defined over S and fuzzy sets over A. This fuzzy relation is the actual rule base of the fuzzy behavior.

Mobile robots are typically equipped with several sensor modalities which may include range sensors, tactile/contact sensors, encoders, and vision systems. The applications discussed in this chapter mainly employ the relatively inexpensive among these, i.e., infrared proximity detectors, sonar, and optical encoders. Given these sensor modalities, the usual procedure for fuzzy behavior synthesis consists of first defining linguistic terminology for the behavior inputs and outputs, partitioning the sensor space and actuator space using appropriate fuzzy sets, and formulating fuzzy rules that satisfactorily govern the desired response of the behavior in all practical situations. This is the same procedure that is used for fuzzy controller synthesis. It is an iterative procedure of trial-and-error which, in practice, involves fine-tuning of the shapes of membership functions used to express uncertainty in inputs and outputs, as well as modifications to the fuzzy rule base. For relatively complex control problems this synthesis procedure can turn out to be quite lengthy depending on the availability of expert knowledge or intuition. This serves to weaken the scalability of one of the strongest attributes of fuzzy logic applications in control—fast development time. Moreover, it is not always obvious what the best combinations of fuzzy input-output resolution, membership functions, and rules should be for a given system behavior. Hence, the likelihood of obtaining an "optimal" fuzzy behavior as a result of trial-and-error synthesis is low.

The establishment of more systematic approaches to fuzzy controller synthesis in the absence of an expert, or sufficient knowledge of the problem domain, is currently an open problem. Various attempts have been made to address this issue. These include the determination of fuzzy membership functions and rules by optimization or search using genetic algorithms (GAs) [8], and by learning using neural networks [9]. When using these techniques to determine rule bases for fuzzy systems, it is often necessary to encode the candidate rules in a numerical form (e.g., binary numbers) suitable for processing by the GA or the neural network, and subsequently decode them into the appropriate linguistic terminology. In Section 1.7 the application of the *genetic programming* (GP) paradigm [10] as a systematic approach to fuzzy behavior synthesis is described. This evolutionary computational method is a more direct approach which manipulates the actual linguistic terminology of the problem, avoiding any intermediate numerical transformations. Its main distinction from the GA is the increase in complexity of the structures undergoing adaptation, i.e., parse trees of hierarchical computer programs rather than numeric strings.

1.4 RELATED WORK

Some of the earliest attempts at applying fuzzy logic to the control of mobile robot vehicles were made by Uragami et al. [11] and Sugeno and Murakami [12]. Uragami and colleagues implemented a fuzzy program to control a simple inchworm robot to simulate a person wandering through a town. They concluded that robots could be used to explore spatial regions, with humans issuing commands in the form of fuzzy instructions. Later, Sugeno and Murakami conducted successful experiments in embedded fuzzy control of a model car for parking based on an operator's control actions. This early work was done before the initial reports on the subsumption architecture became available, and thus, was not influenced by the behavioral decomposition proposed by Brooks.

Following the introduction of the subsumption architecture and the emergence of the behavior control paradigm, a host of groups recognized the advantages to be gained by incorporating fuzzy logic into the framework of behavior control for mobile robots. Saffiotti et al. [13] have developed fuzzy behaviors for complex navigation tasks demonstrating the robustness of fuzzy control in blending reactive and goal-oriented behavior. Pin and Watanabe [14] developed qualitative reasoning schemes for autonomous navigation in unknown environments with emphasis placed on embedded control using VLSI fuzzy chips. A heterogeneous network of fuzzy controllers for reactive behavior-based control was implemented by Goodridge and Luo [15]. In this network, control actions are generated by outputs of independent fuzzy controllers that are linked together through a qualitative rule base. Li [16] emphasizes weighting of reactive behaviors, achieved by implicit mechanisms of fuzzy inference, as an improvement in efficiency over priority-based arbitration. Tunstel and Jamshidi [17,18] have proposed strategies for fuzzy behavior-based mapping and fuzzy spatial map representation for navigation. Research is also being pursued in the area of motion control for path execution[19]. These are but a few of the research activities in fuzzy behavior control of mobile robots. The approaches of each of these research activities are mutually similar and have some things in common with architectures described in this chapter. In Section 1.6, we describe an architecture that exploits the hierarchical nature of robot behavior and generalizes fuzzy set theoretic ideas to facilitate coordination of distributed fuzzy behaviors.

1.5 CONTROL ARCHITECTURES

Behavior controllers that are based on fuzzy-rule based systems can be configured in a number of ways. The alternatives are governed by issues such as the fuzzy set resolution selected for system variables and the complexity of decision making, or reasoning, demanded by the task environment. The fuzzy set resolution of the system variables (inputs) determines the total number of rules (i.e., the rule base cardinality) necessary to cover all possible combinations of fuzzy controller inputs. Individual rule outputs for a given rule base produce a control/decision surface, the nonlinearity of which is a measure of the decision-making complexity. Thus, the resolution of the state space and the nonlinearity of the control surface are inter-related with regard to the interpolation necessary to produce desired behavior via approximate reasoning.

A rule base that considers contingencies throughout the entire state space (given the fuzzy sets of the system variables) is said to be *complete*. The cardinality of a complete rule base is given by the expression:

$$R_c = \prod_{i=1}^{n} L_i \tag{1-2}$$

where n is the number of controller inputs and L_i is the number of linguistic labels assigned to the ith input, which is the size of the ith term set. When all inputs have an equal number of linguistic labels, l, the cardinality grows exponentially, i.e., $L_i = l,\ \forall i \in [1,n] \Rightarrow R_c = l^n$. If R_c is relatively small (say, ≤ 20), then it is feasible to realize the fuzzy behavior controller as a *monolithic*, or single-rule base controller. Otherwise, alternative rule structures may be in order. *Hierarchical* rule structures are a viable alternative for dealing with autonomous behavioral systems that require many rules and/or complex decision-making.

1.5.1 Monolithic Rule Structures

Most of the fuzzy controllers applied to industrial products and reported in the research literature utilize the monolithic rule base structure. This includes much of the related work mentioned in Section 1.4 as well as applications reported herein. All of the precepts that govern the desired behavior of the system are encapsulated as a single collection of *if-then* rules. In most instances, the rule base is designed to carry out a single control policy or behavioral goal. It is also possible to implement additional control policies within a single rule base due to the fusion of information and conflict resolution achieved by aggregation and defuzzification processes respectively. Nevertheless, it becomes more difficult to design monolithic rule bases that implement global behavior which consists of multiple interacting goals.

In order for an autonomous mobile robot to operate in dynamic environments, it must be capable of achieving multiple goals whose relative priorities may change with time. Therefore, its controller should be designed such that a number of task-achieving behaviors can be realized and integrated to achieve different control objectives. This requires the formulation of a large and complex set of fuzzy rules. In this situation a potential limitation to the utility of the monolithic fuzzy controller becomes apparent. During any given control cycle all rules are sequentially processed. Clearly, the cardinality of the rule base has a direct influence on real time performance. As alluded to above, this presents no problem for real time control of systems governed by relatively small rule bases. However, more elaborate controllers implemented as monolithic rule structures can potentially suffer from degraded real time performance.

1.5.2 Hierarchical Rule Structures

We have seen that the cardinality of monolithic rule bases increases exponentially with the number of system variables. An alternative rule structure was proposed [20] that represents a rule base as a hierarchy of rules for which the cardinality increases only linearly with the number of system variables. An alternative approach is based on combining sensory fusion with the hierarchical rule structure [21]. When it is possible to fuse some system variables (e.g., as a set of linear combinations) such that a reduced set of inputs can be fed to the fuzzy controller, further reductions in overall rule base cardinality can be achieved by applying the hierarchical structure to this reduced set. In fact, the cardinality increases in a piecewise-linear fashion and at a significantly slower rate than the linear increase reported in [20]. Hierarchical rule structures have also been proposed for fuzzy control of dynamic systems with interacting goals [22]. Thus, it is possible to overcome the aforementioned limitation of the monolithic rule structure and facilitate practical implementations of more complex behavior by employing hierarchical rule structures.

1.5.3 Hardware Realizations

At some point in the development of mobile robot control architectures, the software must be interfaced with the physical robot hardware for validation of simulation results or final deployment in the target environment. This constitutes the physical embodiment that instantiates the intelligent mobile robot. There are a number of realization options for physically embodied mobile robots given an assortment of computational resources, communications devices, sensors, and actuators. In this chapter the options will be generally classified according to whether the control mode is semi-autonomous or autonomous. By semi-autonomous we mean that the main information processing responsible for the robot's intelligence is resident on some remote processor(s) and not carried aboard the vehicle. In addition, the remote processor communicates with the robot via either physical tether or radio frequency (RF) signals. Autonomous control is meant to refer to situations in which all computational resources are carried onboard.

1.6 HIERARCHY OF DISTRIBUTED FUZZY BEHAVIORS

This section describes an intelligent control architecture that employs an alternative hierarchical rule base structure. It enables the distribution of intelligence among a number of task-achieving fuzzy behaviors. This is facilitated by the inherent hierarchical nature of behavior. The network of distributed behaviors is hierarchical in the sense that overall robot behavior is decomposed into a bottom-up hierarchy of increased behavioral complexity in which behavioral activity at a given level is dependent upon behaviors at the level(s) below. A collection of *primitive* behaviors reside at the lowest level and are encoded as fuzzy rule bases with distinct control policies governed by fuzzy inference. A primitive behavior is a simple, self-contained behavior that serves a single purpose while operating in a reactive fashion. It exists in a state of solipsism, and alone, its capability would be insufficient for typical navigation tasks. Such primitive behaviors are building blocks for more intelligent *composite* behaviors. That is, they can be combined synergistically to produce more intelligent behavior(s) suitable for accomplishing goal-directed operations.

A possible hierarchy for indoor navigation might be arranged as in Figure 1.1. This figure implies that goal-directed navigation can be decomposed as a behavioral function of goal-seek (collision-free navigation to some location) and route-follow (assuming some direction is given perhaps in the form of a path plan). These behaviors can be further decomposed into primitive behaviors as shown, with dependencies indicated by the adjoining lines. wall-follow and avoid-collision are self-explanatory. The doorway behavior implies one that can guide a robot through narrow openings in walls. The go-to-xy behavior will direct a robot to navigate along a straight line trajectory to a particular location. The interconnecting circles between composite behaviors and the primitive level

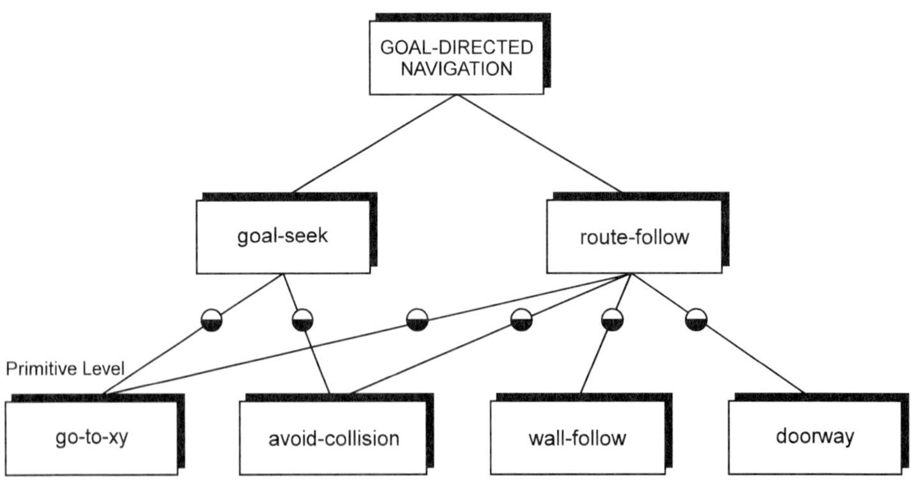

Figure 1.1 Hierarchical decomposition of mobile robot behavior.

represent activation thresholds of associated primitive behaviors. These are described in more detail below.

Note that the decomposition of behavior for a given mobile robot system is not unique. Consequently, suitable behavior repertoires and associated hierarchical arrangements are arrived at following a subjective analysis of the system and the task environment. This idea of a behavior hierarchy is supported by a host of similar conceptual models of motivational control of animal behavior [23,24,25].

1.6.1 Coordination of Multiple Behaviors

Behavior coordination is achieved by introducing a scheme for weighted decision making and behavior selection. The mechanism for weighted decision making is embodied in a concept called the *degree of applicability* which is a measure of the instantaneous level of activation of a behavior. Fuzzy rules of composite behaviors are formulated to include weighting consequents which govern the degree of applicability of primitive behaviors at a lower level. We refer to these as *applicability rules.* Let B_c be a composite behavior comprised of n_p primitive behaviors. Then the degree of applicability, α_p, of primitive behavior p (p = 1, 2, …, n_p) is specified in the consequent of applicability rules of the form:

$$\text{IF } x \text{ is } \tilde{A}_i \text{ THEN } \alpha_p \text{ is } \tilde{D}_i \tag{1-3}$$

where $\tilde{A}_i$ is defined as before. $\tilde{D}_i$ is a fuzzy set representing the linguistic value (e.g. "high") of the degree of applicability of primitive behavior p to the situation prevailing during the current control cycle. It is defined over the unit interval [0, 1]. This feature allows certain robot behaviors to influence the overall behavior to a greater or lesser degree as required by the current goal. It serves as a form of adaptation since it causes the control policy to dynamically change in response to goals and sensory input. Thus, coordination is accomplished by *modulating* behavioral activities using meta-rules that provide a form of the ethological concepts of inhibition and dominance observed in animal behavior. Behaviors with partial applicability ($\alpha < 1$) can be said to be inhibited, while behaviors with maximal applicability ($\alpha = \max_i \alpha_i$) can be said to dominate.

As described here, the degrees of applicability are analogous to neuronal activation levels. Effects similar to neuron threshold activation can easily be implemented with the use of α-cuts of fuzzy sets (this α is not to be confused with the degree of applicability). An α-cut of a fuzzy set, A, defined over a universe U, is a crisp set, A_α, containing all the elements of U with membership grade in A greater than or equal to α [26]. To utilize α-cuts for implementing thresholding behavior activation at the rule level, we consider the rule strength of an applicability rule. If the α-cut of the fuzzy set resulting from evaluation of the rule's consequence is null, then the rule recommends that the level of activation for the behavior is zero. Thus, the rule recommends that the behavior be activated (to degree $\tilde{D}_i$) only if the rule strength equals or exceeds a threshold α.

Alternatively, at the rule base level, primitive behavior p is activated if and only if α_p, resulting from evaluating a *set* of applicability rules, equals or exceeds the threshold.

Coordination of fuzzy behaviors is done in the framework of fuzzy logic theory—via operations on fuzzy sets [27]. In order to briefly describe the process we will concentrate on a single composite behavior, B_c, which can be decomposed into primitive behaviors β_1 and β_2 ($n_p = 2$). We do this without loss of generality since the following description also applies when $n_p > 2$, and across additional composite behaviors. Consider a given control cycle during which B_c dictates that the applicabilities of β_1 and β_2 are say, $\alpha_1 = 0.25$ and $\alpha_2 = 0.75$ respectively. Fuzzy rules of each applicable primitive behavior are processed yielding output fuzzy sets, $\tilde{o}_1$ and $\tilde{o}_2$. These output fuzzy sets are equivalent to the result produced by rule base evaluation in conventional fuzzy controllers *before* applying the defuzzification operator. Following the consultation of individual primitive behaviors, each fuzzy behavior output is weighted by the corresponding degree of applicability. Thus, β_1 and β_2 are activated to degree α_1 and α_2. Here, the behavior activation is accomplished via scalar multiplication of $\tilde{o}_1$ and $\tilde{o}_2$ by the appropriate degrees of applicability. The resulting fuzzy sets are then aggregated using an appropriate t-conorm operator (max is usually used in fuzzy logic control), and defuzzified to yield a crisp output that is representative of the intended coordination of behavior. In this procedure, multiplication by α expresses the relative applicability of a primitive behavior to the current situation, while the scalar α represents the weight of the behavior in the aggregated control decision. Operators other than multiplication can be used to achieve a similar effect. Yager [28] refers to such operators as importance transformations and suggests a general class of them for both t-norm and t-conorm aggregations. This coordination procedure is a generalization of the idea of *rule* weighting in a single rule base to *rule base* weighting among multiple rule bases. Control recommendations from each applicable behavior are considered in the final decision. In the general case, the resultant control action can be thought of as a consensus of recommendations offered by multiple experts.

1.6.2 Behavior Conflict Resolution

In the framework of fuzzy logic theory, conflict resolution is handled implicitly by the mechanisms of conventional fuzzy inference and defuzzification. That is, conflicting fuzzy outputs from individual rules are aggregated and defuzzified (resolved) to yield nonfuzzy control actions. In a similar manner as for coordination, we use fuzzy set theory to generalize this concept to resolve conflicts among conflicting *rule bases*, i.e., behaviors. The manner in which this is done depends on whether there is partial or full conflict among competing primitive behaviors [27]. Partial conflict occurs whenever there is an intersection between the supports of any output fuzzy sets among the competing behaviors. In this case, the fuzzy behavior outputs are coordinated as described above. If the intersection is null, full behavior conflict has occurred and can be resolved using a number of techniques. A common solution is to select a designated *default* action, or to randomly select one action from the set of conflicting behavior outputs. A control action could also be selected based on some criteria such as minimum required control effort. More flexible approaches based on special fuzzy logic defuzzification methods have been proposed in the literature. Yager introduces a method that utilizes nonmonotonic conjunction coupled with the random generation defuzzification method [28] to achieve defuzzification under constraints. Pfluger et al. [29] also proposed a solution based on their centroid of largest area defuzzification strategy.

1.7 FUZZY BEHAVIOR EVOLUTION VIA GENETIC PROGRAMMING

When there is insufficient knowledge about the problem domain, the controller design procedure often consists of the imposition of available knowledge followed by some computational means of improving or optimizing the design. Computational techniques modeled after the biological evolution process are becoming increasingly popular as an alternate means of systematic design of fuzzy systems. The issue of just how much design should be imposed, and how much should be learned or evolved, is unresolved at this time. When a tradeoff can be determined, the genetic programming paradigm is a suitable approach to apply to evolution of fuzzy behaviors or the rule based infrastructure of intelligent systems.

In GP, the population is comprised of computer programs that are hierarchical compositions of the elements of specified sets of *functions* and *terminals* (arguments). Programs in a given population generally vary in size and complexity. They participate in a simulated evolutionary process wherein the population evolves over time in response to selective pressures induced by their relative fitnesses in a particular problem domain. The search space is contained in the set of all possible programs that can be composed recursively from elements of the function and terminal sets. From here on, we will refer to the programs undergoing adaptation as behaviors. A comprehensive introduction to GP is provided in [10].

The artificial evolution of rules using GP for mobile robot control has not received a great deal of coverage in the literature. Koza [10] has applied GP to the evolution of nonfuzzy robot subsumption behaviors for wall following and box-pushing, while others have considered corridor following [30] and path planning [31]. Stimulated by Koza's results, Brooks [32] has advocated using GP to evolve nonfuzzy behavior-based programs for physically embodied robots. The approach proposed herein differs from the others in its focus on evolving fuzzy behaviors as opposed to deterministic behaviors. Other successes at evolving fuzzy behaviors have been reported using GAs [33] and reinforcement learning approaches [34]. For the purpose of evolving fuzzy behaviors the function set can be designated as:

$$F = \{\text{ANT, CONSQ, f\_AND, IF-THEN, f\_OR}\} \tag{1-4}$$

with f_OR taking a variable number of arguments (equal to the number of rules) and the remaining functions each taking two arguments. The function ANT represents a fuzzy proposition in the antecedent of a fuzzy rule. It returns a numerical value in the closed interval [0, 1] representing the membership value, or degree of truth, of the proposition. Note that if a rule contains only one proposition in its antecedent, the membership value represents the rule strength. CONSQ is defined in a similar manner for output linguistic variables and fuzzy sets except that it returns the output fuzzy set(s) designated in the rule consequence. The f_AND function is simply the fuzzy intersection operator of fuzzy set theory. It performs the conjunction of two fuzzy propositions yielding a numerical value for the rule strength. The implementation restricts f_AND to occur only in rule antecedents. Therefore, its two arguments can be return values of either ANT or a recursive call to itself. The function representing a rule is IF-THEN. Its first argument is the rule strength returned by either ANT or f_AND; its second argument is the fuzzy set returned by CONSQ. Finally, the function f_OR serves as a fuzzy aggregation operator. It occupies the root node of every tree in the population of behaviors. Each rule that fires in a fuzzy behavior returns a fuzzy set scaled by the rule strength as the result of the rule consequence. f_OR operates on the output fuzzy sets by taking their fuzzy union to produce a fuzzy set representing the overall output of the behavior. This overall output fuzzy set is the return value of an individual fuzzy behavior in the population. Center-of-gravity defuzzification is used to yield a real number for the behavior control signal. The terminal set, T, is made up of the input and output linguistic variables and the corresponding fuzzy sets associated with the behavior being evolved. Thus, we see that elements of F and T are taken from the linguistic terminology of the problem. Therefore, there is no need to convert between numerical and symbolic representations when using GP to evolve a fuzzy behavior.

1.7.1 Syntactic Constraints and Structure-Preserving Operators

In many GP applications, programs may have any composition of elements from the combined set, $F \cup T$, occupy the nodes of the tree with the only restriction being that the root must be a function and the leaves must be terminals. This is not the case here. To evolve rules for a fuzzy behavior, the GP implementation imposes strong constraints on the syntax of a behavior that are defined by special rules of construction.

A behavior rule base that could potentially evolve from the designated function set and terminal set can be expressed as a rooted, point-labeled tree with pre-ordered branches. An example of a syntactically valid behavior of two rules is depicted in Figure 1.2 along with its interpretation as a linguistic rule base. One can imagine how arbitrary placement of functions and terminals in this tree could lead to severe syntactic violations. Valid behaviors must conform to the syntactic rules of construction listed in Table 1.1.

This syntactic structure, and the rules of construction, are similar to those used in GP application to neural network design [35], but the constraints are stronger here. All behaviors in the initial population are randomly created using these rules, but descendant populations are created by the reproduction, crossover, and mutation operators. The offspring of behaviors modified by crossover and mutation must also conform to the syntactic structure. Structure-preserving crossover is achieved by randomly selecting any non root node as the crossover point in the first parent, and restricting the crossover point in the second parent to be a randomly selected point of the same type. The crossover is completed in the usual way, i.e., by swapping the subtrees at (and including) the crossover points of the two parents. Structure-preserving mutation is done by randomly selecting a non root point in a rule base tree, discarding the selected point and the nodes

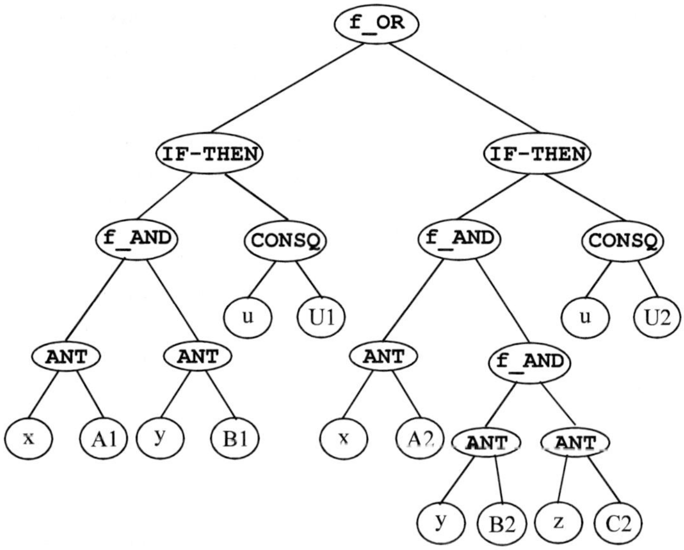

R1: IF x is A1 and y is B1 THEN u is U1

R2: IF x is A2 and y is B2 and z is C2 THEN u is U2

Figure 1.2 Behavior rule tree satisfying syntactic constraints.

Table 1.1 Syntax rules for fuzzy-rule base evolution

1.	f_OR must occupy the root of the tree and cannot occur at non-root points.
2.	Only IF-THEN is allowed at the level immediately below the root.
3.	A left-child of IF-THEN can only be ANT or f_AND.
4.	A right-child of IF-THEN can only be CONSQ.
5.	A child of f_AND can be either ANT or f_AND.
6.	A child of ANT can only be input linguistic variables and input fuzzy sets.
7.	A child of CONSQ can only be output linguistic variables and output fuzzy sets.

below it, and replacing the discarded portion with a randomly generated (but syntactically valid) subtree at that point. The root node is protected from both crossover and mutation.

Amidst all of the constraints on the syntax and structure of candidate fuzzy behaviors, there is room for *some* flexibility. In the creation of the initial population, the number of rules (number of arguments to f_OR) in each behavior is assigned to be a random integer in the prespecified interval, $[R_{\min}, R_{\max}]$. The value for $R_{\min}$ is chosen as a lower bound on the size of a rule base that the control engineer feels may be sufficient to control the system. The upper bound can be chosen such that $R_{\max} \geq R_c$. This feature of the implementation is important for ensuring diversity in the population of behaviors, and also for increasing the potential for evolving a behavior of smaller size than necessary for completeness. It is well-known to practitioners of fuzzy control that a number of dynamic systems exist that can be controlled using fewer rules than dictated by the value of R_c for the fuzzy rule base.

1.8 MOBILE ROBOT BEHAVIOR CONTROL APPLICATIONS

Results of several instances of fuzzy behavior control development are presented in this section. The robots considered are wheeled mobile robots intended for use indoors (as opposed to legged, airborne, submersible, or wheeled articulated mobile vehicles intended for natural outdoor environments). Figure 1.3 shows a family portrait of the robots used in our laboratory to date. As collision avoidance is a fundamental behavioral requirement for any practical mobile robot, we address it first. A collision avoidance behavior developed by trial-and-error synthesis was tested to evaluate the fuzzy behavior under imposed conditions of crude sensing and minimal observability. A second application considers the issues involved when developing fuzzy behaviors for embedded control. In particular, a wall following behavior was developed for a small mobile robot controlled by an embedded microcontroller. Next, we describe a simulation demonstrating the hierarchical fuzzy behavior approach to goal seeking. Finally, an application of GP to evolution of a steering behavior is described.

1.8.1 Collision Avoidance Behavior

A semi-autonomous control scheme was implemented using a monolithic rule base structure. The mobile robot is a Hero-1 educational robot, a tricycle-type vehicle on which the front wheel is actuated for both driving and steering. It is equipped with a

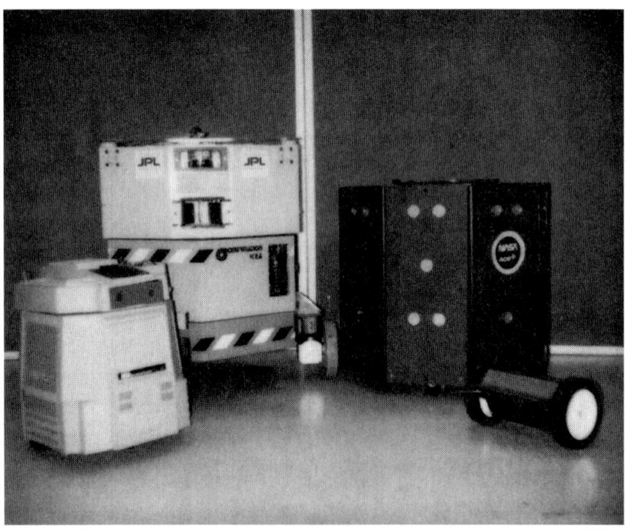

Figure 1.3 Mobile robot testbeds (left to right): Hero-1, Cybermotion Navmaster, UNM LOBOt, LOBOt Jr.

single sonar transmitter-receiver pair (in front) and an embedded Motorola 6808 microprocessor; it was also augmented with two short-range (< 1 foot) infrared proximity detectors on either side. Any information about the robot's environment was provided solely by these range sensors. There was neither state feedback (such as position, heading, or velocity), nor an available map. The robot was controlled via tether by an Intel 386-based PC which performed control calculations and issued motion commands in response to sonar and proximity data received from the 6808.

The collision avoidance behavior is a three-input, two-output behavior. The inputs are frontal range and left and right proximities to obstacles; the outputs are the steering direction and vehicle speed. Four linguistic labels (VeryNear, Near, Far, VeryFar) were used for frontal range, and two labels (Clear, NotClear) each were used for left and right proximity inputs. A total of 16 rules were used for this behavior which is complete per Equation (1-2). The interested reader can find further details on the fuzzy behavior design parameters, hardware, and software in [36]. Four types of experiments were run to test the behavior's performance in situations likely to be encountered indoors: (a) traversal through straight and narrow passageways, (b) head-on approach towards obstacles, (c) partially blocked hallways with free space on one side of the obstacle, and (d) dead-end situations. Obstacles consisted of empty boxes and office furniture. Representative paths executed in each of these scenarios are depicted in Figure 1.4 (a)-(d), respectively.

In experiment (a) the robot successfully traversed the narrow corridor along a fairly smooth path. When approaching obstacles in experiment (b), the robot exhibited a very smooth deceleration before stopping short of colliding with static obstacles. As might be expected, a more abrupt halt was observed when dynamic obstacles (e.g., humans) were encountered. In partially blocked hallways, experiment (c), the robot always avoided collision but its behavior effected a tendency to wander about in the neighborhood of the obstacle. This can be attributed to the fact that the robot was not provided with a particular goal (perhaps on the other side of the obstacle), thus causing it to wander aimlessly. Experiment (d) resulted in similar behavior. The robot could successfully escape from dead-end situations, but with no *particular* place to go it would sometimes wander into the same dead-end again. Trial runs of the robot through the laboratory and hallways demonstrated that the robot was capable of rudimentary collision avoidance limited by the lack of state feedback. The robot wandered about reactively using the minimal sonar and IR proximity data to successfully avoid impact with obstacles.

1.8.2 Embedded Wall Following Behavior

Embedded controller realizations are generally subject to stronger constraints than their PC-based counterparts regarding speed, memory, and the environment in which the embedded system operates. Up to a certain point the realization procedures are quite similar. The point of departure is reached after development of a suitable set of rules. For the embedded controller, the behavior must also be appropriately *encoded* for the target processor, and then *downloaded* to the chip for execution. To give an idea of the steps involved, we describe a procedure for realizing an embedded wall following behavior. The target system is LOBOt Jr., a small mobile robot equipped with an embedded microcontroller based on the Motorola MC68HC11 chip. It is driven by a differential

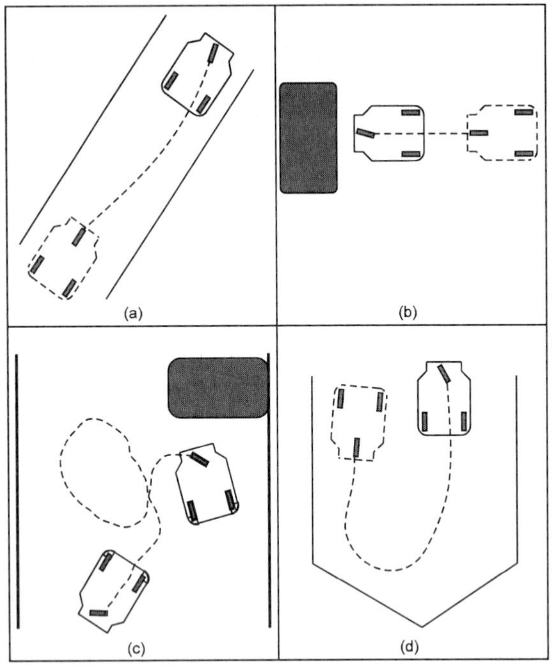

Figure 1.4 Instances of collision avoidance paths.

direct drive system using two independent DC motors. Four optical range sensors are used by the behavior—two on each side, and two in front. Since all control circuitry and computation are embedded in the robot, it is self-contained and travels without the need for any commonly used data links (e.g., tether, RF link, infrared beacon, etc.) to external processors such as a PC. For downloading of fuzzy behaviors, and communicating with the microcontroller during testing, the robot must be connected to a PC via a standard RS232 serial link. Wall following behavior inputs are frontal, left, and right obstacle ranges; outputs are the left- and right-wheel speeds. A monolithic rule base structure is used in which the universe partition for each input consisted of three linguistic labels (Near, Medium, Far). Thus, $R_c = 27$ for the wall following behavior. Several rules prescribed identical consequents for a given combination of two antecedents regardless of the linguistic value of the third antecedent. So in the interest of conserving memory, a rule reduction to 19 rules was performed by eliminating redundant rules.

The presence of a wall in sensor range activates the behavior which simply attempts to align the robot with the wall at a specified distance. Once aligned, the robot follows the wall attempting to maintain alignment by compensating for lateral drifts. The rule base and the generation of the processor-specific code were done using a software product called FIDE (Fuzzy Inference Development Environment) which provides the capability to convert fuzzy inference engines into real time assembly code compatible with the MC68HC11. A typical development procedure is illustrated in Figure 1.5. The code generated by FIDE was assembled into machine code and downloaded to the embedded

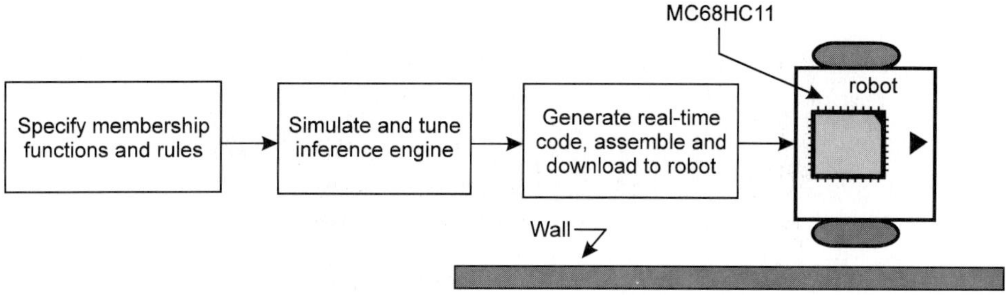

Figure 1.5 Development procedure for embedded fuzzy control.

controller. The machine code utilized less than 1.5 kilobytes of memory. The robot was able to follow walls in real time without apparent difficulty, thus supporting the feasibility of embedded realization of practical fuzzy behavior controllers.

1.8.3 Hierarchical Goal seeking Behavior

In order to demonstrate the fundamental operational aspects of the hierarchical behavior approach, we consider only the composite behavior – goal-seek. Its place in the overall hierarchy is illustrated in the left portion of Figure 1.1 which shows that its effect arises from synergistic interaction between go-to-xy and avoid-collision. These primitive behaviors (and others shown) have been independently developed and tested in simulation to predict their individual performances in indoor spaces with various obstacle arrangements.

The simulated mobile robot is modeled after LOBOt, a custom-built base driven by a 2-wheel differential and supported by two additional passive casters. The sensor suite includes optical encoders on each driven wheel and 16 ultrasonic transducers arranged primarily on the front, sides, and forward-facing obliques. The simulated "world" is a hypothetical indoor layout not unlike a warehouse or office building. Figure 1.6 shows a representative result of sensor-based goal seeking navigation, i.e., no map is employed. The robot icon is displayed every 5 seconds in the figure. The initial pose of the robot is $(x\ y\ \theta) = (10.0\ 5.0\ -\pi/2)$, and the goal is located at $(x, y) = (1.0, 5.0)$.

Recall that behaviors at the primitive level of the hierarchy are only capable of exhibiting their particular primitive roles. If acting alone, avoid-collision would merely display cyclic collision-free motion in the immediate vicinity of the robot's initial location, while go-to-xy would display a taxic reaction that propels the robot toward the goal regardless of obstacles in its path. The path shown in the figure resulted from adaptive coordination of interactions between the two primitive behaviors.

Behavioral interaction during the run is shown in Figure 1.7 as a time history of the degrees of applicability of each primitive behavior. The interaction dynamics shows evidence of brief bouts of competition (overlapping oscillations) and cooperation with varying levels of dominance. During most of the task, avoid-collision has the dominant influence over the robot's motors due to the close proximity of walls and obstacles. The first bout of competition corresponds to the robot's exit from the start room; a longer bout occurs as it enters the doorway just outside of the goal room. While entering the goal

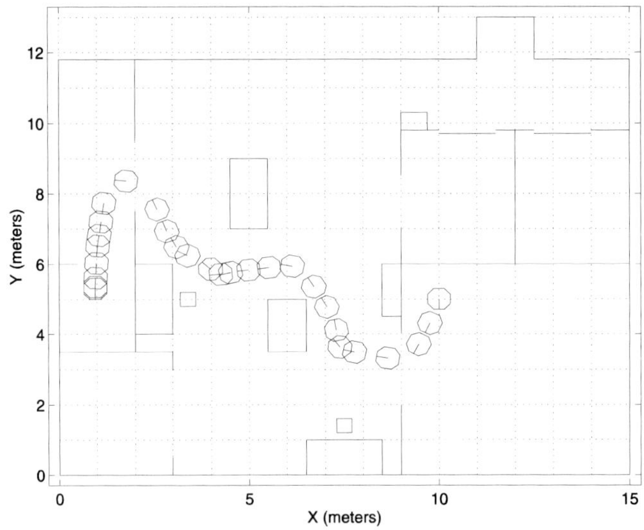

Figure 1.6 Simulation of goal seeking behavior.

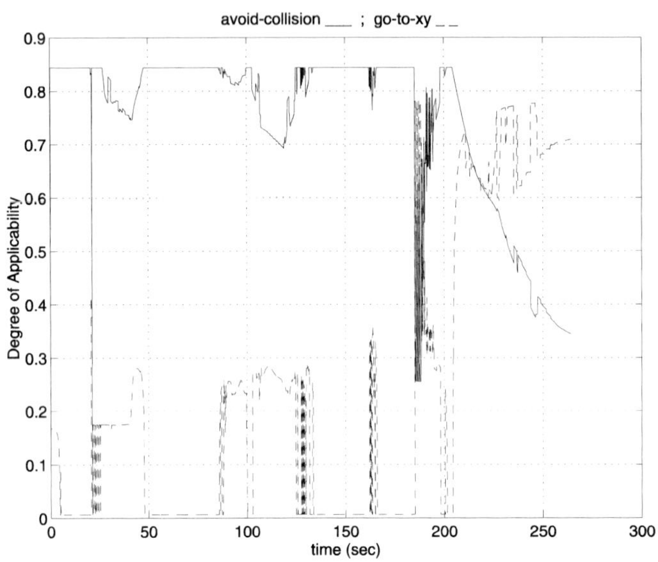

Figure 1.7 Behavior modulation and interaction during goal seeking.

room, the applicability of go-to-xy increases and eventually dominates during the final approach. Elsewhere, the applicabilities vary continuously, reflecting the levels of activation recommended by the behavior control system.

1.8.4 Evolution of Steering Behavior

GP has been applied to evolve a fuzzy behavior to properly steer a mobile robot for path following in the plane. The control problem was formulated by Hemami [37] for low-speed (< 2 m/*s*) tricycle-model vehicles. Hemami derived a state space kinematic model in which the state vector consisted of presumably measurable position and orientation errors associated with path following. The position error is the deviation of the robot's center from the nearest point on the desired path. The orientation error is the angular deviation of the robot's heading from the tangent to the desired path. The steering angle of the front wheel is the corrective control action that causes the error states to decay to zero, thus forcing the robot to follow the desired path. We have used the model to simulate the Hero-1 mobile robot during the evolution of the steering behavior. Full details of this application are documented in [38].

A two input, one-output fuzzy behavior was evolved using predefined membership functions for the linguistic state errors. The membership functions were taken from an existing steering behavior, with a complete rule base of 25 rules, designed by trial-and-error synthesis. For this application, the elements of the terminal set for GP consisted of the state variables, control variable, and the fuzzy I/O linguistic labels. During the GP evolution, each behavior in the current population is evaluated to determine its fitness as a candidate behavior for steering control. This evaluation is achieved by trial simulations of the robot's motion from a number of initial conditions (eight in this case) until either the zero error state is reached or a preset timeout period expires. The behavior fitness function was chosen to be the sum, over the eight trial simulations, of the Euclidean norms of the error state vector at the end of each trial. Lower fitness scores are associated with better behaviors; a perfect score is zero. In a representative run, the best behavior was found on the seventh generation; it had 21 rules and a fitness of 1.58. This compared favorably with the complete rule base, hand derived by trial-and-error, which had 25 rules and a fitness of 1.96.

For performance comparison, Figure 1.8 (a)-(c) illustrates the position and orientation errors, and the control effort for the evolved and hand-derived behaviors. The initial condition is a position error of 0.8 meters and an orientation error of –0.9 radians. The rise time of the evolved behavior is fastest and results in an overshoot of the goal state. Its response time in reaching the goal state, however, is practically the same as for the hand-derived behavior. Although the evolved behavior learned its steering control rules using only eight initial conditions, it was able to generalize when started from initial conditions throughout the error state space. This is shown in the phase portrait of Figure 1.8(d) which reveals that the origin is a stable node of the behavior. It is possible to further improve the performance of evolved fuzzy behaviors by tailoring the fitness function to emphasize desired performance attributes [38]. Behaviors governed by fewer rules are also possible. The smallest rule base discovered by GP had 11 rules and a fitness of 1.99. However, the control signal was unacceptable due to fast transient oscillations of the steering angle.

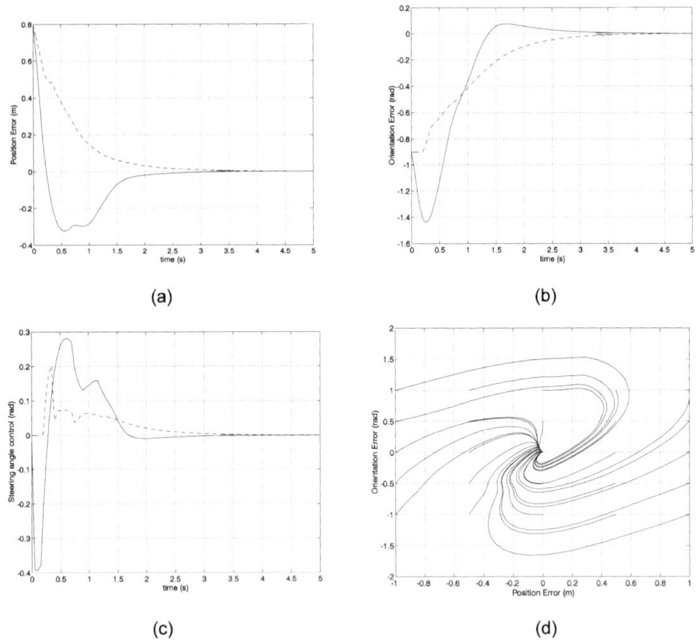

Figure 1.8 Performance comparison between evolved (—) and hand-derived (- -) steering behaviors.

1.9 CONCLUSIONS

We have seen how fuzzy behaviors can be synthesized, evolved, and realized in hardware as semi-autonomous and autonomous embedded systems. Some primitive behaviors do not require precise sensing and detailed models of the environment in order to exhibit satisfactory levels of intelligence. On the other hand, some behaviors may be just complex enough that successful design is best achieved through computational evolution. In such cases, GP is a viable option. It manipulates the linguistic variables directly associated with the behavior, and is capable of discovering highly fit behaviors that require fewer rules than necessary for completeness.

Our discussion of fuzzy behavior control for single monolithic behaviors also applies to the incremental development of behaviors for the hierarchical architecture. Once individual fuzzy behaviors are developed, goal-directed navigation is possible as a hierarchical arrangement coordinated using the concept of "degree of applicability." By using fuzzy reasoning, behavior coordination and conflict resolution are enhanced since all control recommendations contribute to the overall control decision. The practical utility of the hierarchical approach lies in its decomposition of overall behavior into sub-behaviors that are activated only when applicable. When sensor profiles satisfy the conditions for activation of a single behavior, or several, there is no need to process rules from behaviors that do not apply (as is done in the conventional rule base architecture).

Processing rules from irrelevant behaviors would result in unnecessary consumption of computational resources and possible introduction of "noise" into the decision-making process. In the hierarchical fuzzy behavior approach, there are no conceivable circumstances under which the totality of rules in the entire system will be consulted during a given control cycle.

With the various approaches to introducing fuzzy logic into the framework of behavior control, the paradigm is further maturing in its sophistication. The use of ideas such as rule hierarchies will contribute to our ability to develop intelligent control schemes for complex behavioral systems that mandate large rule base cardinalities. For many years, ethologists have developed theories and models to explain aspects of animal behavior. Results of their work are quite useful for developing intelligent *robot* behavior. While they continue to focus on analysis of behavior from the outside, we concentrate on synthesis from the inside. Perhaps we will arrive at a midpoint with some unified understanding of intelligence.

1.10 REFERENCES

1. Aminzadeh, F. and Jamshidi, M. (eds.). *Soft Computing with Fuzzy Logic, Neural Networks and Distributed Artificial Intelligence.* Prentice Hall, 1994.

2. Mamdani, E. H. and Assilian, S. "An Experiment in Linguistic Synthesis with a Fuzzy Logic Controller." *International Journal of Man-Machine Studies.* 1975, pp. 1-13.

3. Zadeh, L. A. "Fuzzy Sets." *Information and Control.* Vol. 12. 1965, pp. 338-353.

4. McFarland, D. J. *Feedback Mechanisms in Animal Behavior.* Academic Press, 1971.

5. Brooks, R. A. "A Robust Layered Control System for a Mobile Robot." *IEEE Journal of Robotics and Automation.* Vol. RA-2, No. 1, 1986, pp. 14-23.

6. Anderson, T. L. and Donath, M. "Animal Behavior as a Paradigm for Developing Robot Autonomy." *Journal of Robotics and Autonomous Systems.* Vol. 6. 1990, pp. 145-168.

7. Gat, E., Desai, R., Ivlev, R., Loch, J. and Miller, D. "Behavior Control for Robotic Exploration of Planetary Surfaces." *IEEE Transactions on Robotics and Automation.* Vol. 10, No. 4, August 1994, pp. 490-503.

8. Homaifar, A. and McCormick, E. "Simultaneous Design of Membership Functions and Rule Sets for Fuzzy Controllers Using Genetic Algorithms." *IEEE Transactions on Fuzzy Systems.* Vol. 3, No. 2, May 1995, pp. 129-139.

9. Huang, S. and Nelson, R. M. "Artificial Neural Networks Used as a Rule Selector for Fuzzy Logic Controllers." *ASME Conference on Computers in Engineering*. 1993, pp. 445-454.

10. Koza, J. R. *Genetic Programming: On the Programming of Computers by Means of Natural Selection*. MIT Press, 1992.

11. Uragami, M., Mizumoto, M. and Tanaka, K. "Fuzzy Robot Controls." *Journal of Cybernetics*. Vol. 6. Hemisphere Publishing Corporation, 1976, pp. 39-64.

12. Sugeno, M. and Murakami, K. "An Experimental Study on Fuzzy Parking Control Using a Model Car." In *Industrial Applications of Fuzzy Control*. (Sugeno ed.). Elsevier Science Publishers, pp. 125-138, 1985.

13. Saffiotti, A., Ruspini, E. H., and Konolige, K. "Blending Reactivity and Goal-Directedness in a Fuzzy Controller." *IEEE International Conference on Fuzzy Systems*. 1993, pp. 134-139.

14. Pin, F. G. and Watanabe, Y. "Navigation of Mobile Robots Using a Fuzzy Behaviorist Approach and Custom-Designed Fuzzy Inferencing Boards." *Robotica*. Cambridge University Press, Vol. 12, No. 6, Nov.-Dec. 1994, pp. 491-504.

15. Goodridge, S. G. and Luo, R. C. "Fuzzy Behavior Fusion for Reactive Control of an Autonomous Mobile Robot: MARGE." *IEEE International Conference on Robotics and Automation*. pp. 1622-1627, 1994.

16. Li, W. "Fuzzy-Logic-Based Reactive Behavior Control of an Autonomous Mobile System in Unknown Environments." *Engineering Applications of Artificial Intelligence*, Special Issue: *Control Applications of Fuzzy Logic*. Elsevier Science Ltd., pp. 521-531, October 1994.

17. Tunstel, E. and Jamshidi, M. "Fuzzy Logic and Behavior Control Strategy for Autonomous Mobile Robot Mapping." *Third IEEE International Conference on Fuzzy Systems*, pp. 514-517, June 1994.

18. Tunstel, E. "Fuzzy Spatial Map Representation for Mobile Robot Navigation." *10th Annual Symposium on Applied Computing*, ACM, pp. 586-589, February 1995.

19. Yen, J. and Pfluger, N. "A Fuzzy Logic Based Robot Navigation System." *AAAI Symposium on Applications of Artificial Intelligence to Real-World Autonomous Mobile Robots*, pp. 195-199, October 1992.

20. Raju, G. V. S., Zhou, J. and Kisner, R. A. "Hierarchical Fuzzy Control." *International Journal of Control*. Vol. 54, No. 5, 1991, pp. 1201-1216.

21. Bruinzeel, J., Jamshidi, M. and Titli A. "A Sensory Fusion-Hierarchical Real time Fuzzy Control Approach for Complex Systems." Technical Report No. 95347. LAAS-CNRS, Toulouse, France, August 1995.

22. Berenji, H., Chen, Y. Y., Lee, C. C., Jang, J. S. and Murugesan, S. "A Hierarchical Approach to Designing Approximate Reasoning-Based Controllers for Dynamic Physical Systems." *6th Conference on Uncertainty in Artificial Intelligence*, pp. 362-369, 1990.

23. Tinbergen, N. *The Study of Instinct*. Oxford University Press, 1951.

24. Baerends, G. P. "A Model of the Functional Organization of Incubation Behaviour in the Herring Gull." *Behaviour, An International Journal of Comparative Ethology*, Suppl. 17. E. J. Brill, 1970, pp. 261-310.

25. McFarland, D. J. and Sibly, R. M. "The Behavioral Final Common Path." *Philosophical Transactions of the Royal Society of London*. B 270. 1975, pp. 265-293.

26. Klir, G. J. and Folger, T. A. *Fuzzy Sets, Uncertainty, and Information*. Prentice Hall, 1988.

27. Tunstel, E. "Coordination of Distributed Fuzzy Behaviors in Mobile Robot Control." *IEEE International Conference on Systems, Man and Cybernetics*, pp. 4009-4014, October 1995.

28. Yager, R. R. and Filev, D. P. *Essentials of Fuzzy Modeling and Control*. John Wiley & Sons, 1994.

29. Pfluger, N., Yen, J. and Langari, R. "A Defuzzification Strategy for a Fuzzy Logic Controller Employing Prohibitive Information in Command Formulation." *IEEE International Conference on Fuzzy Systems*, pp. 717-723, 1992.

30. Reynolds, C. W. "Evolution of Corridor Following Behavior in a Noisy World." *From Animals to Animats 3: Third International Conference on Simulation of Adaptive Behavior*. MIT Press, August 1994, pp. 402-410.

31. Handley, S. "The Genetic Planner: The Automatic Generation of Plans for a Mobile Robot via Genetic Programming." *IEEE International Symposium on Intelligent Control*, pp. 190-195, 1993.

32. Brooks, R. A. and Mataric, M. J. "Real Robots, Real Learning Problems." In *Robot Learning* (Connell, J. H., and Mahadevan, S. eds.). Kluwer Academic Publishers, 1993, pp. 193-213.

33. Leitch, D. and Probert, P. "Genetic Algorithms for the Development of Fuzzy Controllers for Autonomous Guided Vehicles." *Second European Congress on Intelligent Techniques and Soft Computing EUFIT 94*, pp. 464-469, 1994.

34. Bonarini, A. P. "ELF: Learning Incomplete Fuzzy Rule Sets for an Autonomous Robot." *First European Congress on Intelligent Techniques and Soft Computing EUFIT 93*, pp. 69-75, 1993.

35. Koza, J. R. and Rice, J. P. "Genetic Generation of Both the Weights and Architecture for a Neural Network." *IEEE International Joint Conference on Neural Networks*, pp. 397-404, 1991.

36. Martinez, A., Tunstel, E. and Jamshidi, M. "Fuzzy Logic Based Collision Avoidance for a Mobile Robot." *Robotica*. Vol. 12, No. 6, Nov.-Dec. 1994, pp. 521-527.

37. Hemami, A. "Steering Control Problem Formulation of Low Speed Tricycle-Model Vehicles." *International Journal of Control*, Vol. 61, No. 4, 1995, pp. 783-790.

38. Tunstel, E. and Jamshidi, M. "On Genetic Programming of Fuzzy Rule based Systems for Intelligent Control." To appear in *International Journal of Intelligent Automation and Soft Computing*, Vol. 2, No. 2, 1996.

2

FUZZY CONTROL APPROACH FOR THE DESIGN OF ACTIVE AND SEMI-ACTIVE SUSPENSION (COMPARISON WITH OPTIMAL CONTROL AND VARIABLE STRUCTURE CONTROL)

A. Titli
INSA/LAAS CNRS
FRANCE

S. Boverie
SIEMENS AUTOMOTIVE S.A.
FRANCE

The design of active or semi-active suspensions for modern cars is subject to real-time contradictory requirements concerning on the one hand ride comfort and on the other hand road holding and handling.

So, a systems engineering approach is needed; it is based on sophisticated (but not overly so) models and advanced automatic control methods.

However, due to the limitation of design methods for nonlinear systems, the design of the suspension is done on the base of a linear multivariable model; the performance is tested by simulation on more realistic models including nonlinearities and delays. In this chapter, we present, through the use of a full car model, the design of semi-active and active suspensions with an optimal controller, variable structure controller, fuzzy controller, and neuro-fuzzy controller.

The efficiency of the approaches and the performance of the different controllers are compared with a general nonlinear package in the environment of EASY 5/W™, a product of Boeing Computer Services. MATLAB™ is also used for the resolution of some algebraic matrix equations like Riccati and Lyapunov control equations.

Experiments of these three approaches conducted on a bench test at the end this chapter.

2.1 INTRODUCTION

The use of automatic control techniques is increasing for the design of automotive systems devoted to the control of the chassis and engine. For the chassis control, we can mention: traction control system (ASR), anti-blocking system (ABS), steering control, load leveling, tire pressure control, engine mount control, active suspension, and semi-active suspension.

On this last point, we can refer to the design of suspension (adaptive, semi-active, active) for which we know how to build a model and how to define the objectives of the controller to reach a compromise between contradictory requirements like ride quality and road holding and handling.

To solve this problem we used several automatic control theories and techniques—the proportional plus integral plus derivative (PID) approach—and optimization techniques: optimal-suboptimal, and H^∞ robust methods.

In this chapter, we consider three categories of systems:

1. optimal active and semi-active system linear quadratic regulator (LQR)

2. variable structure system (VSS) theory (sliding mode)

3. fuzzy active and semi-active suspension

for the same problem, in order to take into account, some technological constraints on the actuators, which have two kinds of control laws: discrete control laws (e. g., hard/soft control laws) and continuous control law (continuous change of suspension characteristics) according to control strategy developed by one of the different approaches.

The corresponding controllers improve, spectacularly, the dynamic behavior of the car in terms of passengers comfort and security, specified by:

1. Attenuation of the car body accelerations (heave, pitch, and roll) in the frequency range of human acceleration sensitivity.

2. Maximization of the security by maintaining the best contact between wheels and road.

3. Good properties of robustness and perturbation rejection of the controllers against changes of body mass and system damping and stiffness.

In the rest of the chapter, we present a mathematical model on which optimal and VSS controller are designed. After that fuzzy and neuro-fuzzy controllers are introduced and all the results obtained by simulation and on a test bench are compared.

2.2 MATHEMATICAL APPROACH

2.2.1 Modeling

The objective of this modeling is to find a full car model which gives a good representation of the effects of the suspension components on the general dynamical behavior of the vehicle; this model is used to develop different approaches to design a new system of modern suspensions, active or semi-active; the forces acting between car body and unsprung masses come from three components: damper-spring-active or semi-active actuator (Figure 2.1 - Figure 2.3).

A) Description of the Model: The Full Car model is supposed to be linear, Multiple Input/Multiple Output (MIMO), time invariant with 7 degrees of freedom (7 DOF), body inflexible: 4 degrees of freedom for the unsprung masses (vertical oscillations, vertical displacement of each unsprung mass) and 3 degrees of freedom for the body mass (vertical, lateral, and longitudinal oscillations, heave, pitch, and roll).

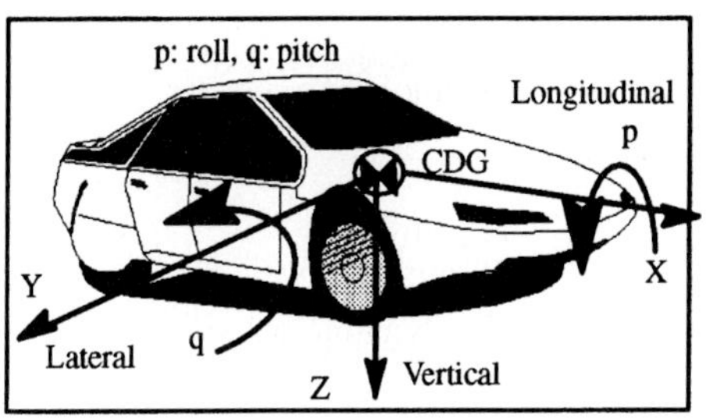

Figure 2.1 Full Car system.

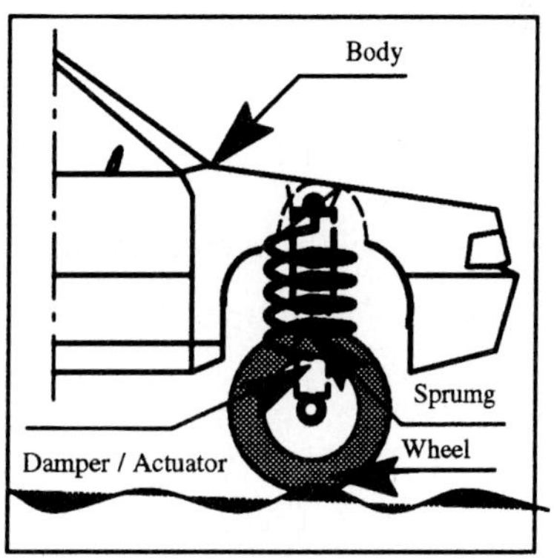

Figure 2.2 Physical system: Quarter car.

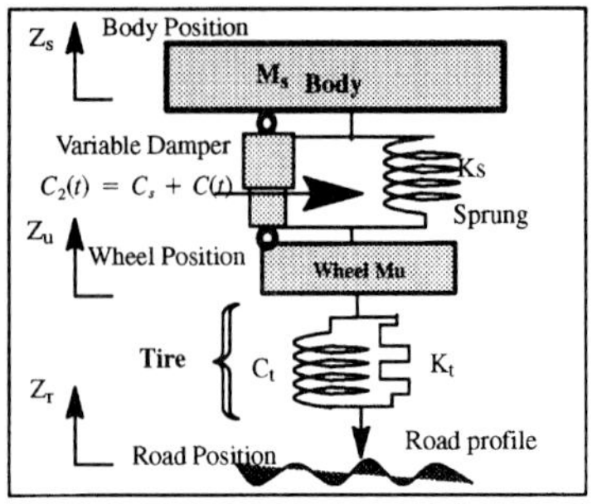

Figure 2.3 Quarter Car Model.

By application of the mechanical laws to the physical system, we can derive the linearized equations of motion for the car body and the unsprung masses; we separate the total equations of motion for this mechanical oscillatory multibody system in two parts:

a) The dynamical equations of motions for the car body: heave, roll, and pitch at the center of gravity G.

b) The equations of dynamical vertical oscillations for the unsprung mass at the four corners: vertical displacement of each wheel.

B) Definition of State Variables: We define the 14 state variables as follows:

X_1: Car body heave, roll and pitch: $X_1 \in R^3$
X_2: Car body heave, roll and pitch rates: $X_2 \in R^3$
X_3: Tire deflection vector: $X_3 \in R^4$
X_4: Absolute velocity vector of unsprung masses: $X_4 \in R^4$

$X = (X_1\ X_2\ X_3\ X_4)^T$ is the global state variable which is used for the LQR and VSS control approaches. The control variable U is the active suspension force.

2.2.2 Design Of Optimal Controller

Optimal control is developed in two ways:

- without constraints by resolution of the Riccati matrix algebraic control equation for active suspension control problem.
- with structural constraints for semi-active suspension control problem.

This problem will be solved by using the Lagrange multipliers to obtain optimality conditions with respect to a quadratic performance index. This approach is more difficult, but the obtained solution leads to a very interesting new control law for the suspension which satisfies conditions on passivity of the actuator and saturation of the control laws.

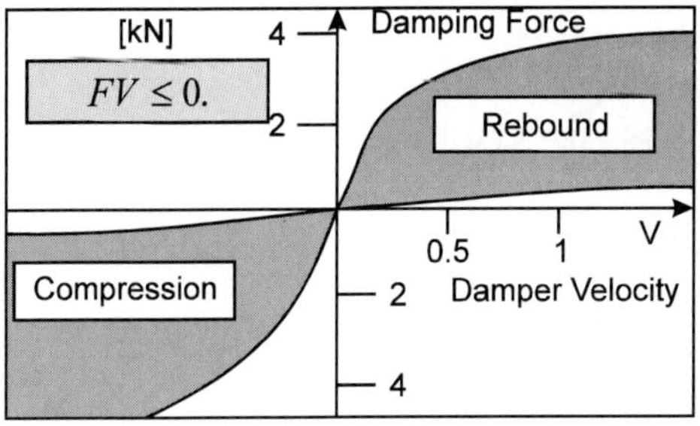

Figure 2.4 Characteristics of variable damper.

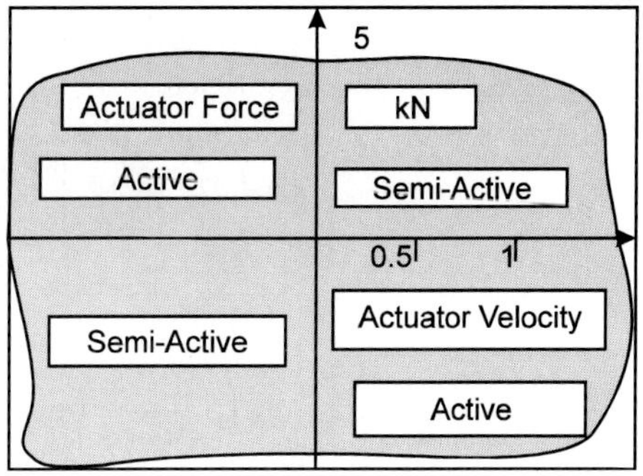

Figure 2.5 Active force generator.

This law controls the damping in the shock absorber between two limits (hard and soft) by continuous variation of the damping, according to the information given by the sensors. This kind of control is superior to the active one because:

1. The simplicity of the control law.

2. The small energy required for signal processing and sensing. Figures 2.4 and 2.5 give the performance characteristics of variable damper and force generator.

Let us begin with the design of active suspension.

A) Active Suspension Controller: All the physical goals in mind during the design of the suspension can be aggregated into a scalar mathematical criterion of the form:

$$J = \int_0^{\infty} \left(X^T QX + 2U^T SX + U^T RU \right) dt \tag{2-1}$$

The full state feedback law (by the LQR method) is obtained as follows:

$$U = F_a = -R_0^{-1}\left(B^T P + S\right)X \tag{2-2}$$

with the P solution of the Riccati algebraic equation:

$$PA^* = A^{*T} P - PBR_0 B^T P + Q^* = 0 \tag{2-3}$$

where $A^* = A - BR_0^{-1}S$ and $Q^* = Q - S^T R^{-1} S.$

The Figure 2.6 shows the bloc diagram of the closed loop system.

B) Semi-Active Controller: Here, we consider the application of semi-active forces for the 4 suspensions which means we try to find the optimal damping control laws for the

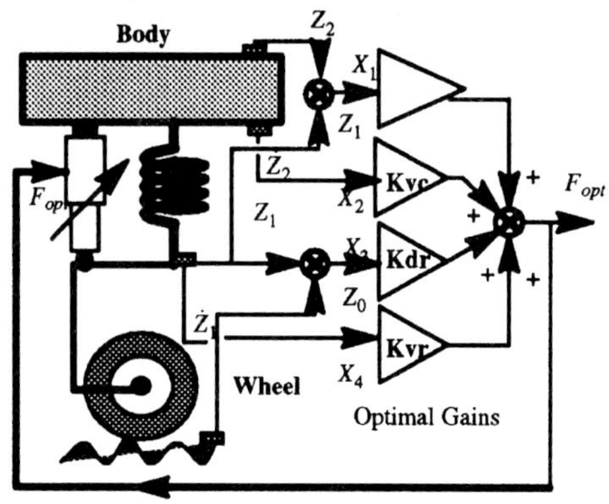

Figure 2.6 Closed loop optimal active suspension.

suspension assuming that damping rate can be changed only between two bounds, according to the technological realization of the actuator.

$$0 \le c_{2i} \le C_{2i}^{\max} \qquad i = 1, 2, 3, 4 \tag{2-4}$$

So, the damping matrix will be considered as the sum of two matrices: one constant matrix corresponding to the static damping and one time-varying matrix given by the optimal control laws, so:

$$C_2 = diag\left[c_{21}^0\, c_{22}^0\, c_{23}^0\, c_{24}^0\right) + diag\left[b_{21}(t)\, b_{22}(t)\, b_{23}(t)\, b_{24}(t)\right) = c_{2^0} + c_v \tag{2-5}$$

(*diag* is the diagonal matrix)

By rewriting the equations of motion for the Full Car model (assuming $U = 0$; semi-active suspension), we get:

$$\dot{X} = AX + \Psi(X)C_v \tag{2-6}$$

where Ψ is function of the state variables and C_v is the variable term in the damping matrix.

We can show that Ψ is given by:

$$\Psi(X) = -B\left(\Sigma^T X_2 - X_4\right) = -BX_{24} \tag{2-7}$$

The performance index becomes ($U = 0$),

$$J = \int_0^\infty \left(X^T Q X\right) dt \tag{2-8}$$

The weighting matrix Q can be divided into two matrices: the first one related to the static damping and the second one to the variable damping, i.e.,

$$Q = Q_0(C_v = 0) + Q_t \tag{2-9}$$

where Q_t is given by:

$$Q_t = \begin{bmatrix} 0 & \Sigma K_2 R^* C \Sigma^T & 0 & \Sigma K_2 R^* C \\ \Sigma C R^* K^2 \Sigma^T & \Sigma H \Sigma T & -\Sigma C R^* K_2 & -\Sigma H \\ 0 & -K_2 R^* C \Sigma^T & 0 & K_2 R^* C \\ -C R^* K_2 \Sigma^T & -H \Sigma^T & C R^* K_2 & H \end{bmatrix} \tag{2-10}$$

with $H = C_0 R^* C + C R^* C_0 + C R^* C$ and $C_s = (b_1\ b_2\ b_3\ b_4)^T$. It can be shown that the performance index can be rewritten as:

$$J = \int_0^\infty \left(X^T Q_0 X + 2 C_s^T S(X) X + C_s^T R^* C_s \right) dt \tag{2-11}$$

where $S(X) = -X_{24} S_0$ with $S_0 = S(C_s = 0)$, $R^*(X) = X_{24} R^* X_{24}$.

In general, we assume that for physical reasons the modulated damping vector C_s has the following maximum values:

$$C_{\max} = \left(C_{m1} \quad C_{m2} \quad C_{m3} \quad C_{m4} \right)^T \tag{2-12}$$

The problem is then to find the optimal control vector C^* in order to minimize the performance index (2-11), subject to the constraints (2-4).

We use the calculus of variations to solve this problem the Lagrangian of the problem becomes:

$$\begin{aligned} L \equiv\ & X^T Q_0 X + 2 C^T S(X) X + C^T R(X) C \\ & + 2 P^T A X + \Psi(X) C - \dot{X} - 2\alpha^T C + 2\beta^T (c - C_{masx}) \end{aligned} \tag{2-13}$$

where $\alpha^T = (\alpha_1, \alpha_2, \alpha_3, \alpha_4)$, $\beta^T = (\beta_1, \beta_2, \beta_3, \beta_4)$. α and β are the Lagrange multiplier vectors for the state and damping constraints; using the calculus variation, an optimal damping factor c^* can be derived.

C) Experiments: Experiments of this approach have been conducted in two ways: a) by simulation using linear and non linear models of the system including delays, hysteresis, dynamics of the sensors and actuators. b) by real experiments on a test bench. They consist on a frequency analysis where signals applied to the system (road profile) are sinusoidals at 1Hz (frequency of the body) and 10Hz (frequency of the wheel) and on a

Table 2.1 Vehicle inertia data

Effective Weight of Wheel	Front	50.0 kg
	Rear	40.0 kg
Radius of wheel		0.2 m
Weight of sprung mass		1600.0 kg
Principle inertias of sprung mass	I_{xx}	500.0 kg m^2
	I_{yy}	3000.0 kg m^2
	I_{zz}	3000.0 kg m^2
Center of gravity of sprung mass	from front axle	1.32 m
	from rear axle	1.5 m
	above ground	0.5 m
	from left / right side	0.75 m

Table 2.2 Suspension and tire data.

		Front	Rear
	Spring stiffness	20.0 kN m^{-1}	27.0 kN m^{-1}
	Damper coefficient	1.4 kN s m^{-1}	2.0 kN s m^{-1}
	Anti-roll bar Stiffness	20.0 kN m^{-1}	
	Vertical stiffness	250.0 kN m^{-1}	250.0 kN m^{-1}
	Lateral stiffness	66.0 kN m^{-1}	70.0 kN m^{-1}
Steering Stiffness		196.0 kN rad^{-1}	

time analysis where the input signal is a 5cm step. We present here only the results this time analysis.

The data corresponding to a real car are given in the Tables 2.1 and 2.2.

Some results are given in Figure 2.7, and Figure 2.9. The improvements obtained by optimal control with respect to passive suspension are obvious.

2.3 DESIGN OF VSS CONTROLLER

A) The Method: The principle of the VSS approach is as follows. Consider an n-dimensional system and an $(n-1)$-dimensional hyperplane S given by $S(X) = 0$, where X is the system state and S represents the switching boundary. Then using a control of the form $U = U_{\max}$ if $S(X) > 0$ and $U = U_{\min}$ if $S(X) < 0$, sliding mode is obtained: once the system trajectory reaches S, it is constrained to remain on S, therefore with reduced-order dynamics determined by the switching surface $S(X) = 0$. The condition for sliding can be expressed as:

$$S\dot{S} < 0 \ \text{ near } S(X) = 0 \tag{2-14}$$

In this section, we apply the control by sliding mode to the multivariable (MIMO) suspension system; in this case, for each local control law i (i = 1, 2, 3, 4), there is an actuator which switches between two values (minimum and maximum) by using sliding mode property.

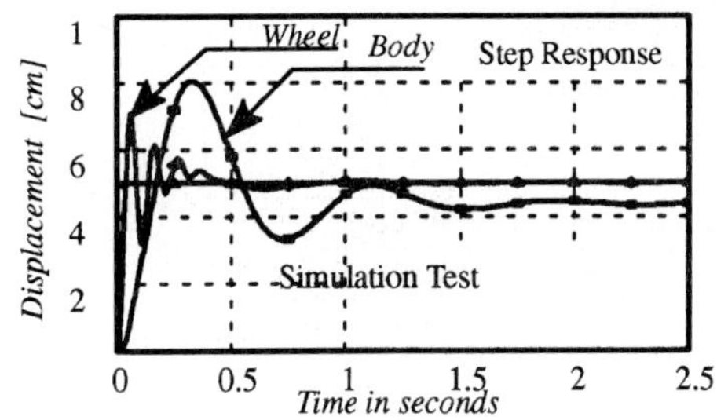

Figure 2.7 Performance of passive standard suspension.

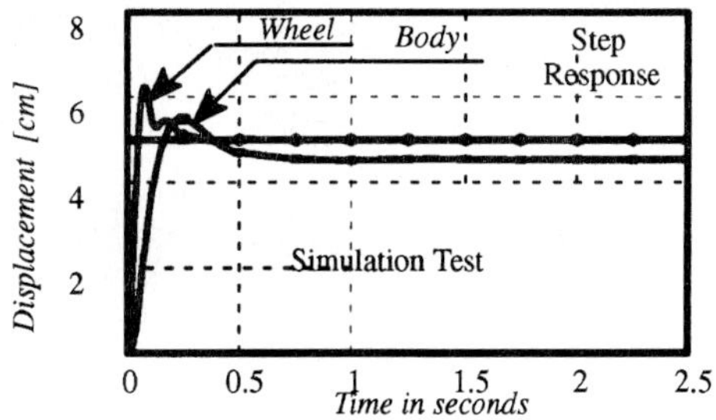

Figure 2.8 Performance of optimal-semi-active suspension.

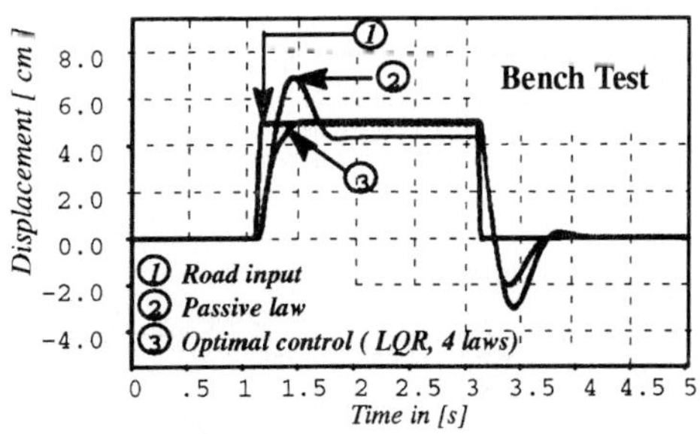

Figure 2.9 Body displacement for passive and LQR system (semi-active system).

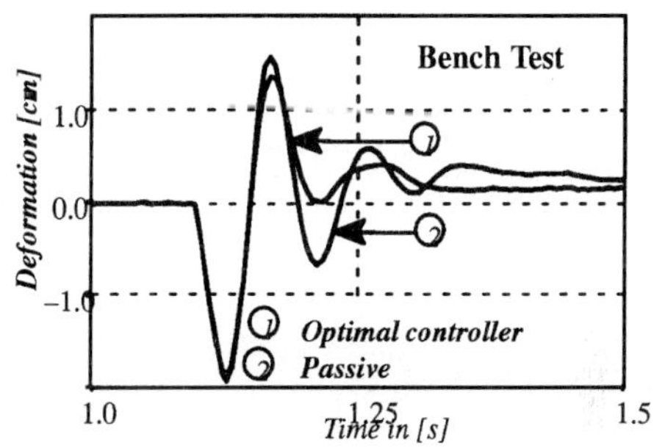

Figure 2.10 Tire deformation for passive and LQR system (semi-active system).

The design of the logic control law will follow the different steps:

1. Definition of MIMO Proportional Integral (PI) control law

2. Definition of equivalent vector control law

3. Design of sliding hyperplane using a pole assignment technique

All the mathematical details of these steps are rather complicated and for the sake of space, we refer to [15] for a complete description of the method. The corresponding control structure is given in Figure 2.11.

B) *Experiment*: In the same conditions as for optimal control, some experimental results are given below as an illustration of the efficiency of the method for the semi-active suspension.

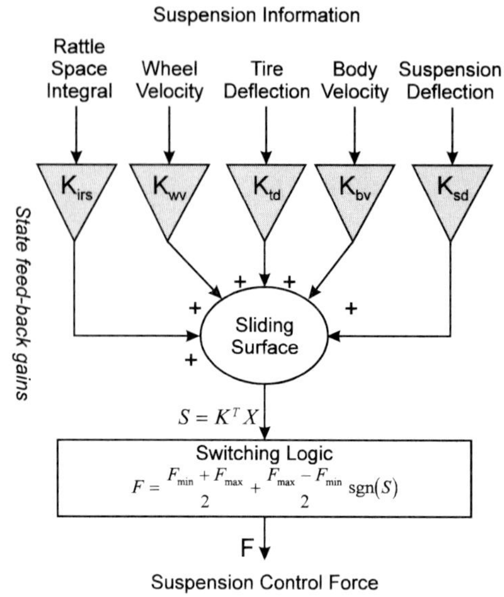

Figure 2.11 Variable structure controller for a Quarter Car.

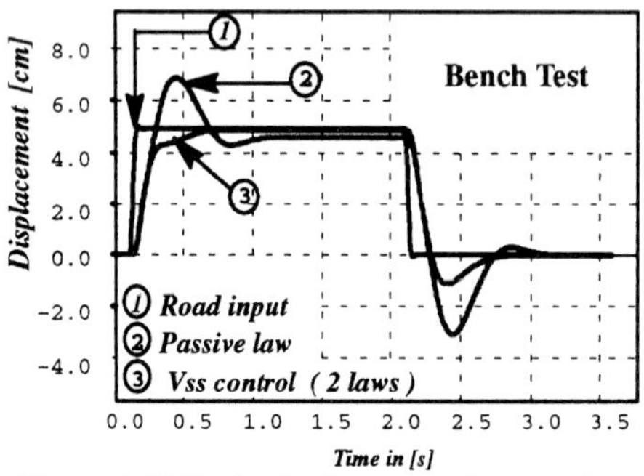

Figure 2.12 Body displacement for passive and VSS system (semi-active system).

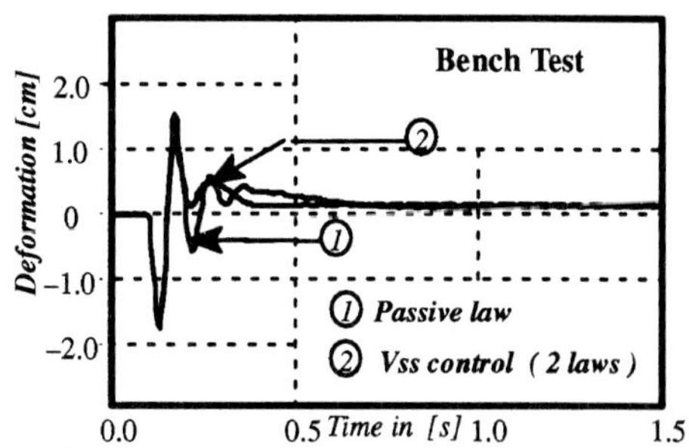

Figure 2.13 Tire deformation for passive and VSS system (semi-active system).

2.3 KNOWLEDGE-BASED APPROACH

2.3.1 Motivation

Based on classical mechanical laws and a few hypotheses, linear mathematical models (from Quarter Car to the Full Car) can be obtained. These models have been used here and in other papers to design semi-active and active suspensions, applying mathematical control theory and specific techniques like: optimal control, variable structure control, adaptive control, robust control, etc.

These approaches, based on a linear model, do not take into account some constraints like a) nonlinearities of physical phenomena (dry friction for example), b) saturation on actuators, c) discrete values of the control forces delivered by actuators, d) need for

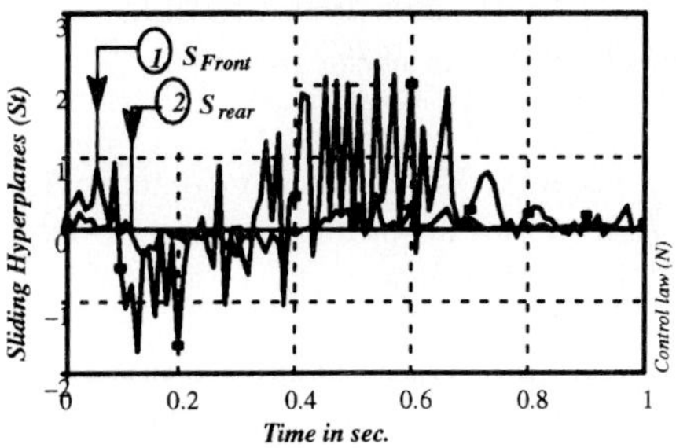

Figure 2.14 Sliding mode hyperplanes for front and rear semi-active-suspension system with VSS controller.

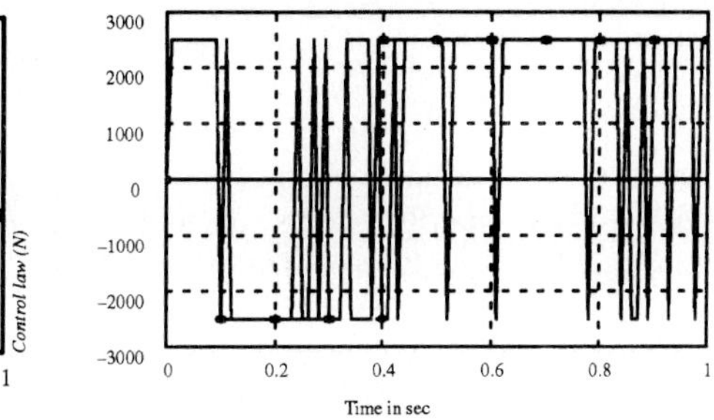

Figure 2.15 Sliding mode control force time history, showing switching and soft and hard actuator cycles.

robustness of the closed loop system, e) noise on measurements, and f) approximate measurements.

Here we want to develop a robust nonlinear controller based on the expertise of automotive engineers and not on mathematical models sometimes difficult to obtain or with limited validity. For this, we use fuzzy logic to design a fuzzy controller, a kind of knowledge-based controller.

Moreover, it is easy to see by the open loop dynamical analysis that the physical system has two different dynamics, one corresponding to the wheel hop (the fast part of the system) and the other one corresponding to the car body (the slow part of the system). We try to take advantage of this dynamical property to design two parallel fuzzy controllers coordinated by a fuzzy supervisor.

2.3.2 Design of Fuzzy Controller

Fuzzy logic was introduced by Zadeh in 1965 to give a possible mathematical representation of vagueness and approximation. A fuzzy set X (for example, X = "comfortable") has a membership function μ whose values are in the interval 0 to 1. The grade of membership functions relating to the fuzzy set X is defined for each element x, x being a possible element of various fuzzy sets with different membership grades.

These fuzzy sets are used to express the human expertise in a set of rules (rule base) like:

IF (the suspension deflection is Negative Big) THEN (the control action is Positive Big)

High similarity becomes apparent when comparing the Sliding Mode Controller (SMC) with a boundary layer to a FC (Fuzzy Controller) whose rules have been derived from a phase plane analysis. An analysis of the two controllers leads to the following conclusion: the fuzzy controller, in the above mentioned sense, is an extension of sliding mode with a boundary layer.

Then, the control delivered by FC can be represented by:

$$U = -K_{FC}(y, \dot{y}) \cdot \text{sgn}\left(\frac{S}{\Delta S}\right) \tag{2-15}$$

where S is the sliding hyperplane, and K_{FC} is a nonlinear function of y (outputs of the system). This analogy helps us build the rule base according to the following recommendations with respect to the normalized phase plane:

R1) U is negative above the switching hyperplane and positive below it.
R2) $|U|$ increases as the distance grows between the actual state and the switching hyperplane
R3) $|U|$ increases as the distance grows between the actual state and the hyperplane perpendicular to the switching hyperplane,
R4) y, $\dot{y}$ that fall out the phase space are covered by the maximum values $|U|_{max}$ with the alternative sign of U.

Table 2.3 shows an example of this kind of rule base.

The different steps in the running of the fuzzy controller are (see Figure 2.17): a) fuzzification of the physical variables by projection on fuzzy sets describing qualitative (linguistic) concepts, b) fuzzy implication for the evaluation of the truth value of the different rules, c) fuzzy reasoning by application of the generalized modus ponens d) rule aggregation for the combination of the contribution of different rules, e) defuzzyfication to set of real value for the control applied to the physical system (see Figure 2.16 and Figure 2.17).

A) Fuzzy Slow Comfort Controller

1. Controller Purpose: The purpose of this controller is to improve the ride comfort which is characterized by the human sensitivity to the acceleration in the frequency bandwidth (0.1 to 10 Hz).

e \ e	NB	NM	NS	ZE	PS	PM	PB
NB	NB	NB	NB	NB	NM	NS	ZE
NM	NB	NB	NB	NM	NS	ZE	PS
NS	NB	NB	NM	NS	ZE	PS	PM
ZE	NB	NM	NS	ZE	PS	PM	PB
PS	NM	NS	ZE	PS	PM	PB	PB
PM	NS	ZE	PS	PM	PB	PB	PB
PP	ZE	PS	PM	PB	PB	PB	PB

NB: Negative Big
NM: Negative Medium
NS: Negative Small
PB: Positive Big
PM: Positive Medium
PS: Positive Small

Table 2.3 Example of fuzzy controller design using tow input variables *e*, *e* and one output variable *U* (Mac-Vicar Wheelan rule base).

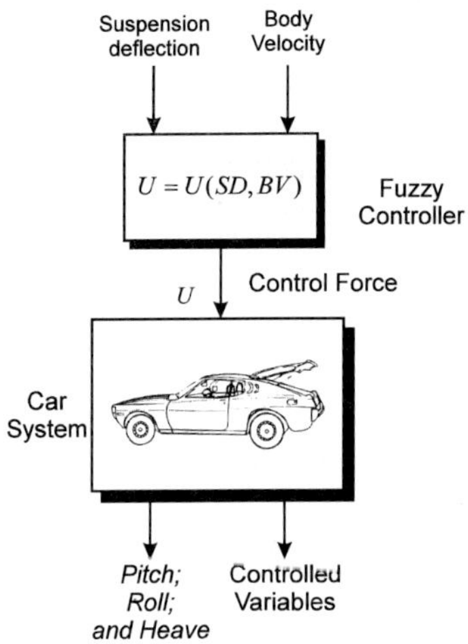

Figure 2.16 Basic fuzzy logic control system.

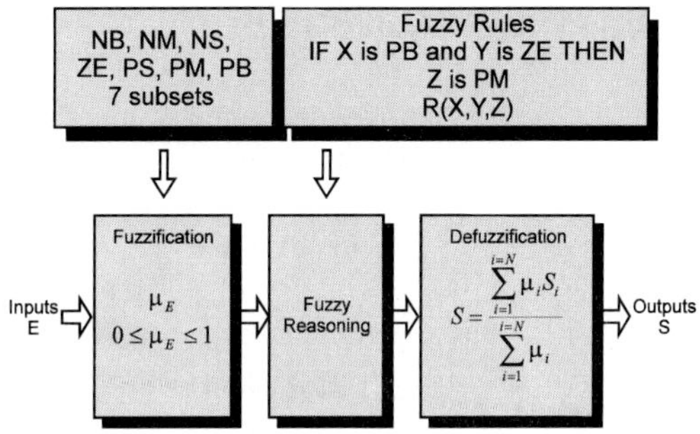

Figure 2.17 Details of a fuzzy controller.

2. Controller Inputs and Output: This controller tries to control the "slow" part of the system in order to optimize the ride comfort It is a PI non linear type controller using as inputs 1) suspension deflection, 2) integrated body acceleration (absolute velocity of the car body) and the output gives a control force acting between body and wheel as output for the actuator system (active or semi-active). See Figures 2.18 and 2.19.

B) Fuzzy Handling Fast Controller

1. Controller Purpose: The controller is used to have the best property in terms of road holding and handling by maintaining the contact between the road and the wheel in all operating conditions in order to ensure the ride safety for the passengers. Handling performance of a Quarter Car is difficult to define, but a measure of road holding is the contact tire/ground which is characterized by the tire deformation and excursion of the wheel.

2. Controller Inputs and Output: This controller tries to control the "fast" part of the system. It is a Proportional Derivative (PD) nonlinear type controller using as inputs: 1) tire deflection, 2) wheel absolute vertical velocity and as output the control force acting between wheel and body (see Figure 2.21 and Figure 2.22).

C) Fuzzy Supervisor: The design of two local fuzzy independent controllers is not enough to pilot the vehicle suspension with the maximum of benefit for the control of ride comfort and the control of handling and holding, because we use only one actuator to optimize the two aspects of the problem. So, we need a fuzzy weighting to be applied to each local controller. The determination of fuzzy weighting factors is done by a fuzzy supervisor. This supervisor uses as inputs the following global vehicle information:

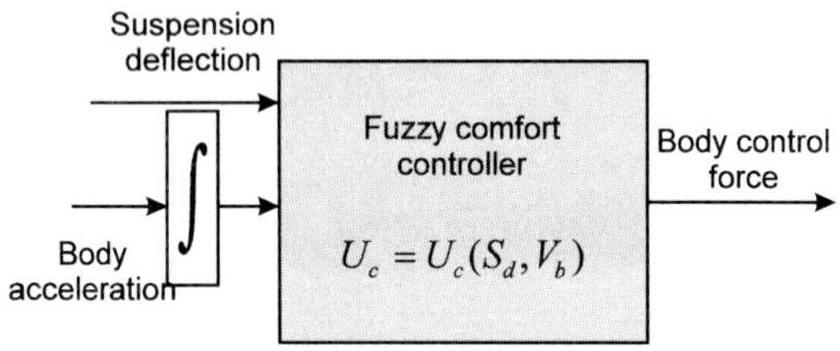

Figure 2.18 Fuzzy comfort controller.

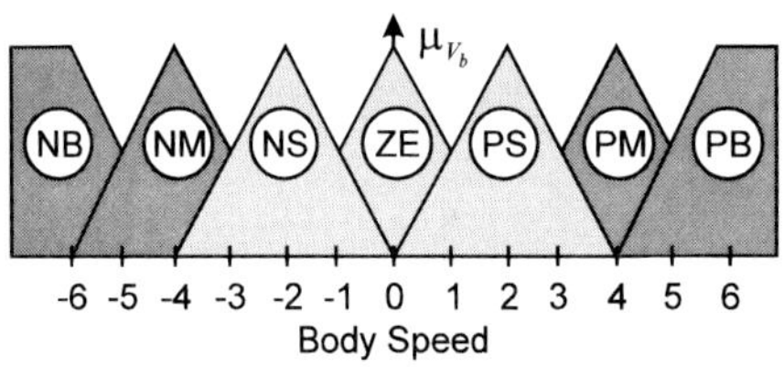

Figure 2.19 Membership functions for body speed on a quantized universe of discourse.

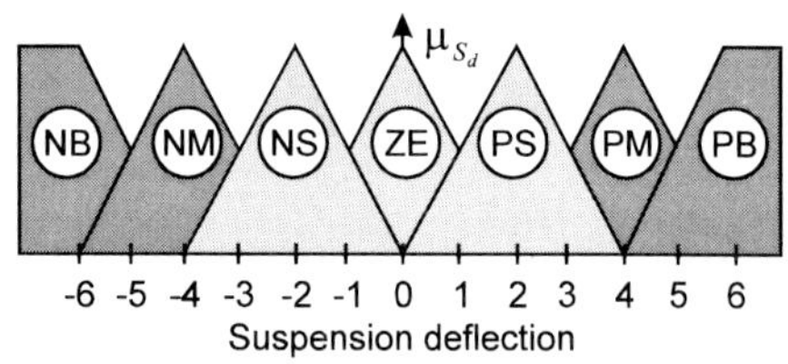

Figure 2.20 Membership functions for suspension deflection.

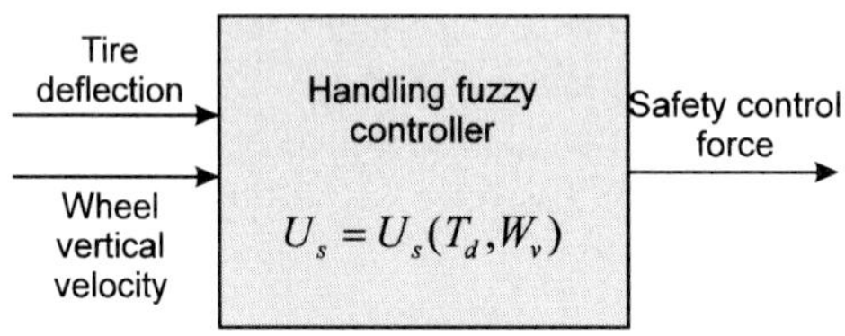

Figure 2.21 Handling fuzzy controller.

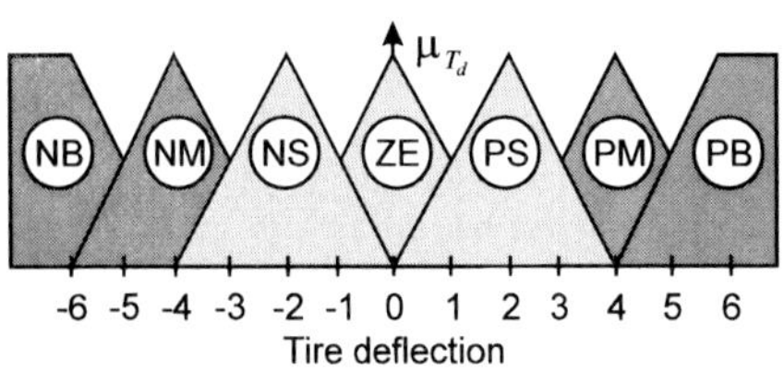

Figure 2.22 Membership functions of tire deflections.

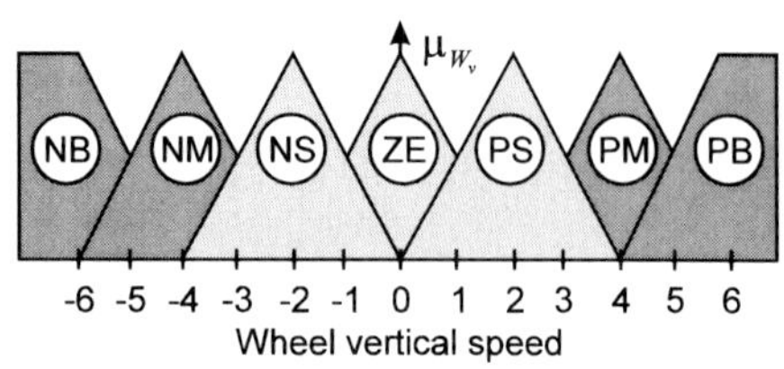

Figure 2.23 Membership functions of wheel vertical speed.

vehicle speed V, vehicle acceleration or deceleration A, braking pressure F, steering angle θ and body height H. (see Figure 2.24, Figure 2.25 and Figure 2.26).

D) Experiments: Always with the same data as for optimal and VSS controls, some illustrative results obtained on the test bench are given in figures 2.27 and 2.28 and seem satisfactory.

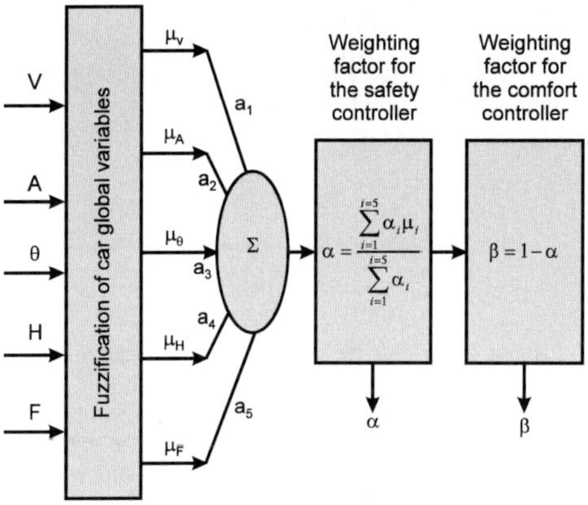

Figure 2.24 Fuzzy suspension supervisor.

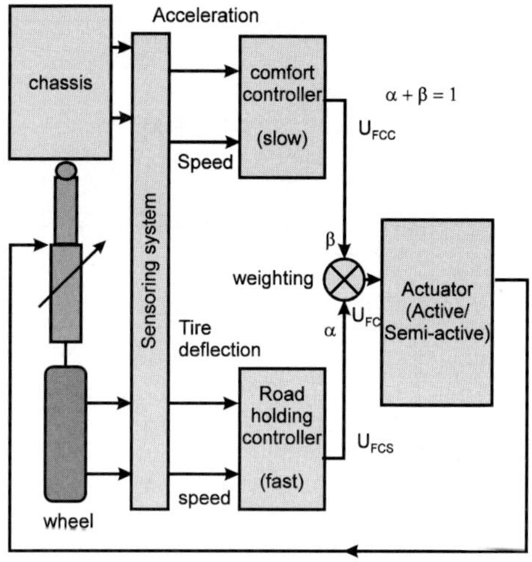

Figure 2.25 Active and semi-active suspension controller.

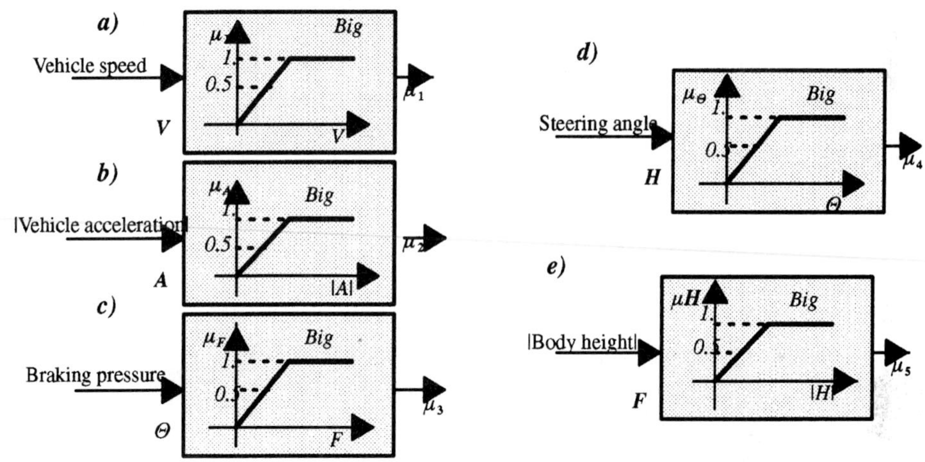

Figure 2.26 Membership functions for global vehicle variables.

2.3.3 Optimization of the Fuzzy Controller

It is well known that if fuzzy controllers are efficient and robust, they present a lot of parameters to tune in order to obtain the best performance: content of the rules, distribution and shape of the membership functions of the different used fuzzy sets, choice of the operators used in the development of the approximate reasoning, choice of the deffuzification technique, to mention a few of them. One way to overcome these difficulties is to reduce the set of parameters and to consider special fuzzy controllers. For

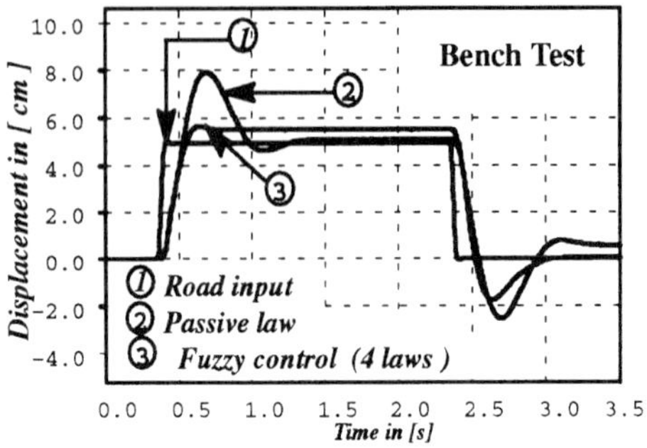

Figure 2.27 Body displacement for passive and fuzzy system (semi-active suspension).

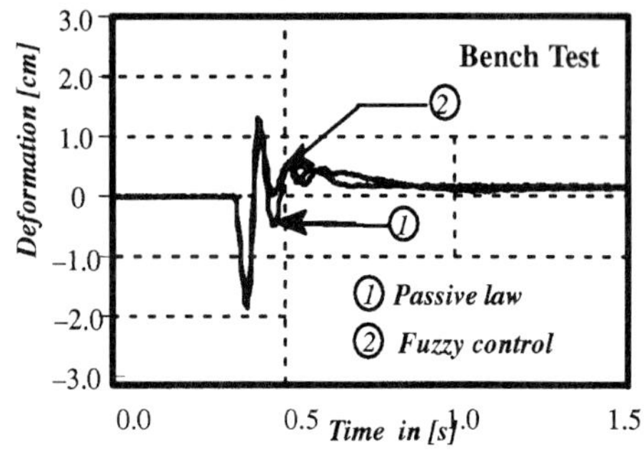

Figure 2.28 Tire deformation for passive and fuzzy system (semi-active system).

example, we will consider here a Sugeno-type fuzzy controller where the only parameters to tune are the singletons in the conclusion of the rules.

Each controller is defined by rules such:

$$\text{If } x_1(t) \text{ is } A_1^k \text{ AND } x_2(t) \text{ is } A_2^k \text{ THEN } F(t) \text{ is } \gamma^m$$

where A_j^k represents the membership function coding the linguistic term k associated with the variable j and γ^m represents the corresponding output. According to [11, 12], we assimilate such an output to a crisp value. For instance, the third linguistic term for the output, i.e., $m = 3$ (*negative small*, for instance) is given by $\gamma^3 = -2$.

In our approach, fuzzy controllers are realized by using isosceles triangles for membership functions, and Sugeno's method for inference and defuzzification; therefore, fuzzy output is a derivable function with respect to each input and each parameter.

The parameters shaping the triangle A_j^κ are the center α_j^κ, and the base β_j^κ. Membership values, ith rule strength, and fuzzy output (force) are computed, respectively, as follows:

$$A_j^k(x_j) = \max\left[1 - \frac{2|x_j - a_j^k|}{\beta_j^k}, 0\right] \tag{2-16}$$

$$\mu_i = \prod_i A_j^k(x_j) \tag{2-17}$$

The output of the controller is given by:

$$F = \frac{\sum_i \mu_i \gamma_i}{\sum_i \mu_i} \quad \text{if } \sum_i \mu_i \neq 0$$
$$F = 0 \qquad \text{if } \sum_i \mu_i = 0 \tag{2-18}$$

where x_j represents a generic variable and γ_i the output of the rule i.

2.3.3.1 Optimization of the Fuzzy Local Controllers

The parameters defining each controller are adapted using a gradient descent technique.

Those parameters are adapted in an independent way for each subsystem. Since the optimization process is the same for each controller, in order to simplify the presentation, we will restrict the following analysis to the comfort controller. We perform optimization with respect to discrete time (T) with period ΔT.

Let ω_c be a generic parameter (α_j^k, β_j^k or γ^k) of the comfort controller and η the iteration parameter. The optimization algorithm will be

$$(\omega_c)_{T+1} = (\omega_c)_T - \eta\left(\frac{\delta E_c}{\delta\omega_c}\right)_T \tag{2-19}$$

Once defined the comfort controller square error at time T is given by:

$$E_c(T) = \frac{1}{2}\left(x_1^2(T) + x_2^2(T)\right) \tag{2-20}$$

Its derivative with respect to the parameter ω_c is:

$$\frac{\delta E_c}{\delta\omega_c} = \left(\frac{\delta E_c}{\delta x_1}\frac{\delta x_1}{\delta F} + \frac{\delta E_c}{\delta x_2}\frac{\delta x_2}{\delta F}\right)\frac{\delta F}{\delta\omega_c} \tag{2-21}$$

What is unknown in the absence of an analytical model of the process is the Jacobian $\delta x_j / \delta F$. When using a gradient-based technique, a perfect knowledge of the Jacobian is not required. Since substituting $\delta x_j / \delta F$ by its sign ± 1 just alters the amplitude of the variation but not its direction, the gradient method can still give an adequate value of the control parameters [14].

In the last expression, derivatives of the error with respect to any input are easy to calculate and Jacobian values may be approximated by their sign. Since parameters ω_c do not depend on F_s, we may write:

$$\frac{\delta F}{\delta\omega_c} = \alpha\frac{\delta F_c}{\delta\omega_c} \tag{2-22}$$

When deriving Euation (2-18), we finally found the expression of derivatives for centers and bases of membership functions and crisp outputs given respectively by:

$$\frac{\delta F_c}{\delta \alpha_j^k} = \frac{1}{\sum_i \mu_i} \left(\sum_{i \in I_j^k} \mu_i \gamma_i - F_c \sum_{i \in I_j^k} \mu_i \right) \frac{2 \operatorname{sgn}\left(x_j - \alpha_j^k\right)}{A_j^k\left(x_j\right) \beta_j^k} \qquad (2\text{-}23)$$

$$\frac{\delta F_c}{\delta \beta_j^k} = \frac{1}{\sum_i \mu_i} \left(\sum_{i \in I_j^k} \mu_i \gamma_i - F_c \sum_{i \in I_j^k} \mu_i \right) \frac{1 - A_j^k\left(x_j\right)}{A_j^k\left(x_j\right) \beta_j^k} \qquad (2\text{-}24)$$

$$\frac{\delta F_c}{\delta \gamma^k} = \frac{1}{\sum_i \mu_i} \left(\sum_{i \in I_j^k} \mu_i \right) \qquad (2\text{-}25)$$

where I^k represents the set of rules in which γ^k appears and I_j^k the set of rules in which α_j^k and β_j^k appear.

These sets of derivatives, very similar to the one presented in [13], have been discussed in [11] and [12]. Their originality resides in the fact that just the linguistic terms are optimized and not all the parameters appearing in the rules (this justifies the presence of I^k and I_j^k).

2.3.3.2 Simulation Results and Comments

Each input variable is associated with a linguistic term taken among seven possibilities, so each fuzzy controller is defined by up to 49 rules. In our simulations, we use all the rules and the whole controller is composed of a 98-rule set. Crisp output is also defined by seven crisp values. The gradient descent is applied to each parameter every 0.002 seconds.

Simulation results show the ability of our controller to improve the suspension response when applying a single step in input (the road profile).

Figures 2.29 through 2.32 show the system response (the body position *B* and the wheel position *W*) when applying a 5cm step during one second. We may observe the improved behavior of the optimized systems.

It is important to remark that when using a passive suspension, the oscillations of state variables are 30% wider and the period 5 times longer.

Figures 2.33 and 2.34 show the behavior of some parameters defining the fuzzy controller (the center position corresponding to the fuzzy set *Zero* and the value of the crisp output *Positive Small*) during the adaptation procedure.

We performed several simulations with different values of α, changing the mass and adapting only one controller at a time. Results have shown a good robustness when optimizing at least the comfort controller.

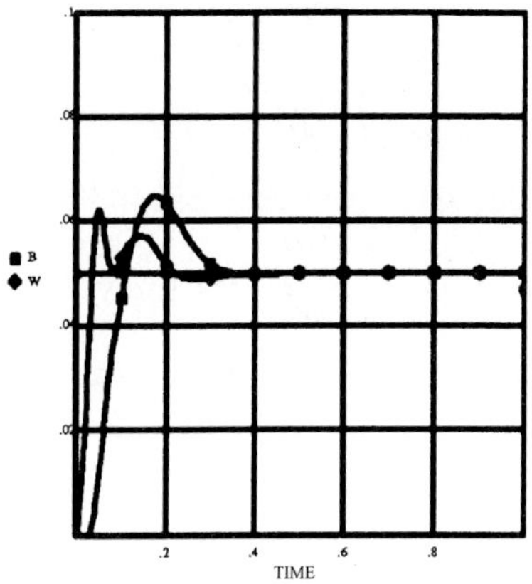

Figure 2.29 No optimization.

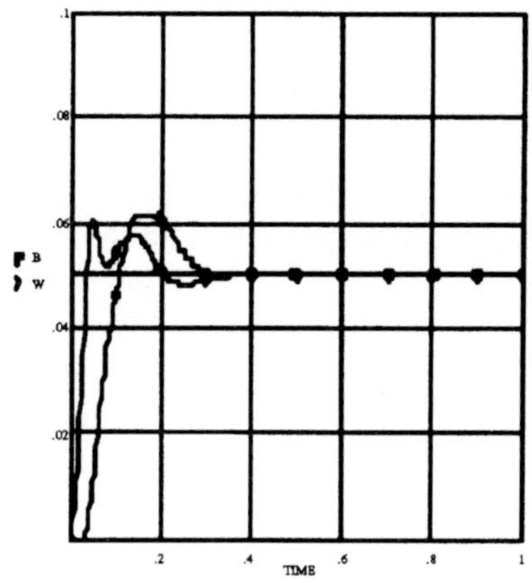

Figure 2.30 Optimization of both controllers.

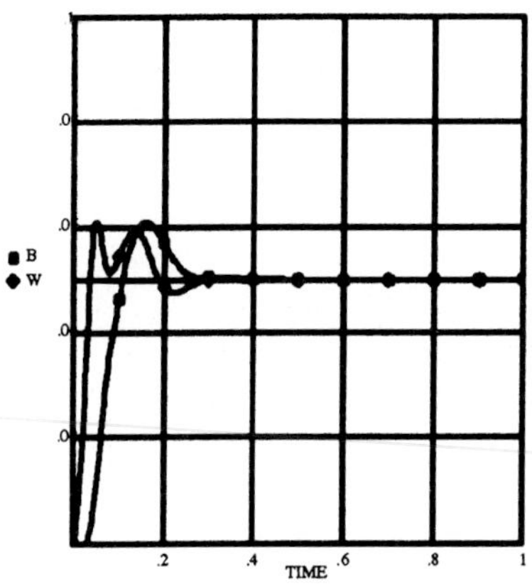

Figure 2.31 Only comfort optimization.

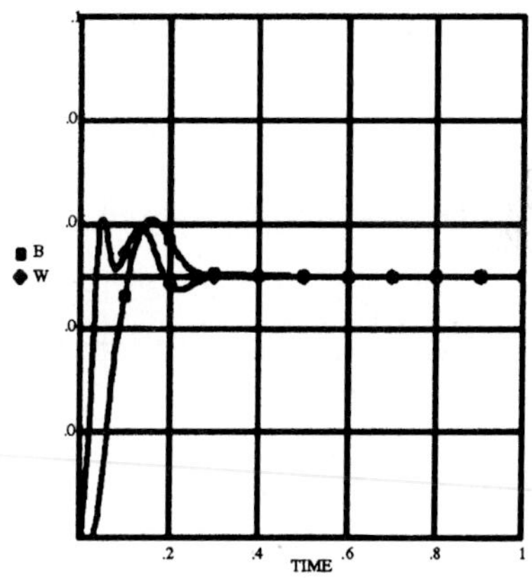

Figure 2.32 Only safety optimization.

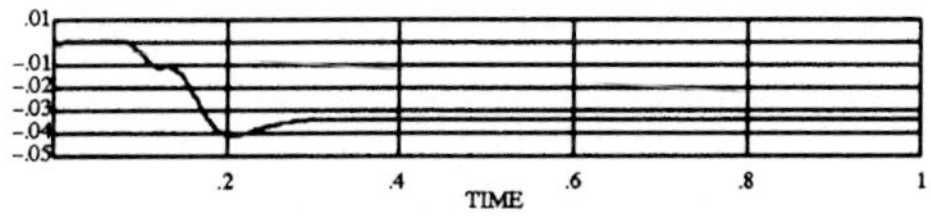

Figure 2.33 Optimization of the center for the fuzzy set ZERO.

Figure 2.34 Optimization of one crisp output.

2.4 NEURO-CONTROLLER

2.4.1 Design

Finally in this section, a neural network is introduced to give adaptive properties to the system in order to catch slow parameter variations in the system (weight, aging of the damper, etc.). In this case, the neural controller is initialized by the fuzzy controller defined previously (Figure 2.35)

After a learning phase using the classical back propagation algorithm, with the data obtained by simulation and the fuzzy controller, the following structure of the neuro-controller has been defined (Figure 2.36).

This previous learning phase is developed off-line. Now, the neural controller is inserted in a closed loop and a learning procedure is maintained active in order to get adaptiveness and robustness to the global control structure.

In order to tackle the safety and comfort criteria, we define two control errors. The first one is linked to the wheel behavior Δu_w and the second one to the body behavior Δu_b. As for the supervised fuzzy controller, we weight the control error in order to determine the error Δu to be used for the back propagation procedure. The structure of the neural adaptive controller is given Figure 2.37.

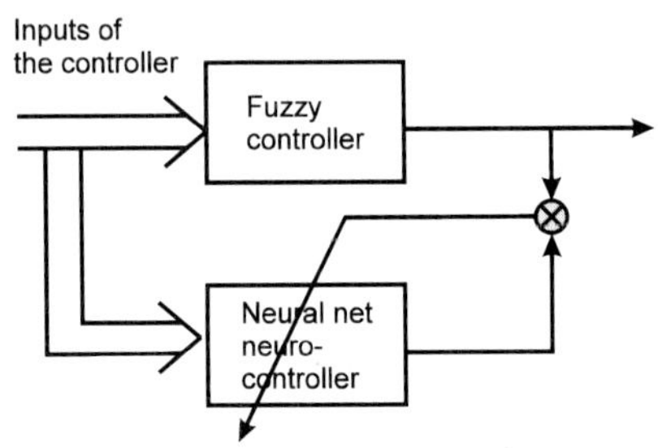

Figure 2.35 Learning the fuzzy controller.

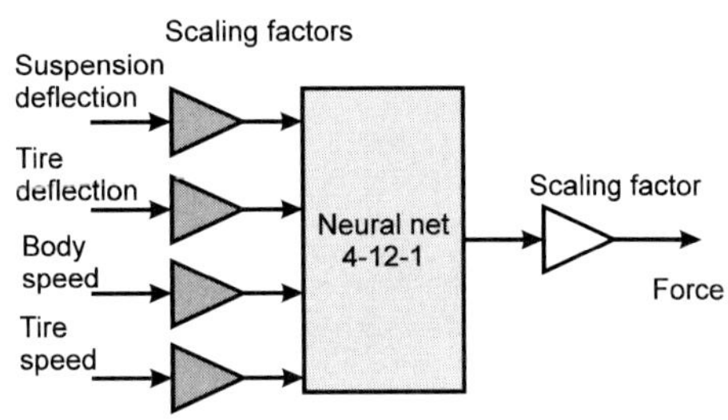

Figure 2.36 Structure of the neuro-controller (4 neurones on the input layer, 12 neurones on the hidden layer, 1 on the output layer).

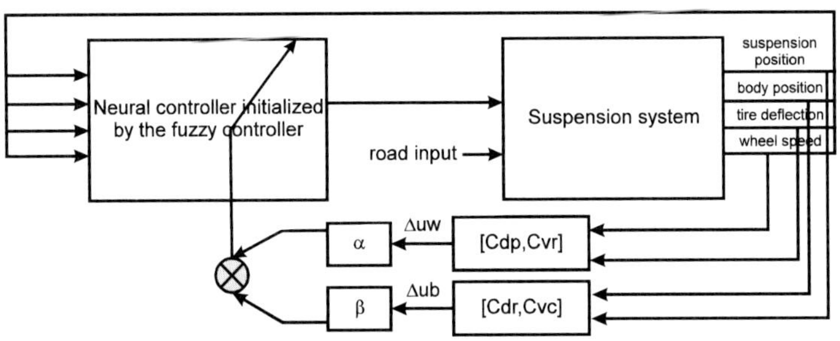

Figure 2.37 Neural adaptive controller.

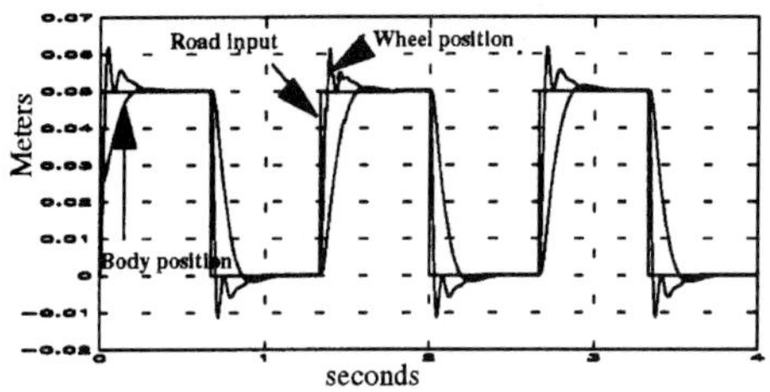

Figure 2.38 Response of the fuzzy controller for a 5cm step input (simulation).

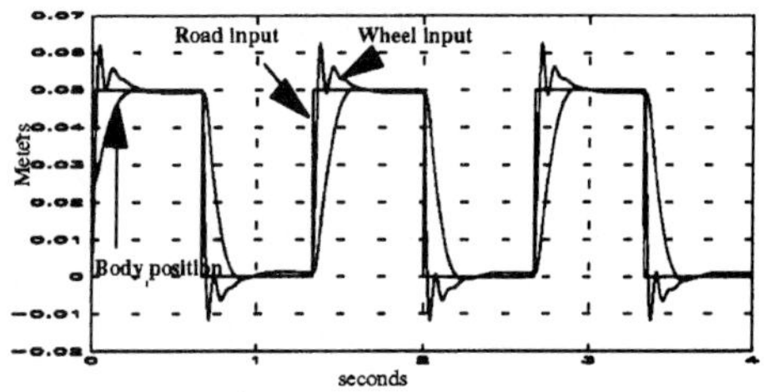

Figure 2.39 Response of an adaptive neural-controller for a 5cm step input (simulation).

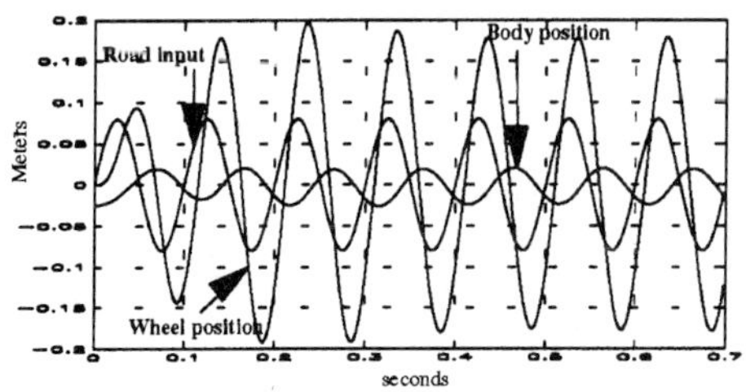

Figure 2.40 Response of a fuzzy controller for a sinus excitation (simulation).

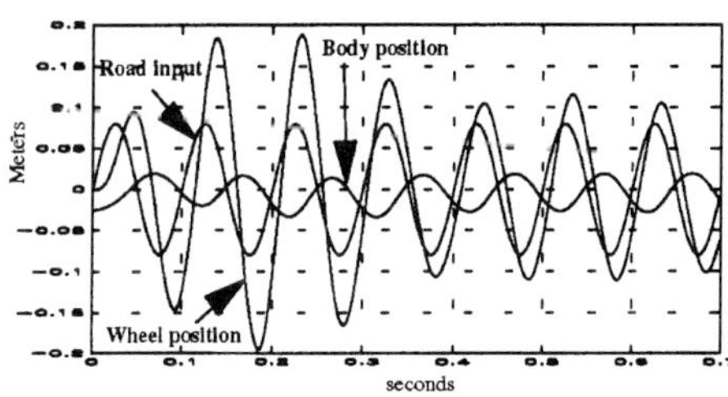

Figure 2.41 Response of an adaptive neural-controller for a sinus excitation (simulation).

2.4.2 Experiments

In a first step, we have checked that the fuzzy controller has been copied well and that the learning phase has been efficient. We can see Figures 2.38 and 2.39 the response to a 5cm step input, both for the fuzzy controller and the neural controller. The results are really in good agreement.

After that, in order to check the adaptiveness, and then the robustness of the neuro-controller, we have applied to the wheel a sinusoidal signal of 10 Hz with a simultaneous change of 100% both on the mass of the body and on the damping ratio of the shock absorber, in a semi-active suspension.

The results given Figures 2.40 and 2.41 show that, thanks to the on-line learning phase of the neuro-controller, this last one is very robust and adapts by itself to the changes mentioned above.

2.5 COMPARISON OF THE DIFFERENT APPROACHES

A systematic analysis of the results obtained on a test bench with the different approaches, has been done both in frequency and the time domain. It is easier to report here on the time domain analysis with the following tables (Tables 2.4 and 2.5) corresponding to the response to a 5cm step.

Finally, we can summarize all these results in this last table which gives some characteristics of the different techniques.

Table 2.4 Comparison of the different approaches.

		Passive	Optimal	VSS	Fuzzy	Neuro
Body	1st overshoot	40%	2.7%	No overshoot	14%	as good as fuzzy plus robustness
	Settling time	705 ms	235 ms	210 ms	150 ms	
	Static error	−11.4%	2.5%	−1%	11%	
Wheel	1st overshoot	33%	28%	27%	25%	
	Settling Time	405 ms	205 ms	250 ms	290 ms	
	Static error	2%	3%	3%	4%	

2.6 CONCLUSIONS AND PERSPECTIVE

This work has been oriented toward the design of semi-active suspension using different tools at our disposal from the automatic control field.

To take into account some constraints on the actuator and some specificity of the problem, we have considered three control techniques: optimization under constraint, variable structure control, and fuzzy control.

These three approaches have been tested by simulation on realistic nonlinear models of the process and on a real test bench. Results are really satisfactory. We can summarize the characteristics and performances of the controllers as follows:

1. Better control of body trajectory (for ride comfort) and wheel dynamic loads (for road handling) by optimal, VSS, and fuzzy semi-active and active suspensions.

2. LQR gives practically same performance but caution is required in respect to robustness of the optimal controller.

3. VSS controller is efficient for actuators that have two states (hard and soft); fuzzy controller is convenient for all actuators.

Table 2.5 Global comparison of the four approaches

	Optimal	VSS	Fuzzy	Neural
Performances	high	high	high	high
Robustness	low	medium	high	very high
Portability (flexibility)	low	medium	high	high
Critical aspects for the design	Def. of the performance criterium	Definition of the sliding surface	Writing of the rule base	Learning phase
Best suited type of suspension	Continuous active suspension	Semi-active suspension with discrete laws	Semi- or fully-active suspension	Semi- or fully-active suspension

4. VSS and fuzzy controllers are better in terms of robustness and for taking into account the actuator constraints.

5. Bench test results confirm simulation results for the controlled system.

6. Better performances would be obtained with high-performance actuator.

Concerning fuzzy control, without using any mathematical model, but taking advantage of the physical characteristics of the system to be controlled, we have designed a fuzzy controller composed of two subcontrollers (one oriented toward a security goal, the second toward a comfort goal), supervised by a simple fuzzy system using global information on the car.

In special case the fuzzy controller has been easily tuned using optimization techniques. Robustness and adaptability of the fuzzy controller have been increased by introducing an on-line learning phase on a neural network representing the basic fuzzy controller.

These methodological results have been validated on simulations and also on a suspension test bench. The corresponding techniques seem valid and efficient solutions for real problems.

2.7 ACKNOWLEDGMENTS

The authors want to express their thanks to all of their students who have worked on this project: S. Roukieh, E. Dayre, Y. Lespine, and V. Gorrini.

2.8 REFERENCES

1. Zadeh, L. A. "Fuzzy Logic", *IEEE Computer*. May 1992, pp. 83-93.

2. Holmblad, L. P. and Østergaard, J. J. "Control of a Cement Kiln by Fuzzy Logic". In *Fuzzy Information and Decision Processes*, North Holland, 1982, pp. 389-399.

3. Buhler, H. "Réglage par Mode de Glissement." *Presses Polytechniques Romandes*. Lausane, 1986.

4. Mira, C. "Systèmes asservis non linéaires." *Traité des Nouvelles Technologies-Série Automatique*. Hermès, 1990.

5. King, J. and E. H. Mamdani. "The Application of Fuzzy Control Systems to Industrial Process," *Automatica*. Pergamon Press, vol. 13, 197, pp. 235-242.

6. Roukieh, S. and Titli A. *On the Model-Based Design of Semi-Active and Active Suspension for Private Cars*. Technical paper, Vol. 2, pp. 305-318. London, FISITA, 1992.

7. Tseng, T. and Hrovat, D. "Some Characteristics Of Optimal Vehicle Suspensions Based on Quarter-Car Models". *29th Conference on Decision and Control (IEEE)*, Honolulu, HI. December 1990.

8. Yamashita, M., Fujimori, K., Uhlik, C., Kawatani, R. and Kimira, H. "H^{∞} Control Of Automotive Active Suspension". *29th Conference on Decision and Control (IEEE)*, Honolulu, HI. December 1990.

9. Yue, J., Hedrick, K. and Butsuen, T. "Alternative control laws for Automotive Active Suspension". *Proceedings, American Control Conference*, Atlanta, GA, 1988.

10. Falb, Athans, Optimal Control, McGraw-Hill, 1966.

11. Bersini, H. and Gorrini, V. "FUNNY (FUzzY or Neural Net) Methods for Adaptive Process Control". *Proceedings of the First European Congress on Fuzzy and Intelligent Technologies*. pp. 55-61, 1993.

12. Bersini, H., Nordvik, J. P. and Bonarini, A. "A Simple Direct Adaptive Fuzzy Controller Derived from its Neural Equivalent." *Proceedings of the Second IEEE International Conference on Fuzzy Systems*, pp. 345-350, 1993.

13. Nomura, H., Hayashi, I. and Wakami, N. "A Learning Method of Fuzzy Inference Rules by Descent Method." *Proceedings of the First IEEE International Conference on Fuzzy Systems*, pp. 203-210, 1992.

14. Renders, J.-M., Bersini, H. and Saerens, M. "Adaptive Neurocontrol: How Simple and Black-box Can It Be?" *Proceedings of the Tenth International Workshop on Machine Learning*, Amherst, pp. 260-267,1993.

15. Roukieh, S. "Contribution des Commandes Optimale, à Structure Variable et par Logique Loue à la Synthèse de Suspensions Automobile". Ph.D. Thesis, LAAS INSA Toulouse, France, 1993.

2.9 TECHNICAL REFERENCES

16. "MATLAB User's Manual for PC and workstation Computer." The Mathworks, Inc, 1990.

17. "EASY/W Users Guide Manual for Workstation Computer." The Boeing Computer Services Programming Product, 1991.

3

FUZZY MODELING AND MODEL-BASED CONTROL FOR NONLINEAR SYSTEMS

R. Babuška and H. B. Verbruggen
Delft University of Technology
THE NETHERLANDS

Based on a fuzzy model of a process, a nonlinear control system can be designed. Unlike conventional fuzzy control, this approach does not require (but can incorporate) heuristic knowledge and eliminates the usual trial-and-error tuning of the controller. The fuzzy model of the process is obtained directly from process measurements using an identification method based on fuzzy clustering. The method estimates simultaneously antecedent membership functions and consequent parameters of a Sugeno–Takagi fuzzy model with linear rule consequents that represent local linear models of the process. Based on the identified Sugeno–Takagi model, a Mamdani model with singleton consequents can be derived. A simple feedforward control law is derived by inverting this fuzzy model. For processes with constraints, a predictive control approach based on the fuzzy model improves the performance significantly. An application to a nonlinear pressure control system is described and simulation and real-time control results are presented.

3.1 INTRODUCTION

The increasing technological demands and performance requirements call for complex production and manufacturing systems that in turn require sophisticated control systems. Contemporary production is characterized by frequent changes in product throughput and product mix, causing large variations in operating points and operating conditions. To satisfy the product quality requirements in a flexible production environment, advanced control techniques that can deal with uncertainty and nonlinearities need to be introduced.

Conventional modeling and control approaches based on differential and difference equations are often insufficient, mainly due to the lack of precise, formal knowledge about the process to be controlled. While conventional control theory typically considers nominal operating conditions and small variations from these conditions, industrial practitioners widely agree that a design that delivers excellent performance under nominal conditions may produce poor results when the operating conditions vary. Start-up and shut-down of a continuous production facility are examples that require a careful coordination of many variables with patterns different from what are present under normal operating conditions.

Although most of the conventional and modern control theory is based on linear systems, there is a general agreement that design and analysis of controllers for complex industrial processes cannot be based solely on linear methods [1]. Nonlinear behavior of most industrial processes can be approximated by linear models only locally and additional information must be supplied in order to design reliable and well-performing controllers. An important source of such information is the knowledge of plant operators, control engineers and process designers. This knowledge is usually present both in qualitative and quantitative forms, and knowledge-based methods have been developed for its use in control.

Among various knowledge-based control techniques, fuzzy control has gained considerable interest both in the research community and also in industry. Since the first industrial application to a cement-kiln control in the beginning of 1980's [2], fuzzy control has been applied to various systems in the chemical process industry [3,4], consumer electronics [5,6], automatic train operation [7], and many other fields. The theory of fuzzy control has been extensively studied too [8,9,10].

The main idea of fuzzy control is to use the operator's knowledge verbalized as a collection of if–then control rules, that are directly encoded into a control algorithm. An advantage of this approach is that no explicit model of the system under consideration is required, which can significantly reduce the development time if sufficient knowledge is available or is easy to obtain. In practice, however, much effort must be spent on knowledge acquisition, since a complete rule base is required for proper functioning of the fuzzy controller. If no sufficient knowledge is available, the design must rely on the tuning phase, which may be a tedious and time-consuming trial-and-error procedure. Moreover, in an industrial environment, an on-line trial-and-error controller tuning is often not acceptable for safety, economical, and environmental reasons. Extensive simulation tests must be carried out before the fuzzy controller can be applied, and for realistic simulations a good model of the plant is needed. The problem is that a fuzzy control approach might have been chosen just because a plant model was not available.

Another shortcoming of all design methods that are not based on a process model is that the dynamic properties of the control system such as stability or robustness cannot be analyzed in a rigorous way. This prevents these techniques from entering applications where such analysis is required. Finally, it has been observed that human control skills may be difficult to verbalize since the operator's control strategy can be based on various control principles simultaneously, combining feedforward, feedback, and predictive strategies in a complex, time-varying fashion, and that an operator may not be able to explain why a particular control action is chosen. Experience from knowledge acquisition also shows that the rules provided by different operators are often contradicting [11].

Another approach to fuzzy control design is to incorporate human knowledge into the development of a *fuzzy model* of the plant (in combination with data, first principles, and other pieces of information). Then a nonlinear controller can be designed based on this fuzzy model. An advantage of this approach is that the heuristic design and tuning of a fuzzy controller are replaced by a more structured process of modeling and identification, in which all available knowledge and data are combined and the resulting model can be easily validated. The controller is then designed using established techniques and the dynamic properties of the closed loop, such as stability, can be analyzed [12]. Based on a model, a nonlinear control system may be developed that integrates low-level control, supervision, planning and diagnostic tasks.

This chapter describes in detail the procedure of building a fuzzy model from process data and the subsequent controller design based on this model. The material presented is a revision and expansion of our previous papers [13,14,15] and is organized as follows: In Section 3.2, after reviewing the basic structure of fuzzy models considered in this text, we proceed with a detailed description of the identification method based on fuzzy clustering. Section 3.3 then describes two controller design techniques: the first is an inversion of a fuzzy model and the second is based on the concept of predictive control. In Section 3.4, we apply the methodology to modeling and control of the pressure in a laboratory fermentation tank. The chapter concludes with a summary of advantages and drawbacks of the presented approach.

3.2 FUZZY MODELING

The area of fuzzy modeling encompasses several kinds of models that use various concepts from fuzzy sets theory to construct abstract models of systems [9]. Typically, fuzzy sets provide partitioning of the domains of the problem (input, output and state variables) into overlapping regions. A fuzzy model describes the system by establishing relations between the input and output fuzzy regions in the form of if–then rules or by using a simple model for each region. The overlap of the regions ensures a smooth transition among the rules or a weighted combination of the outputs of several local models.

3.2.1 The Structure of Fuzzy Models

Various classes of fuzzy models exist, depending on the form of the rules and the way the rules are organized in the rule base. The identification and control design methods described in this chapter employ two types of fuzzy models, namely the Mamdani model and the Sugeno–Takagi model. In both cases, the fuzzy rule base corresponds to a static nonlinear map f that can be used to model a dynamic system, for instance in an input-output NARX (Nonlinear AutoRegressive with eXogenous input) form:

$$y(k+1) = f\left(y(k), y(k-1), \ldots, y(k-n+1), u(k), u(k-1), \ldots, u(k-m+1)\right) \tag{3-1}$$

Here $y(k), \ldots, y(k-n+1)$ and $u(k), \ldots, u(k-m+1)$ denote the lagged model outputs and inputs respectively and n, m are integers related to the model order. The lagged inputs and outputs are called the regressors. The NARX structure applies directly also to Multi-Input, Single-Output (MISO) systems and to systems with transport delays. To simplify the notation, in this chapter the rules will be written for Single-Input, Single-Output (SISO) models without delays. In the application section, we present also a result for a MISO process.

A *Mamdani* fuzzy model consists of rules of the following form:

$$\begin{aligned} R_i: &\textbf{If } y(k) \text{ is } A_{i,1} \textbf{ and } y(k-1) \text{ is } A_{i,2} \textbf{ and,} \ldots\ y(k-n+1) \text{ is } A_{i,n} \textbf{ and } u(k) \text{ is } B_{i,1} \\ &\textbf{and } u(k-1) \text{ is } B_{i,2} \textbf{ and,} \ldots,\ u(k-m+1) \text{ is } B_{i,m} \textbf{ then } y(k+1) \text{ is } C_i \end{aligned} \tag{3-2}$$

where R_i denotes the ith rule, $A_{i,1}, \ldots, A_{i,n}, B_{i,1}, \ldots, B_{i,m}$, and C_i and are fuzzy sets that usually have some linguistic meaning such as *Zero*, *Small*, *Big*, etc. The overall output of the rule base is computed using the well-known Mamdani inference algorithm, see e.g., [8]. In a variant of the Mamdani model, the consequent fuzzy sets C_i are reduced to singletons and can be represented as real numbers c_i. In the sequel, this model will be called the *singleton model*.

A *Sugeno–Takagi* fuzzy model [16] combines logical rule antecedents as in (3-2) with numerical consequents that are known functions of the input variables:

$$\begin{aligned} R_i: &\textbf{If } y(k) \text{ is } A_{i,1} \textbf{ and } y(k-1) \text{ is } A_{i,2} \textbf{ and,} \ldots\ y(k-n+1) \text{ is } A_{i,n} \textbf{ and} \\ &u(k) \text{ is } B_{i,1} \textbf{ and } u(k-1) \text{ is } B_{i,2} \textbf{ and,} \ldots,\ u(k-m+1) \text{ is } B_{i,m} \\ &\textbf{then } y(k+1) = g_i\left(y(k), y(k-1), \ldots, u(k), u(k-1), \ldots\right) \end{aligned} \tag{3-3}$$

The overall model output is computed as a weighted mean of the consequent functions:

$$y(k+1) = \frac{\sum_{i=1}^{K} \beta_i \cdot g_i(\cdot)}{\sum_{i=1}^{K} \beta_i} \tag{3-4}$$

where β_i is the degree of fulfillment of the ith rule antecedent and K is the total number of rules; see [16] for details. The same defuzzification formula can be used for the singleton model (3-2), substituting the singleton domain values for the values of C_i. The Sugeno–Takagi model often employs linear consequent functions, e.g.,

$$\begin{aligned} R_i\text{: } &\textbf{If } y(k) \text{ is } A_{i,1} \textbf{ and } y(k-1) \text{ is } A_{i,2} \textbf{ and,} \ldots\ y(k-n+1) \text{ is } A_{i,n} \textbf{ and} \\ &u(k) \text{ is } B_{i,1} \textbf{ and } u(k-1) \text{ is } B_{i,2} \textbf{ and,} \ldots,\ u(k-m+1) \text{ is } B_{i,m} \\ &\textbf{then } y(k+1) = \sum_{j=1}^{n} a_{i,j}\, y(k-i+1) + \sum_{j=1}^{m} b_{i,j}\, u(k-i+1) + c_i \end{aligned} \tag{3-5}$$

where $a_{i,j}$, $b_{i,j}$, c_i are parameters for the ith rule which can be easily estimated from measurements.

Since both the Mamdani and Sugeno–Takagi fuzzy models can approximate any smooth function to any degree of accuracy (see [17] for a proof), (3-2) and (3-3) can approximate any observable and controllable modes of a large class of discrete-time nonlinear systems [18]. Yet, the way they approximate functions is different as schematically shown in Figure 3.1. Besides the general approximation capabilities, another important property of fuzzy models is their ability to integrate different kinds of information, namely knowledge expressed as if–then rules, local linear(ized) models in terms of linear functions (3-5) and numerical data.

In this chapter, methods are presented for constructing a Sugeno–Takagi model from process measurements and for converting the identified Sugeno–Takagi model into a singleton model.

3.2.2 Constructing Sugeno–Takagi Models from Data

This section describes a method for estimating a Sugeno–Takagi model from the process data, using fuzzy clustering in the product space of the model input and output

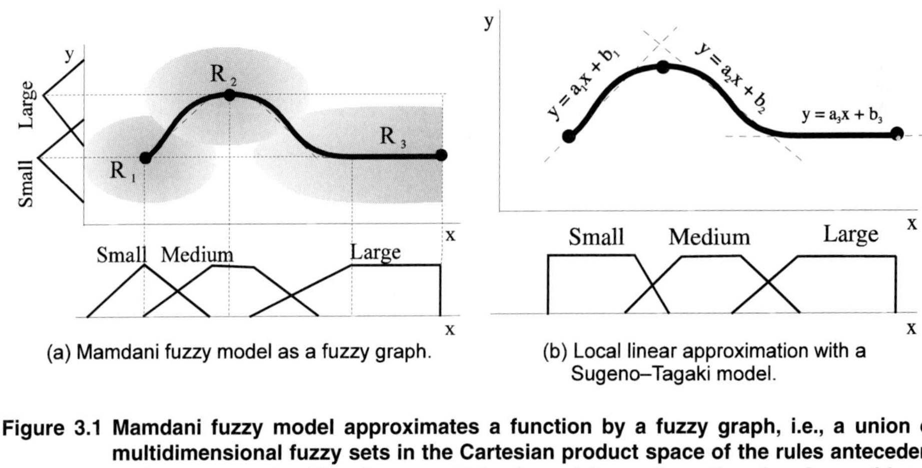

(a) Mamdani fuzzy model as a fuzzy graph.

(b) Local linear approximation with a Sugeno–Tagaki model.

Figure 3.1 Mamdani fuzzy model approximates a function by a fuzzy graph, i.e., a union of multidimensional fuzzy sets in the Cartesian product space of the rules antecedent and consequents. The Sugeno—Takagi model, on the other hand, provides a smooth combination of local linear models.

variables. The identification algorithm requires the structure of the model (inputs, outputs, order of the system) and an initial number of clusters to be fixed beforehand. Then a clustering algorithm combined with a cluster merging procedure finds a partition of the product space of model variables fuzzy regions where the system behavior can be approximated locally by linear functions. Finally, the antecedent membership functions are extracted by projecting the clusters and the consequent parameters are estimated using least squares. These steps will now be described in more detail.

Identification by fuzzy clustering: the main idea. This approach is based on a geometric interpretation of the NARX model (3-1). In the product space of the regressors and the model output

$$R^{(n+m+1)} = \underbrace{Y\times\cdots\times Y}_{n \text{ times}}\times \underbrace{U\times\cdots\times U}_{m \text{ times}}\times Y \tag{3-6}$$

the nonlinear mapping f defines a subspace $R^{(n+m)}$, generally a nonlinear hypersurface. Identification of the system using the Sugeno–Takagi model can be regarded as an approximation of this hypersurface by a collection of locally valid hyperplanes. The rule antecedents determine the validity regions of these hyperplanes and the defuzzification formula (3-4) provides a smooth interpolation among the individual rules.

The antecedent membership functions and the consequent parameters must be estimated such that a good fit to the identification data is obtained. Various methods proposed in the literature treat the optimization of the membership functions and of the consequent parameters separately, see e.g., [16–19]. In our view, both tasks are highly interdependent and should be solved simultaneously.

The optimization problem can be conveniently defined as fuzzy clustering with hyperplanar prototypes. The data in the product space of the regressors and the model output partitioned into clusters, such that the sum of distances of the data points from each cluster prototype weighted by the membership degrees is minimized. The available clustering algorithms fall in two main groups:

1. Cluster prototypes are a priori defined as hyperplanes (fuzzy linear varieties) or hyperellipsoids (fuzzy c-elliptotypes), see [20]. The distance function measures the distance of data points from the prototypes. Cluster prototypes are points and the distance measure adapts during the clustering process to match the shape of the clusters, using for instance the approach of Gustafson and Kessel [21].

2. Our experience shows that despite the higher computational demands, for function approximation the Gustafson–Kessel (GK) algorithm exhibits better convergence properties than the fuzzy linear varieties and the fuzzy c-elliptotypes. In order to make the description self-contained, we briefly review the GK algorithm below.

The Gustafson–Kessel algorithm. Assume that a set of input–output process measurements is available: $u(j)$, $y(j)\}$, $j = 1, 2,\ldots, N$. Considering the nonlinear regression model (3-1) and fixing the structure of the model (orders n, m) we form a matrix of data for fuzzy clustering consisting of the regressors and the model output, i.e., $X = \{x_j \mid j = 1, 2,\ldots, N\}$,

$$x_j = \left[y(j), y(j-1),\ldots, y(j-n+1), u(j), u(j-1),\ldots, u(j-m+1), y(j+1)\right]^T$$

The total dimension of the data space will be denoted d, where $d = n + m + 1$. Further, denote a $V = \{v_j \mid j = 1, 2, \ldots, K\}$ a K-tuple of cluster prototypes (centers) in R^d, $v_i = [v_{i,1}, v_{i,2}, \ldots, v_{i,d}]^T$. Matrix $U = \{\mu_{i,j} \mid i = 1, 2,\ldots, K; j = 1, 2,\ldots, N\}$, $\mu_{i,j} \in [0,1]$ is called the fuzzy partition matrix, where $\mu_{i,j}$ represents a membership degree of data element x_j in cluster v_i. The aim of fuzzy clustering is to minimize an objective function defined as:

$$J(X,V,U) = \sum_{i=1}^{K}\sum_{j=1}^{N} \mu_{i,j}^p D^2(x_j, v_i) \tag{3-7}$$

Here $p > 1$ is a real parameter that controls cluster fuzziness, usually $p = 2$ and $D(x_j,v_i)$ is the distance of the data point x_j to the cluster prototype v_i induced by a positive definite matrix M_i:

$$D^2(x_j, v_i) = (x_j - v_i)^T M_i (x_j - v_i) \tag{3-8}$$

M_i can be derived as a normalized inverse of the cluster covariance matrix $F_i : M_i = \det(F_i)^{1/d} F_i^{-1}$. Cluster covariance matrix F_i is computed as:

$$F_i = \frac{\sum_{j=1}^{P} \mu_{i,j}^p (x_j - v_i)(x_j - v_i)^T}{\sum_{j=1}^{N} \mu_{i,j}^p}. \tag{3-9}$$

Algorithm. Given a data set X, number of clusters K, and initial (e.g. random) partition U, the Gustafson–Kessel algorithm minimizes (3-7) by performing the following steps iteratively:

1. Updates cluster prototypes (centers): $v_i = \dfrac{\sum_{j=1}^{N} \mu_{i,j}^p x_j}{\sum_{j=1}^{N} \mu_{i,j}^p}$.

2. Updates cluster covariance matrices using (3-9).

3. Calculates the distances: $D^2(x_j, v_i) = \det(F_i)^{1/d} (x_j - v_i)^T F_i^{-1} (x_j - v_i)$.

4. Updates the fuzzy partition matrix U: $\mu_{i,j} = \frac{D^2\left(x_j, v_i\right)^{-1/(m-1)}}{\sum_{l=1}^{K} D^2\left(x_j, v_i\right)^{-1/(m-1)}}$ if $D^2\left(x_j, v_i\right) = 0$

   for some $i = q$, set $\mu_{q,j} = 1$ and $\mu_{i,j} = 0$, $\forall i \neq q$.

until a specified convergence criterion is satisfied, e.g., $| U_l - U_{l-1} | < \varepsilon$ where $| \cdot |$ is a suitable matrix norm, l is the iteration step and ε is a termination tolerance, typically $\varepsilon = 0.01$. As proven in [21], this algorithm converges to a strict local minimum of (3-7).

Deriving antecedent membership functions. Each cluster found by the GK algorithm corresponds to one rule of the Sugeno–Takagi model. The membership degrees of the fuzzy partition matrix U correspond to the degrees of fulfillment of each rule's antecedent for the data contained in the identification data set. To obtain the membership functions of the antecedents in the conjunctive form (3-3), each cluster (a multidimensional fuzzy set) is projected onto the antecedent variables (regressors $y(k), \ldots, y(k-n+1), u(k), \ldots, u(k-m+1)$). For illustration, see the projection of a two-dimensional fuzzy set C onto two one-dimensional fuzzy sets A and B schematically depicted in Figure 3.2(a). When computing the degree of fulfillment of the rule antecedent, the original multidimensional cluster is approximately obtained as an intersection of the cylindrical extensions of the projected one-dimensional fuzzy sets, as shown in Figure 3.2(b). Some information is lost due to a *decomposition error*. On the other hand, this method is computationally efficient (compared to a nonlinear multidimensional interpolation) and the rule base can be further simplified by neglecting those antecedent variables whose membership functions are close to a universal set and therefore have a negligible influence on the degree of fulfillment of the rule.

Notice that when identifying dynamic systems, the identification data are scattered over the regressor–output product space. The projection of the clusters results in a fuzzy

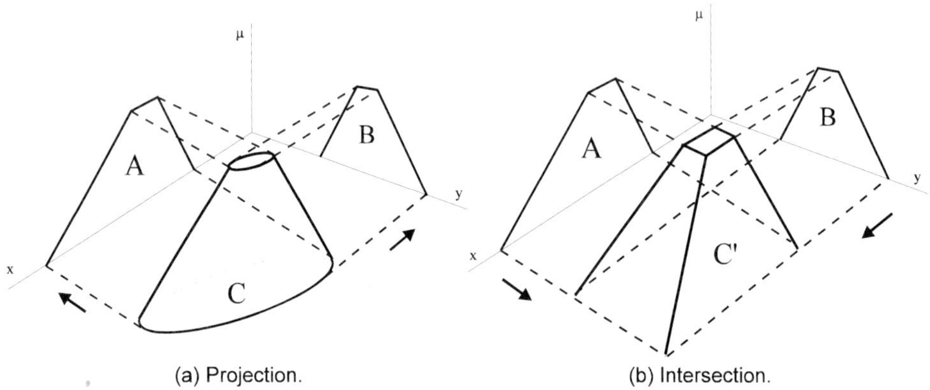

(a) Projection. (b) Intersection.

Figure 3.2 A decomposition error arises by projecting a multidimensional fuzzy set onto the antecedent variables and composing the original fuzzy set by applying intersection in the Cartesian product space.

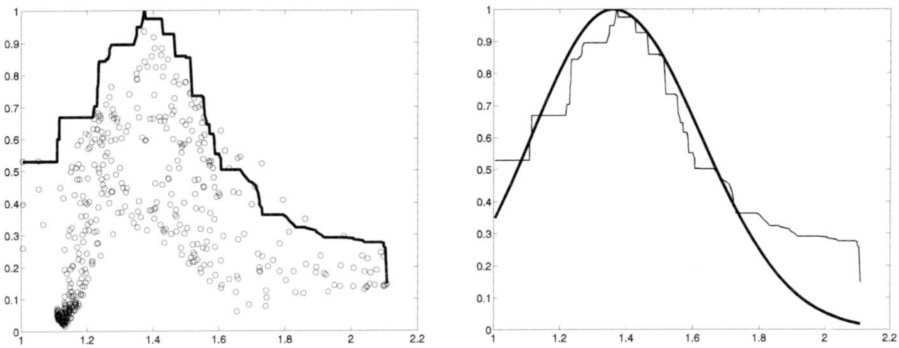

(a) Fuzzy cluster projection and its approximate envelope.

(b) Fitting the envelope with the exponential membership function.

Figure 3.3 Obtaining a parametric membership function from the cluster projection.

set defined pointwise for the given data, where points with high and low membership degrees are interlaced, as shown in Figure 3.3(a). Application of the maximum operator on discrete domains corresponds to extracting an envelope of the membership function, e.g., by low-pass filtering. In order to simplify the inference calculations, it is convenient to approximate the membership function by a suitable parametric function. Typical forms include *trapezoidal*, *exponential*, or *sigmoidal* membership functions. For instance an exponential membership function may be defined as

$$\mu(x, \sigma_1, \alpha_1, \alpha_2, \sigma_2) = \begin{cases} e^{-(\frac{x-\alpha_1}{2\sigma_1})^2} & \text{if } x < \alpha_1 \\ e^{-(\frac{x-\alpha_2}{2\sigma_2})^2} & \text{if } x > \alpha_2 \\ 1 & \text{otherwise} \end{cases}$$

The four-tuple of parameters σ_1, α_1, α_2, σ_2 is numerically optimized to fit the envelope, see Figure 3.3b. We found the exponential shape to be the most suitable approximation of the original cluster shape.

Estimating consequent parameters. There are two ways of estimating the consequent parameters: 1) direct extraction from the cluster prototypical points and the covariance matrices, 2) least square estimate using the identification data set. The first method provides an implicit equation for each hyperplane, corresponding to the geometrical interpretation of the model, i.e. there is no explicit distinction between the model inputs and outputs. The second method provides a set of optimal parameters with respect to the model output and also can reduce the decomposition error in the antecedent membership functions. If the model serves as a numerical predictor, the least squares approach is usually preferred. In the following, we briefly describe both approaches.

Extracting consequents from cluster prototypes. The distance measure (3-8) defines a cluster as a hyperellipsoid with its shape roughly described by the eigenstructure of the cluster covariance matrix F_i. The eigenvectors Φ_{ij} of F_i determine the direction of the hyperellipsoid axes and the length of the axes is given by the eigenvalues of F_i. For flat hyperellipsoids close to hyperplanes, the eigenvector corresponding to the smallest eigenvalue, say Φ_{is}, determines the normal to that hyperplane, see Figure 3.4. This normal together with the cluster prototypical point determines the hyperplane in an implicit form:

$$\left[\left(y(k),\ldots,y(k-n+1),u(k),\ldots,u(k-m+1)\right)^T - v_i\right]\Phi_{is} = 0 \qquad \text{(3-10)}$$

By rewriting (3-10) into the explicit form (3-5), we obtain the parameters , and $a_{i,j}$, $b_{i,j}$ and c_i.

Estimating consequents by least squares. When using linear consequent equations (3-5), the parameters and $a_{i,j}$, $b_{i,j}$, c_i can be estimated by least squares, since (3-4) is linear in these parameters. First, the degree of fulfillment β_i of each rule is computed by combining the individual antecedent terms using a t-norm (conjunction) operator t

$$\beta_i = \mu_{A_{i,1}}(y(k))\,t\ldots t\,\mu_{A_{i,n}}(y(k-n+1))\,t\,\mu_{B_{i,1}}(u(k))\,t\ldots t\,\mu_{B_{i,m}}(u(k-m+1))$$

Further, the degrees of fulfillment are normalized $\overline{\beta}_i = \beta_i / \sum_{i=1}^{K} \beta_i$. With regard to (3-4), the identification of the local models boils down to a standard least squares problem. Denoting ψ a matrix formed from the regressors and a unit column

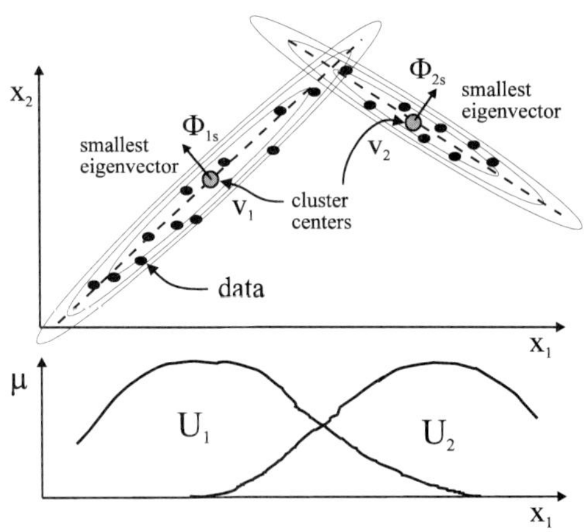

Figure 3.4 When applied to function approximation, the Gustafson–Kessel algorithm produces flat hyperellipsoidal clusters that can be regarded as hyperplanes. The smallest eigenvector Φ_{is} determines the normal vectors to the hyperplanes.

$$\psi = \left[y(k),\ldots,y(k-n+1),u(k),\ldots,u(k-m+1),1\right],$$

Θ_i the parameter vector, $\Theta_i = [a_{i,1},\ldots,a_{i,n},b_{i,1},\ldots,b_{i,m},c_i]$ and Y the model output vector $y(k+1)$, the following set of equations can be written in a matrix form:

$$\left[\overline{\beta}_1\psi,\overline{\beta}_2\psi,\ldots,\overline{\beta}_K\psi\right]\cdot\left[\Theta_1,\Theta_2,\ldots,\Theta_K\right]^T = Y \tag{3-11}$$

A least square solution of (3-11) gives an estimate for the parameters and $a_{i,j}$, $b_{i,j}$ and c_i.

Optimizing the number of rules. The number of clusters (i.e., the number of rules in the resulting Sugeno–Takagi model) must be specified a priori before clustering. This number is related to the type of nonlinearity the system is expected to exhibit. The more rules the model contains, the finer approximation can be obtained, but more parameters must be estimated and their variance is therefore higher. The number of clusters also depends on the dimension of the data matrix (i.e., the model orders) and on the number of measurements available. Few data points cannot be classified in many clusters because in such a case the hyperplanes are not sufficiently determined. Empirical relations among the number of clusters K, the dimension of the data $d(d = m+n+1)$, and the number of data points required have been suggested [20].

If no particular knowledge about the type of process nonlinearities is available, a Compatible Cluster Merging (CCM) procedure can be applied to find an appropriate number of rules [22]. The idea of the method is to set the initial number of cluster sufficiently high (such that the nonlinearities in the system hypersurface can be accurately captured) and to reduce the number of clusters iteratively by merging clusters with similar prototypes. The degree of similarity is calculated from the cluster centers and the eigenvectors of the covariance matrix, using the so-called cluster compatibility criteria [23,22]. Based on the numerical values of the criteria for all pairs of clusters, a fuzzy decision-making procedure determines which clusters should be merged.

The benefit from using the CCM technique is twofold: 1) The number of rules can be optimized without testing all the cluster numbers from two to the specified upper limit, which significantly reduces the CPU time; 2) by decreasing the number of clusters gradually, using CCM, for a given number of clusters, a better solution can be found than when initializing the partition matrix at random; see [22] for details.

Another approach to rule-base simplification has been developed in [24], using a measure of similarity among the membership functions obtained by cluster projections. Similar membership functions are replaced by one common function, reducing the number of linguistic terms needed and thus also the number of rules. In this way, a more transparent and understandable rule base is obtained. Experimental results show that both the CCM and the similarity-based approach lead to good results.

The entire identification procedure can be summarized as follows:

1. Design the identification experiments and collect a set of representative process measurements.

2. Choose the model structure, i.e., the model inputs, orders n, m and delays.

3. Fix the initial number of clusters, find clusters in the product space of the model inputs and outputs using the GK algorithm.

4. Merge the compatible clusters, repeat clustering until no more clusters can be merged.

5. By projecting the clusters on the antecedent variables, generate the membership functions.

6. Estimate the consequent parameters using least squares.

7. Validate the model.

The design of the experiment is an important initial step for any identification method, since it determines the information content of the identification data set. As opposed to linear techniques, pseudo-random binary excitation signals are not suitable for nonlinear identification in general and for fuzzy clustering in particular. Although the choice of the excitation signal may be problem-dependent, the input data should preferably excite the system in the entire range of the signals both in amplitude and in frequency. The pseudo-random binary signal is not suitable since it contains only two amplitude levels.

The last step, model validation, is even more important, since it should decide whether a model is acceptable or not. The validation of fuzzy models has several facets, namely a standard validation through numerical simulation and a comparison with the process data, analysis of the linear consequent models (stability, step responses, gains, time-constants, nonminimum phase behavior, etc.) and analysis of the coverage of the input space by the combinations of the antecedent membership functions. For an incomplete rule base, additional rules can be provided based on prior knowledge or local linearization of first-principles models.

3.2.3 Deriving the Mamdani Model

An identified Sugeno–Takagi model can be converted into a Mamdani model with singleton or linguistic consequents, using a method introduced in our earlier work [13]. There are several reasons for this conversion:

- The antecedent partitions of a singleton model cannot be identified directly from data as with a more structured Sugeno-Takagi model.
- The singleton model provides a simple and controllable (possibly linear) interpolation among the rules and does not exhibit the interpolation error inherent to the Sugeno–Takagi mechanism [25]. The inference in the singleton model is also computationally simpler and therefore faster.
- The singleton model has fewer free parameters and therefore lower variance in the parameter estimates.

- A simple control law can be designed by inverting the singleton model, which is not possible with the Sugeno–Takagi model, since both its antecedents and consequents are generally functions of the model inputs.

The transformation of a Sugeno–Takagi model into a singleton model is based on the simple observation that a Sugeno–Takagi model exhibits local linear behavior when a single rule fires, while in a singleton fuzzy model, local linear behavior results from a linear interpolation among the adjacent rules. Notice that the interpolation in the singleton model is linear if the antecedent partition consists of normal fuzzy sets with triangular membership functions and the sum of membership degrees for each domain element equals one. The product operator must be used to represent the logical **and** connective between the individual antecedent variables.

From the above it follows that to achieve approximately the same input–output mapping, the cores of the membership functions of the singleton model must be placed between the cores of the adjacent membership functions of the Sugeno–Takagi model, see Figure 3.5. The top part of this figure shows a function $\alpha = f(x)$(solid line), approximated by three Sugeno–Takagi rules with linear consequents:

$$
\begin{aligned}
R_1&: \text{ If } x \text{ is } A_1 \text{ Then } y = a_1 x + b_1 \\
R_2&: \text{ If } x \text{ is } A_2 \text{ Then } y = a_2 x + b_2 \\
R_3&: \text{ If } x \text{ is } A_3 \text{ Then } y = a_3 x + b_3
\end{aligned}
$$

The antecedent membership functions A_1, A_2 and A_3 defined on the domain X are shown below. The dashed lines in the upper figure denote the local linear consequent models.

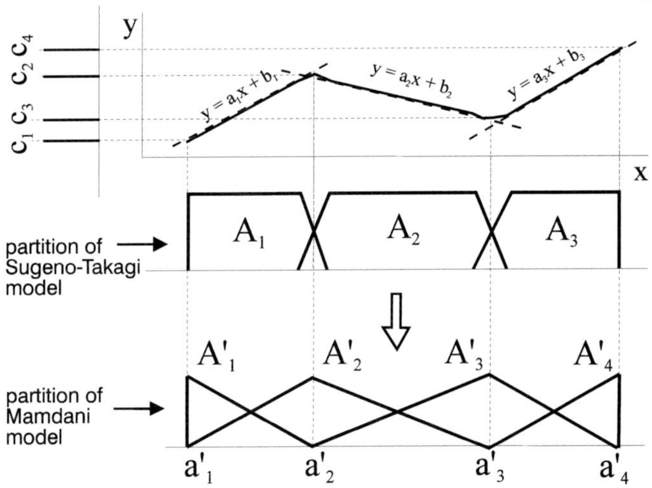

Figure 3.5 Cores of the triangular membership functions of the singleton fuzzy model are placed at the points where the intersections of the adjacent fuzzy sets of the Sugeno–Takagi model attain the maximal membership degree.

The function $f(x)$ also can be described using the following four rules with singleton consequents:

R_1: If x is $A_1^{'}$ Then y is c_1 R_3: If x is $A_3^{'}$ Then y is c_3

R_2: If x is $A_2^{'}$ Then y is c_2 R_4: If x is $A_4^{'}$ Then y is c_4

where the antecedent membership functions $A_1^{'}$ to $A_2^{'}$ are derived from the partition of the Sugeno–Takagi model such that their cores $a_1^{'}$ to $a_4^{'}$ are placed at the intersection points of the adjacent membership functions of the Sugeno–Takagi model and at the boundaries of the domain. After the antecedent partitions are derived, the singleton consequents are estimated using the least squares method described in Section 3.2.2, with the parameter vector Θ equal to a unit vector. Applications presented in [13, 26] show the usefulness of this method in terms of improved accuracy of the model and better model validation, since the rules of the Mamdani model can be easily checked using the existing knowledge of operators and experts.

3.3 FUZZY-MODEL-BASED CONTROL

Based on a fuzzy model of the process, a (fuzzy) control law can be synthesized using various methods. The Sugeno–Takagi model is suitable for local application of linear design techniques per rule. The global control action is obtained by interpolation using the antecedent membership functions of the fuzzy model. Stability of the closed-loop system can be analyzed, using e.g. the direct Lyapunov method [27,12]. If the local controllers differ only in their parameters, the global control law is an interpolative form of the well-known *gain-scheduling* mechanism. On the other hand, *heterogeneous control* [28] can employ different control laws in different operating regions. In the latter case, e.g., time-optimal control for dynamic transitions can be combined with PI(D) control near the set-point.

In this chapter, we investigate another approach based on causality inversion of the singleton fuzzy model. The model can be inverted on-line, using an algorithm introduced in our earlier paper [15]. The application example shows that without constraints on the control signal, this approach delivers excellent results. For the constrained case, we combine the controller based on model inversion with a predictive control approach.

3.3.1 Inverting the Plant Model

Assume that a SISO singleton model of the process is available. As shown in the previous section, such a model can be derived from an identified Sugeno-Takagi model. The rules have the following general form:

$$\begin{aligned}&\textbf{If } y(k) \text{ is } A_1 \textbf{ and } y(k-1) \text{ is } A_2 \textbf{ and,}\dots\ y(k-n+1) \text{ is } A_n \textbf{ and } u(k) \text{ is } B_1 \\ &\quad \textbf{and } u(k-1) \text{ is } B_2 \textbf{ and,}\dots\ u(k-m+1) \text{ is } B_m \textbf{ then } y(k+1) \text{ is } c\end{aligned} \tag{3-12}$$

where $A_1,\ldots,A_n$ and $B_1,\ldots,B_m$ are fuzzy sets and c is a fuzzy singleton (to simplify the notation, the rule index will be omitted in this section). Let us introduce an artificial state vector $x(k)$ that contains all the past inputs, the past outputs and the current output (i.e. all the antecedent variables from (3.12) except $u(k)$):

$$x(k)=\left[y(k),\ldots,y(k-n+1),u(k-1),\ldots,u(k-m+1)\right]$$

The corresponding fuzzy sets are composed in one multidimensional state fuzzy set X, applying a t-norm operator on the Cartesian product space of the state variables: $X = A_1\ t \ldots t\ A_n\ t\ B_2\ t \ldots t\ B_m$. After formally substituting U for B_1, rule (3-12) can be written as:

$$\textbf{If } x(k) \text{ is } X \textbf{ and } u(k) \text{ is } U \textbf{ then } y(k+1) \text{ is } c \tag{3-13}$$

Note that transformation of (3-12) into (3-13) is only a formal simplification of the rule base that does not change the order of the model dynamics, since X is a multidimensional fuzzy set. In the following we assume that the if–then relation is modeled by a t-norm. Denote M the number of fuzzy sets X_i, defined for state $x(k)$, and N the number of fuzzy sets U_j, defined for input $u(k)$. The rule base consists of all possible combinations of sets X_i, $i = 1,\ldots, M$, and $j = 1,\ldots, N$ ($M \cdot N$ rules in total), and can be represented as the following table:

	U_1	U_2	$\cdots$	U_N
X_1	c_{11}	c_{12}	$\cdots$	c_{1N}
X_2	c_{21}	c_{22}	$\cdots$	c_{2N}
$\cdots$	$\cdots$	$\cdots$	$\cdots$	$\cdots$
X_M	c_{M1}	c_{M2}	$\cdots$	c_{MN}

(3-14)

Using the product t-norm operator, the degree of fulfillment of the rule antecedent βij is: $\beta_{ij}(k)=\mu_{X_i}(x(k))\cdot\mu_{U_j}(u(k))$ where $\mu_{X_i}(x(k))$ is the membership degree of a particular state $x(k)$ in the fuzzy set X_i and $\mu_{U_j}(u)$ is the membership degree of input $u(k)$ in the fuzzy set U_j. Output $y(k + 1)$ is the weighted average of the rule contributions:

$$y(k+1)=\frac{\sum_{i=1}^{M}\sum_{j=1}^{N}\beta_{ij}(k)\cdot c_{ij}}{\sum_{i=1}^{M}\sum_{j=1}^{N}\beta_{ij}(k)}$$

Inverting the fuzzy model means finding a mapping f^{-1} such that $u(k) = f^{-1}(x(k), y(k + 1))$. If the same t-norm represents the if–then relation and the **and** connective in the rule antecedent, rule (3-13) can be rewritten as:

$$\textbf{If } x(k) \text{ is } X_i \textbf{ then } \left(\textbf{if } u(k) \text{ is } U_j \textbf{ then } y(k+1) \text{ is } c_{ij}\right) \tag{3-15}$$

For a given state $x(k)$, the degree of fulfillment of the first antecedent proposition "$x(k)$ is X_i" can be calculated as $\mu_{X_i}(x)$. Then the consequents c_{ij} of the rules containing identical U_j's, i.e., columns in (3-14), are aggregated (there are M such rules for each U_j):

$$c_j = \frac{\sum_{i=1}^{M} \mu_{X_i}(x) \cdot c_{ij}}{\sum_{i=1}^{M} \mu_{X_i}(x)}, \quad j = 1,2,\ldots,N$$

As a result, a set of rules N for the input $u(k)$ is obtained:

$$\textbf{If } u(k) \text{ is } U_j \textbf{ then } y(k+1) \text{ is } c_j \quad j = 1,2,\ldots,N$$

Since the if-then relation in the Mamdani model is symmetrical (t-norm), each rule can be inverted by simply exchanging the cause and effect, i.e., swapping the antecedent and the consequent. (Note that in the Mamdani model the individual rule-based inference is equivalent to the global inference for both crisp and fuzzy data.) The inverted rule has the following form:

$$\textbf{If } y(k+1) \text{ is } c_j \textbf{ then } u(k) \text{ is } U_j \quad j = 1,2,\ldots,N$$

The entire rule base is invertible if the kernel of each U_j is a single point (card(ker(U_j)) = 1) and $\ker(U_l) \geq \ker(U_m) \rightarrow c_l > c_m$, resp. $\ker(U_l) \geq \ker(U_m) \rightarrow c_l > c_m$, $\forall l, m\ l \neq m$, i.e. the rule base is *monotonic*. Figure 3.6 gives an example of non-invertible rule base, for which extra information is needed to select an appropriate control action, given the desired system output $y(k+1)$. Since the values c_j are singletons, it is necessary to

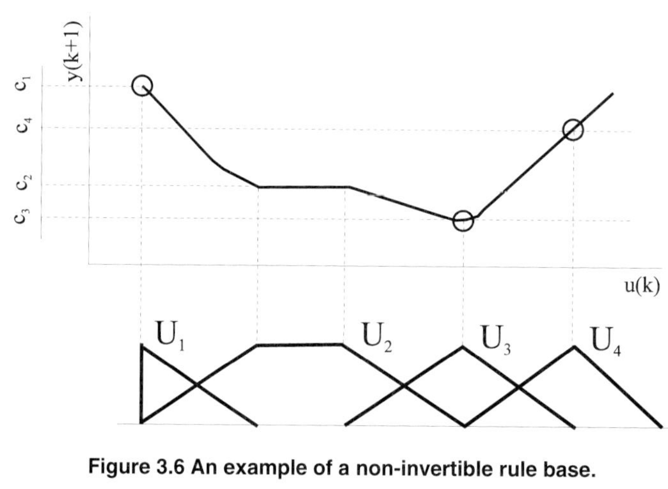

Figure 3.6 An example of a non-invertible rule base.

interpolate between the consequents c_j to obtain $u(k)$. The interpolation is accomplished by constructing a fuzzy partition such that $\ker(C_j) = c_j$ and $\sum_{i=1}^{N} \mu_{C_i}(y) = 1, \forall y$. Note that this partition depends dynamically on the values c_j, and thus on the process state. An inverted plant model yields directly the control action needed to achieve the desired system output. Substituting the reference $r(k+1)$, for the future plant output $y(k + 1)$, we obtain the following fuzzy control law:

$$\textbf{If } r(k+1) \text{ is } C_j \textbf{ then } u(k) \text{ is } U_j \quad j = 1,2,\ldots,N \tag{3-16}$$

Without disturbances, constraints, or modeling errors, (3-16) is a time optimal controller with a zero steady-state error. This is clearly an ideal case; in reality the control actions are usually limited both in the amplitude and the rate, which results in overshoots when using (3-16). This problem can be solved by applying predictive control as shown in the next section. Modeling errors and disturbances must be compensated by a feedback, using e.g., the well-known Internal Model Control (IMC) scheme [29].

3.3.2 Fuzzy-Model-Based Predictive Control

A predictive controller uses an optimization algorithm to calculate a sequence of future controller outputs over a control horizon, such that a specified objective (cost) function is minimized. Most of the objective functions are some modifications of the quadratic function [30]:

$$J = \sum_{i=1}^{H} \alpha_i \left(r(k+i) - \hat{y}(k+1)\right)^2 + \sum_{i=1}^{H_c} \beta_i \Delta u(k+i-1)^2 \tag{3-17}$$

where $\hat{y}$ denotes the predicted process output, r the desired process behavior (reference trajectory) and u the future control signal. H is the prediction horizon and H_c is the control horizon. Vectors α and β define weighting of the output error and the control effort with respect to each other and with respect to time. Additional "hard" constraints on the process variables can be specified as a part of the optimization problem. Only the first control $u(k)$ is applied to the plant and the optimization is repeated at the next time sample $k + 1$ (the receding horizon principle).

Using a fuzzy model as a process model, the most straightforward way of minimizing (3-17) is the use of some numerical optimization technique (Nelder-Mead, sequential quadratic programming, etc.). Such methods, however, require a significant computing power which may be a serious obstacle for the real-time implementation. Besides, convergence to a global minimum is not guaranteed and local solutions may result in undesired control actions. Another solution is to apply discrete optimization over a few predetermined discrete-valued control actions. For unconstrained $\Delta u(i)$, the cores of antecedent fuzzy sets U_j of the fuzzy model (3-15) are typically used, i.e. $u(i) = \{\text{core}(U_j) \mid j = 1,\ldots, M\}$. If the control input change $\Delta u(i)$ is constrained, discrete controls $u(i) = \{u(i-1) - \Delta u_{\max}, u(i-1), u(i-1) - \Delta u_{\max}\}$ are used, where $\Delta u_{\max}$ is a physical limit on $\Delta u(i)$.

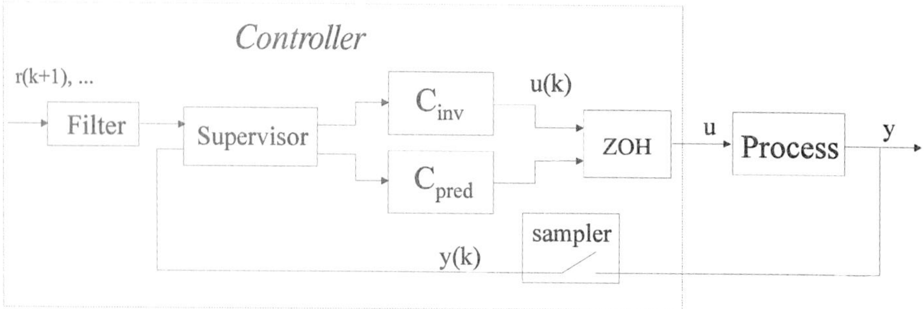

Figure 3.7 Combination of predictive and inverse-model-based control.

In both cases, the optimization problem can be solved by, e.g., dynamic programming with a guaranteed optimality (for the given discretization). Applying discrete control actions near the set points, however, may result in oscillations that need to be avoided. Since the constraints are active during the dynamic transients but not in a close vicinity of the set points, the predictive optimization-based control scheme can be combined with an inverted-model-based control active only around the set points, Figure 3.7 shows the resulting control scheme with a supervisory system that selects the predictive control strategy if (3-16) results in $|\Delta u(i)| > \Delta u_{\max}$ or $u(i)$ saturated, for some $i \in [k, k + H - 1]$ and the inverse model control law (3-16) otherwise. A standard reference shaping filter, usually a first- or second-order linear filter avoids large controller outputs as a reaction to step changes in the reference.

3.4 APPLICATION TO PRESSURE CONTROL

The described identification and control design methods were applied to modeling and control of the pressure in a laboratory fermentor operating at 30° C. The volume of the fermentor tank is 40 l and normally it contains 25 l of water. At the bottom of the tank, air is fed in the water at a flow rate controlled by a local mass-flow controller. The air pressure above the water level is controlled by an outlet valve at the top of the tank. With a constant inlet flow rate, the system has a single input, the valve position, and a single output, the air pressure. The inlet flow rate can be manipulated too; in that case the system has two manipulated inputs and one output. The pressure dynamics are nonlinear because of the underlying physical principles and because of the nonlinear characteristic of the outlet valve. Also the water volume strongly influences the pressure dynamics: with larger volumes, the pressure reacts faster as the valve position changes.

At the above specified conditions, the smallest time constant of the process is about 45s, which allows a sample time of 5s. The controller was implemented on an Intel 486/50Mhz-based computer, connected to the process via A/D and D/A convertors. The fuzzy model identification, control design and simulation programs run under Matlab [31], the routines for real-time control and the run-time inference engine were programmed in PASCAL.

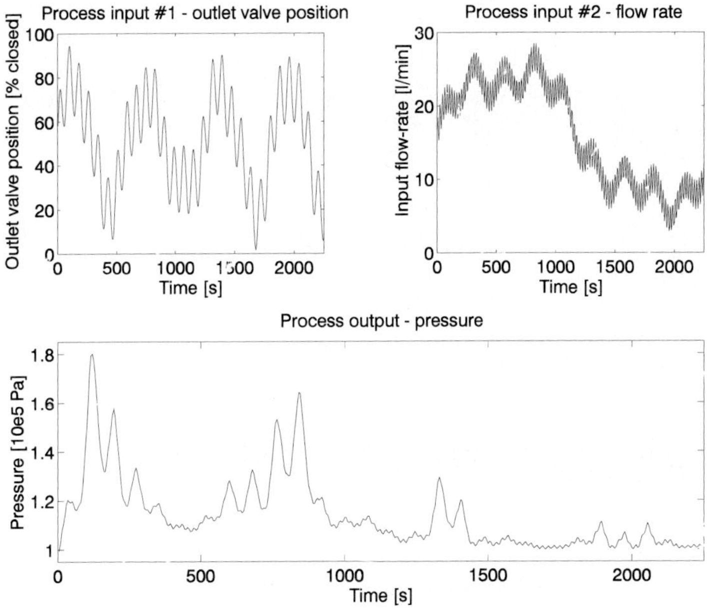

Figure 3.8 Identification data.

3.4.1 System Identification via Fuzzy Clustering

In order to collect the identification data, the system inputs were excited with multi-sinusoidal signals, as shown in Figure 3.8. The upper plots show the input signals -- the outlet valve position and the inlet flow rate the bottom plot shows the corresponding pressure. The process dynamics can be modeled by a nonlinear first-order dynamic system $y(k + 1) = f(y(k), u_1(k), u_2(k))$, where k is discrete time, $y(k)$is the predicted model output at time k, $y(k + 1)$ is the current model output, and $u_1(k)$ and $u_2(k)$ are the model inputs, the outlet valve position, and the inlet flow rate, respectively. From the process measurements $\{y(k), u_1(k), u_2(k)\}$, $k = 1, 2, \ldots, N$, where N is the number of measurements, a data matrix for clustering is formed:

$$\begin{bmatrix} y(1) & u_1(1) & u_2(1) & y(2) \\ y(2) & u_1(2) & u_2(2) & y(3) \\ \cdots & \cdots & \cdots & \cdots \\ y(N-1) & u_1(N-1) & u_2(N-1) & y(N) \end{bmatrix}$$

Initially, the data were clustered in 9 clusters, using the GK algorithm. This number is considered sufficiently high to capture the system nonlinearities. Applying the CCM algorithm resulted first in merging 4 clusters. The GK algorithm was reinitialized with a reduced partition matrix, created by adding up the membership grades of the clusters to

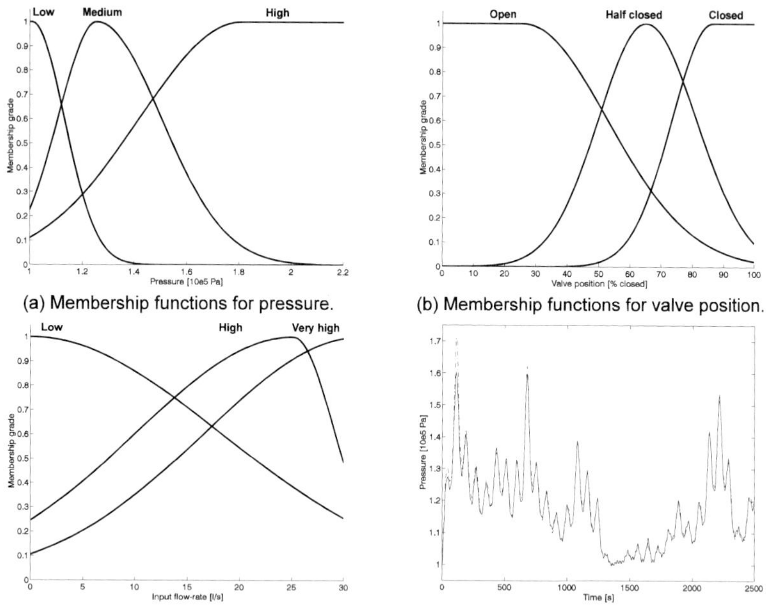

(a) Membership functions for pressure.
(b) Membership functions for valve position.
(c) Membership functions for flow-rate.
(c) Model validation (solid – process, dash-dotted – model).

Figure 3.9 Membership functions and model validation.

be merged. In the next cycle, 2 compatible clusters were merged, giving a final partition with 3 clusters.

Figure 3.9(a) shows exponential membership functions fitted to the projection of the fuzzy partition onto the pressure antecedent variable. Figures 3.9(a) and 3.9(c) show the membership functions for the valve position and the inlet flow rate respectively. After estimating the consequent parameters with the least squares method, the following rule base was obtained:

- If pressure $y(k)$ is *Low* and valve $u_1(k)$ is *Open* and inlet flow rate $u_2(k)$ is *Low*, then pressure $y(k+1) = 0.52y(k) + 0.00005u_1(k) + 0.0032u_2(k) + 0.44$.
- If pressure is $y(k + 1)$ *Medium* and valve $u_1(k)$ is *Half Closed* and inlet flow rate $u_2(k)$ is *High*, then pressure $y(k+1) = 0.76y(k) + 0.00024u_1(k) + 0.0049u_2(k) + 0.04$.
- If pressure is $y(k)$ *High* and valve $u_1(k)$ is *Closed* and inlet flow rate $u_2(k)$ is *Very High*, then pressure $y(k+1) = 0.85y(k) + 0.00074u_1(k) + 0.0053u_2(k) - 0.45$.

Consequent parameters of the individual rules represent stable local linear models valid around the three operating points determined by the cluster centers. Figure 3.9(d) shows validation of the model by simulation from the inputs only. Visual comparison of the process and model outputs as well as the root mean squared error, *rms* = 0.0172, indicate that even this simple fuzzy model can serve as a reliable long-range predictor for this process.

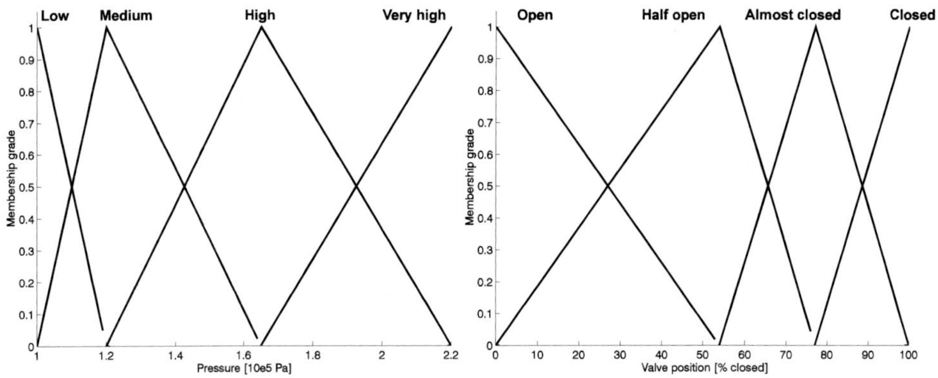

(a) Pressure. (b) Valve position.

Figure 3.10 Antecedent partitions for the singleton model.

3.4.2 Deriving A Singleton Model

In order to design the control law based on the model inversion, we first convert the Sugeno–Takagi model into a singleton model. Here only the SISO system is considered for the control experiments, with the air flow rate kept constant at 22.5 *l*/min. First a Sugeno–Takagi model was estimated for the SISO process. From its antecedent partitions, triangular antecedent membership functions of the singleton model were derived, see Figures 3.10(b) and 3.10(a).

Since there are four membership functions per antecedent variable, the rule base consists of 16 rules, containing all possible combinations of the antecedent labels. The consequents we estimated using least squares.

	Pressure			
Valve	*Low*	*Medium*	*High*	*Very high*
Open	1.06	1.13	1.46	1.70
Half open	1.05	1.19	1.51	2.03
Almost closed	1.07	1.23	1.63	2.12
Closed	1.11	1.33	1.79	2.34

3.4.3 Fuzzy-Model-Based Control

The controller based on the fuzzy model inversion and the predictive controller were first tested in simulation, using an approximate mechanistic model of the process

$$\frac{dP}{dt} = \frac{1000.RT}{22{,}4.V_h} \cdot \left[\Phi_g - \left(\pi R_H^2\right) \sqrt{\frac{2P_0}{\rho_0 K_f} \ln\left(\frac{P}{P_0}\right)} \right]$$

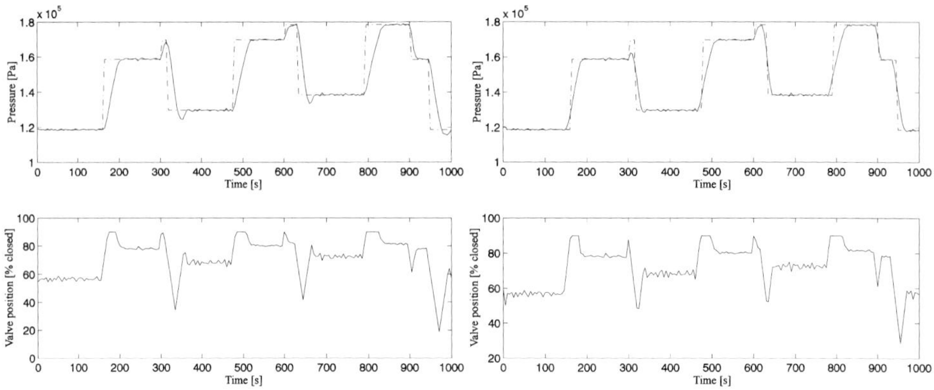

(a) Fuzzy controller based on an inverted fuzzy model.

(b) Combination of predictive and inverse model control.

Figure 3.11 Simulation results.

where R is the gas constant (8.134 $Jmol^{-1}K^{-1}$), T temperature (305 K), V_h gas volume (0.015 m^3), Φ_g gas flow rate ($3.75 \cdot 10^{-4}$ m^3s^{-1}), R_H radius of the outlet pipe (0.0178 m), P_0 the reference pressure ($1.013 \cdot 10^5$ Nm^{-3}), ρ_0 outside air density (1.2 Kgm^{-3}), P pressure in the tank (Nm^{-2}), and K_f a valve friction factor ($Jmol^{-1}$), a nonlinear function of the valve position u and the flow rate Φ_g. The maximum change in the valve position $|\Delta u(k)|$ is limited to 10% per sample. The simulation is quite realistic since a mismatch appears between this mechanistic process model and the fuzzy model used for the controller design.

Based on the current pressure and the desired pressure at the next sample, the inverted model (3-16) computes the exact value of the valve setting needed to bring the pressure to the desired value. Figure 3.11(a) shows the simulation results for several step changes in the reference.

Note that the undershoots, due to the physical constraints on the control signal rate, [see Figure 3.11(a)], are eliminated using the proposed predictive control scheme, [see Figure 3.11(b)]. The prediction horizon is 3 samples. Also the integrated time error at the stepwise set-point changes decreases due to the predictive way of control.

The predictive control scheme gives better results and therefore has been used for controlling the process in real time. A response under nominal conditions is shown in Figure 3.12(a). In Figure 3.12(b), depicting a response with the inlet flow rate increased to 30 l/min, a difference between the process output and the model prediction is observed, since the model does not account for the change in the flow rate. Still the controller is able to compensate this model–plant mismatch and stabilize the process.

3.5 CONCLUSIONS AND DISCUSSION

A method for constructing a fuzzy model of a process from data and designing a nonlinear controller based on this model was presented. As in conventional control, the design has two steps: 1) system modeling and identification, 2) controller synthesis. In

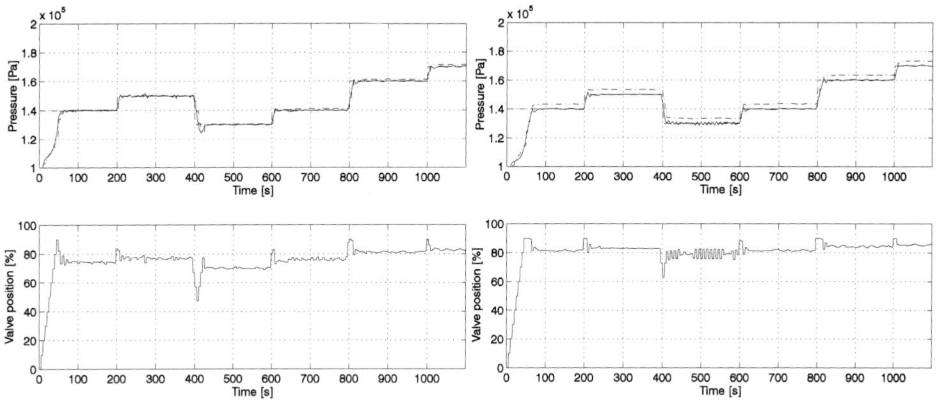

(a) Process response under nominal conditions. (b) Inlet flow-rate increased from 22.5 to 30 l/min.

Figure 3.12 Real-time responses of the fuzzy predictive controller. Dashed line: reference; solid line: process output; dash-dotted line: model output.

contrast to conventional methods, fuzzy models can also accurately approximate highly nonlinear processes and can integrate a priori knowledge with information obtained from the process data. If no prior knowledge is available, the rules and the membership functions can be extracted from process measurements. An identification method based on unsupervised fuzzy clustering combined with a cluster merging technique was presented. This method builds a Sugeno–Takagi model with linear consequent functions describing the process locally as linear ARX models. An advantage of this approach is that from a single set of global process data a collection of local models can be obtained. A drawback is that sufficiently informative process measurements must be available, which is not always the case in industrial practice.

The Sugeno–Takagi model can be converted into a Mamdani model with singleton consequents, suitable for designing a simple controller based on causality inversion of the rules. For invertible plants with a minimum phase behavior and without constraints, this controller yields an optimal control law. In general, a predictive control scheme based on a fuzzy model is more suitable, since it can easily deal with constrained and multivariable systems. In both cases, the controller design is based entirely on the model and no tuning is necessary. The controller performance depends directly on the quality of the model that can be assessed using standard validation techniques.

An illustrative application to pressure control has been presented to show how the method can be applied in practice. The pressure control system was chosen since because of its nonlinearities, it cannot be controlled by a global linear controller and therefore the use of more sophisticated nonlinear control techniques is fully justified. Very good results achieved both in the process modeling and control show that the technique is not only theoretically sound but also practically applicable.

3.6 REFERENCES

1. Shinskey, F.G. *Controlling Multivariable Processes.* Instrument Society of America, Research Triangle Park, NC, 1981.

2. Holmblad, L.P. and Østergaard, J.-J. "Control of a Cement Kiln by Fuzzy Logic". In, *Fuzzy Information and Decision Processes.* (Gupta, M. M. and Sanchez, E. eds.) North-Holland, 1982, pp. 389–399.

3. Santhanam, S. and Langari, R. "Supervisory Fuzzy Adaptive Control of a Binary Distillation Column." *Proceedings of the Third IEEE International Conference on Fuzzy Systems*, Vol. 2, pages 1063–1068, Orlando, FL, June 1994.

4. Tani, T., Utashiro, M., Umano, M., and Tanaka, K. "Application of Practical Fuzzy–PID Hybrid Control System to Petrochemical Plant." *Proceedings of Third IEEE International Conference on Fuzzy Systems*, Vol. 2, pp. 1211–1216, June 1994.

5. Hirota, K., ed. *Industrial Applications of Fuzzy Technology.* Springer Verlag, 1993.

6. Bonissone, Piero P. "Fuzzy Logic Controllers: An Industrial Reality." In *Computational Intelligence: Imitating Life.* (Zurada, Jacek M., Marks, Robert J. II and Robinson, Charles J., eds.). IEEE Press, 1994, pp. 316–327.

7. Yasunobu, S. and Miyamoto, S. "Automatic Train Operation System by Predictive Fuzzy Control. In *Industrial Applications of Fuzzy Control*, (Sugeno, M., ed.). pp 1–18. North-Holland, 1985.

8. Driankov, D., Hellendoorn, H. and Reinfrank, M. *An Introduction to Fuzzy Control.* Springer Verlag, 1993.

9. Yager, R.R. and Filev, D.P.. *Essentials of Fuzzy Modeling and Control.* John Wiley & Sons, 1994.

10. Pedrycz, W. *Fuzzy Control and Fuzzy Systems (second, extended, edition).* John Wiley & Sons, 1993.

11. Ouwerkerk, S.F.J. *Development and Implementation of a Fuzzy Logic Controller for a Chemical Process.* Master's dissertation, Shell Nederland B.V., Delft University of Technology, Delft, The Netherlands, February 1995 (in Dutch).

12. Wang, Li-Xin. *Adaptive Fuzzy Systems and Control, Design and Stability Analysis.* Prentice Hall, 1994.

13. Babuška, R. and Verbruggen, H.B. "A New Identification Method for Linguistic Fuzzy Models." *Proceedings FUZZ-IEEE/IFES'95*, pp. 905–912, Yokohama, Japan, March 1995.

14. Babuška, R. and H.B. Verbruggen. "Identification of Composite Linear Models via Fuzzy Clustering." *Proceedings European Control Conference*, Rome, Italy, September 1995.

15. Babuška, R., Sousa, J., and Verbruggen, H.B. "Model-Based Design of Fuzzy Control Systems." *Proceedings 3rd. European Congress on Fuzzy and Intelligent Technologies*, Aachen, Germany, August 1995.

16. Takagi, T. and Sugeno, M. "Fuzzy Identification of Systems and Its Application to Modeling and Control." *IEEE Transactions on Systems, Man and Cybernetics*. Vol. 15, No. 1, 1985, pp. 116–132.

17. Wang, L.-X. "Fuzzy Systems Are Universal Approximators." *Proc. IEEE Int. Conf. on Fuzzy Systems 1992*, pp. 1163–1170, San Diego, CA, March 1992.

18. Leonaritis, I.J. and Billings, S.A. "Input-output parametric models for non-linear systems." *International Journal of Control*. Vol. 41, 1985, pp. 303–344.

19. Sugeno, M. and Tanaka, K. "Successive Identification of a Fuzzy Model and Its Application to Prediction of a Complex System." *Fuzzy Sets and Systems*, Vol. 42, 1991, pp. 315–334.

20. Bezdek, J.C. *Pattern Recognition with Fuzzy Objective Function*. Plenum Press, 1981.

21. Gustafson, D.E. and Kessel, W.C. "Fuzzy Clustering with a Fuzzy Covariance Matrix." *Proc. IEEE CDC*, pages 761–766, San Diego, CA, 1979.

22. Kaymak, U. and Babuška, R. "Compatible Cluster Merging for Fuzzy Modeling." *Proceedings FUZZ-IEEE/IFES'95*, pp. 897–904, Yokohama, Japan, March 1995.

23. Krishnapuram, R. and Freg, Chin-Pin. "Fitting an Unknown Number of Lines and Planes to Image Data Through Compatible Cluster Merging." *Pattern Recognition*, Vol. 25, No. 4, 1992, pp. 385–400.

24. Setnes, M. *Fuzzy Rule Base Simplification Using Similarity Measures*. Master's dissertation, Delft University of Technology, Delft, The Netherlands, July 1995.

25. Babuška, R., R. Jager, and H.B. Verbruggen. Interpolation issues in Sugeno–Takagi reasoning. *In Proceedings IEEE World Congress on Computational Intelligence*, pages 859–863, Orlando, U.S.A., June 1994.

26. Babuška, R., Verbruggen, H.B., and van Can, H.J.L. "Fuzzy Modeling of Enzymatic Penicillin–G Conversion." Submitted to *IFAC World Congress*, San Francisco, CA, July 1996.

27. Tanaka, K. and Sugeno, M. Stability Analysis and Design of Fuzzy Control Systems." *Fuzzy Sets and System.* Vol. 45, No. 2, 1992, pp. 135–156.

28. Kuipers, B. and Åström, K. The Composition and Validation of Heterogeneous Control Laws. *Automatica.* Vol. 30, No. 2, 1994, pp. 233–249.

29. Braake, H.A.B., Babuška, R., and van Can, H.J.L. "Fuzzy and Neural Models in Predictive Control. *Journal A*. Vol. 35, No. 3, 1994, pp. 44–51.

30. Clarke, D. "Advances in Model-Based Predictive Control." In *Advances in Models-Based Predictive Control* (Clarke, D., ed.). Oxford University Press, 1994, pp. 3–21.

31. Babuška, R. "Fuzzy Toolbox for MATLAB." *Proceedings IMACS Conference MIMS-2 '93*, Brussels, Belgium, April 1993.

4

NEURO-FUZZY STRATEGIES FOR URBAN TRAFFIC NOISE MONITORING

A. Fichera, L. Fortuna, S. Graziani, M. G. Xibilia
University of Catania, ITALY

In this chapter some new techniques of modeling for complex systems are applied to determine the relationship between the noise pollution in urban areas and some related parameters. In particular, a classical empirical model, an optimized fuzzy model, a neural, and a neuro-fuzzy model have been compared by using the same set of measured data. Even if all the new approaches behave better than the classical one, the neuro-fuzzy modeling approach has revealed a good compromise between performance and computational efforts.

4.1. INTRODUCTION

In the last decade new techniques have been applied to develop an *intelligent system* able to solve complex problems in modeling and control. In this chapter some of these techniques have been adopted to derive the model of traffic noise pollution as a function of some meaningful parameters which are easily measurable.

Noise pollution in urban areas degrades the quality of life and therefore threatens health. This problem is drawing increasing attention both from governments and the scientific community [1,2].

The main source of noise is the flow of motor vehicles, moreover some physical parameters, related to the road structure act as modifying quantities [3,4,5]. Many researchers have been involved with the problem of noise pollution in order to understand the phenomenon and describe it through mathematical models with the aim to limit it in urban areas. Many models that use linear relationships to describe the noise pollution phenomenon are available in the literature [3,4,5,6]. These models have been derived on the basis of empirical considerations and do not lead to really satisfactory results.

In this chapter the authors propose new models for the prediction of acoustical noise; the proposed models are nonlinear and have been obtained by using the fuzzy logic approach, neural networks and neuro-fuzzy networks [7,8,9]. The nonlinearity of the models allows us to capture the complex interactions among the variables that govern acoustic noise pollution. The models have been obtained by using a set of measurements, performed on various roads of a typical medium-size European city.

4.2. THE MEASUREMENT SURVEY

The measurements necessary to determine the model of noise pollution with the different approaches mentioned above have been taken in Catania, a medium-sized city (about 400,000 people) in southern Italy.

The outdoor acoustical observations have been performed in sixteen locations placed in residential, commercial and industrial areas. The selected roads can be roughly grouped into two different sets:

- town center roads (6 measurement sites);
- suburban roads (4 measurement sites).

For each measurement a sound level meter, Bruel & Kjær type 2231, plus integrating module BZ7110 and a video tape recorder were used; moreover the sound level meter was directly interfaced with a computer. The measuring equipment was placed in accordance with Italian law [10].

The sound level meter was used to acquire 100 samples per second of the A-weighted sound level L_i. From measured data, the *equivalent continuous sound level*, L_{AeqT}, was obtained by using software hosted in the connected computer. The *equivalent continuous sound level* can be approximated as follows:

$$L_{AeqT} = 10\log\frac{1}{T}\sum_i 10^{\frac{L_i}{10}} \tag{4-1}$$

and represents a typical parameter in noise pollution analysis [11]. In the above expression the following symbols have been adopted:

- $T = 3600$s, the observation time;
- L_i = the (A-weighted) sound level measured over 1 sec (dB).

Some of the parameters to be taken into account to build up the model are as follows:

- the number of flowing vehicles (1/h);

- the nature of the vehicles;
- the speed of the vehicles, (m/s);
- urban area characteristics;
- geometry of the road section (width of the road and height of the buildings), (m);
- pavement type.

Such a model has a very complex structure and therefore is hard to identify or even to understand. Thus a reduced set of parameters among the reported ones has been selected and the noise pollution has been modeled as a function of:

- the equivalent number of vehicles, n_{eq} (1/h);
- the median height of the buildings, h (m), occurring in the considered road section;
- the width of the road, w (m);

The equivalent number of vehicles is defined as follows:

$$n_{eq} = n_{cars} + c_1 \cdot n_{mc} + c_2 \cdot n_{hv} \tag{4-2}$$

where:

- n_{cars} is the number of cars; (1/h)
- n_{mc} is the number of motorcycles (1/h)
- n_{hv} is the number of heavy vehicles. (1/h)

The values $c_1 = 3$ and $c_2 = 6$ are two coefficients suggested in the literature [12]. that allow the calculation of the number of motorcycles and heavy vehicles in an equivalent number of cars, producing the same L_{AeqT} value. Therefore the model assumes the following expression:

$$L_{AeqT} = f\left(n_{eq}, h, w\right) \tag{4-3}$$

For each measurement site, the video camera was used to compute, off-line, the number of vehicles belonging to the above defined classes and therefore n_{eq}. The video camera was also utilized to recognize abnormal phenomena. Table 4.1 reports the ranges of the measured parameters:

From Table 4.1 it can be observed that the town is very noisy. The L_{AeqT} value never goes below 65 dB which is considered the maximum limit for a residential area according to Italian law, and reaches 81 dB, well above the limit. Such high noise levels are due to the large number of vehicles that, combined with the narrowness of the roads, make the traffic very chaotic.

In Figure 4.1 a subset of the measured data is reported.

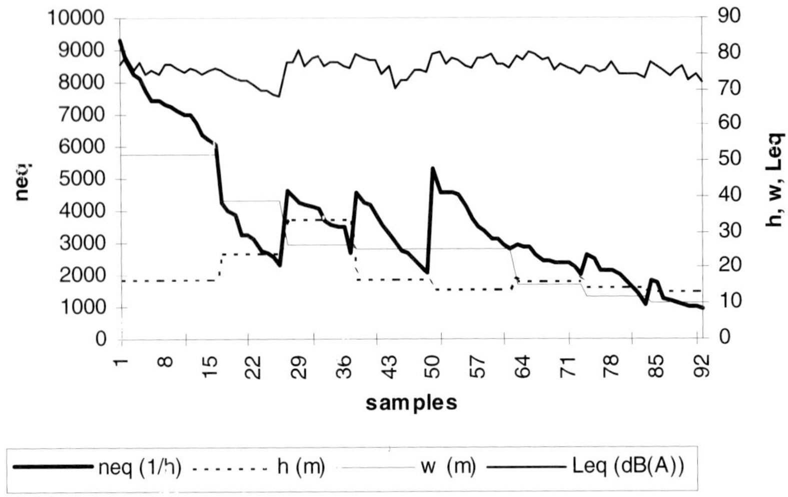

Figure 4.1 A subset of the measured data

Table 4.1 The ranges of the measured parameters				
	L_{AeqT}	n_{eq}	h	w
min	68.3	924	12	10
max	81	10300	33.5	52
average	76.06	3846	17.54	25.52

4.[illegible] THE NOISE POLLUTION CLASSICAL MODELS

Several kinds of models for Equation (4-3) can be obtained depending on the type of functional relationship $f(\cdot)$ [3,5]. Previous efforts have almost exclusively been devoted to determining linear representations. Hence the model assumes the following form:

$$L_{AeqT} = a_0 + a_1 \cdot w + a_2 \cdot h + a_3 \cdot n_{eq} \tag{4-4}$$

The modeling efforts are aimed at determining the parameters a_i. The a_i values are generally identified with suitable optimization methods. The following model (suggested in [3]) is a linear function of the logarithm of some of the considered parameters and will be denoted in the following as the *Josse model*

$$L_{AeqT} = 38.8 + 15\log(n_{eq}) - 10\log(w) \tag{4-5}$$

The linear models are often used because they are easy to handle; on the other hand they allow only a rough modeling of the real phenomena.

For more accurate results nonlinear models should be considered. The primary problem that researchers are faced with in considering nonlinear models is the choice of the particular nonlinearity to be taken into account, i.e., the form of the function $f(\cdot)$ in Equation (4-3). Nowadays some innovative techniques are available to alleviate this problem. Among these, three strategies can be advantageously applied: neural networks, fuzzy logic and neuro-fuzzy networks. These strategies represent valid approaches in noise pollution modeling, as it will be shown in the following paragraph [13,14,15].

4.4 THE FUZZY MODEL FOR THE ACOUSTIC NOISE

Different approaches can be adopted in order to determine a fuzzy model of a complex system. In particular, when a set of I/O data are available, optmization strategies to find the best fitting model can be considered. In the following, the optmization strategy reported in [16] has been considered.

The form of fuzzy inference used in order to build the model is:

$$\text{If } x_1 \text{ is } A_1 \text{ and } x_2 \text{ is } A_2 \text{ and} \ldots \text{ and } x_n \text{ is } A_n \text{ then } y = g(x_1, x_2, \ldots x_n) \tag{4-6}$$

where: y is the variable whose value is inferred; x_1, x_2,..., x_n are the variables that determine the y value; A_1, A_2,..., A_n are fuzzy sets, with piecewise linear membership functions, defined on the x_1, x_2,..., x_n variables; $g(\cdot)$ is a function that implies the value of y; x_i is A_i means x_i belongs to the fuzzy set A_i.

The expression:

$$\text{If } x_1 \text{ is } A_1 \text{ and } x_2 \text{ is } A_2 \text{ and} \ldots \text{ and } x_n \text{ is } A_n$$

is known as the *rule premise*, and the expression:

$$\text{then } y = g(x_1, x_2, \ldots x_n)$$

represents its *consequence*.

In the proposed model the function $g(\cdot)$ is assumed to be a linear one. The generic ith inference, used in a *fuzzy algorithm,* assumes therefore the form shown in Figure 4.2.

For each implication the output value is computed as:

$$y^i = g^i(x_1^0, x_2^0, \ldots, x_n^0) \tag{4-7}$$

where $(x_1^0, x_2^0, \ldots, x_n^0)$ is the actual input vector.

The truth value of this output can be obtained by using a generalization of the modus ponens criterion: *the consequence of a fuzzy rule is as true as its antecedent.* In this case each rule has a truth value

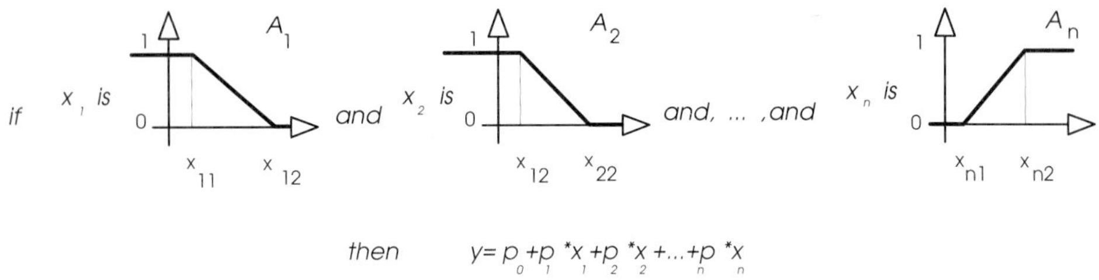

Figure 4.2. A schematic representation of a fuzzy implication.

$$|y = y^i| = \min\left(\mu_1^i(x_1^0), \mu_2^i(x_2^0), \ldots, \mu_n^i(x_n^0)\right) \tag{4-8}$$

where $|\ast|$ means the truth value of proposition *. Finally, the output is computed as:

$$y = \frac{\sum |y = y^i| y^i}{\sum |y = y^i|} \tag{4-9}$$

where the summation is extended to all the rules that make up the fuzzy algorithm.

In the case of the model proposed for noise pollution, the following variables were chosen:

- x_1 = number of equivalent vehicles (1/h);
- x_2 = average building height (m);
- x_3 = road width (m);
- $y = L_{AeqT}$ value (dB).

The fuzzy model described above depends on the fuzzy sets contained in the rule premises and on the polynomials of the associated consequences. A detailed description of the optimization strategy required to determine the fuzzy model is reported in [16]; in this chapter a brief description is reported.

To obtain the fuzzy model, the following steps are required: (a) to establish which variables must be considered in the premises; (b) to decide on the membership functions of each rule, i.e., to fix the values of x_{j1}^i and x_{j2}^i, introduced in Figure 4.2, which characterize the membership function shape; (c) to define the polynomial associated to the *consequence* of each fuzzy rule as in (4-7).

After steps (a) and (b) are performed, the solution of step (c) consists in the determination of those coefficients of the polynomials $g^i(\cdot)$ that minimize a suitable performance index. The performance index is generally assumed to be the root mean square E_r, of the errors between the measured L_{AeqT} and the estimated one:

$$E_r = \sqrt{\sum (y_i - \bar{y}_i)^2} \tag{4-10}$$

where y_i is the measured data and $\overline{y}_i$ is its estimation. The coefficients can be obtained by using the Least Square Method (LSM).

The shape of the fuzzy sets for each rule in step (b) depends on the values x^i_{j1} and x^i_{j2}. An optimization strategy must be used to find those values that minimize the error function E_r, with respect to these parameters. The problem of finding the optimum premise parameters is a nonlinear programming problem and the authors used the Nelder and Mead optimization method [17].

The solution of the whole problem starts with step (a), i.e., to establish how many fuzzy sets (eventually none) must be considered for each variable. In the following, a string of type *Mijk* is adopted to refer to each model, where *i, j*, and *k* are natural numbers related to the complexity of the model; in particular, each number indicates how many fuzzy sets are considered for each variable.

The stated problem is a combinatorial one and a general solution does not exist. Therefore a heuristic method has been used [16]. The search algorithm starts with the analysis of the simplest models. In this case we considered those with two fuzzy sets defined on one variable, e.g., M200 if the first variable is considered.

For each model the optimum premise and consequence parameters are found, as previously discussed, and the error E_r is computed. The model that reaches the minimum E_r value is considered for the generation of more complex ones. Each successive model is obtained either by doubling the number of fuzzy sets defined on a variable or introducing two fuzzy sets for the variables that have not been taken into account in the previous step. The procedure is reiterated until a satisfactory low E_r value is gained.

With the procedure described above, the search tree reported in Figure 4.3 was obtained.

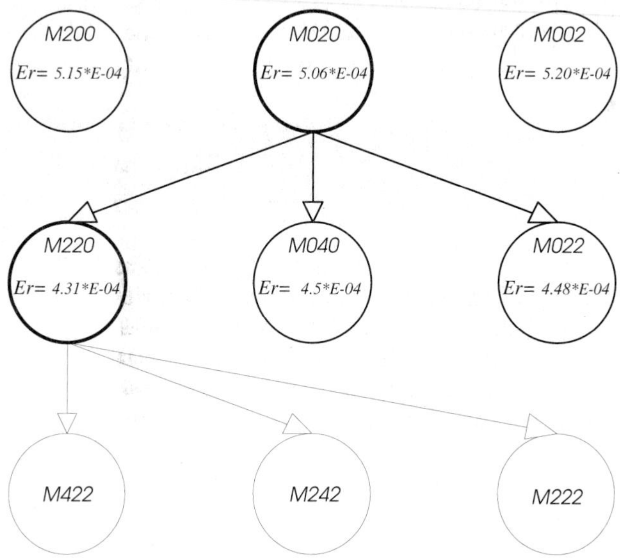

Figure 4.3 The search tree for the fuzzy model.

In Figure 4.3, each circle represents a model. The circles of the same layer refer to models of the same complexity, e.g., the first layer refers to the models that contain two fuzzy sets on one variable and so on. For each layer, the thick circle indicates the model that has the minimum E_r value among the models of the same complexity. This model has been chosen for the next step.

Finally the model obtained was labeled M220. In fact, the models with eight fuzzy sets have been discarded, due to the large number of parameters involved. This number becomes too large when it is compared with the number of the L_{aeqT} measurements (about 350) such a condition makes the problem undefinable and numerical drawbacks may arise.

The chosen model contains two fuzzy sets, labeled *small* and *large*, for the equivalent number of vehicles (n_{eq}) and the average building height (h). No fuzzy set was considered for the road width. However, this last variable was considered in the consequences of the fuzzy rules. The fuzzy model contains four fuzzy rules and is reported in Figure 4.4.

Figure 4.4 reports the values of the parameters that identify the fuzzy sets of the premises and the coefficients of the consequence polynomials.

A schematic example of how the obtained model works is reported in Table 4.2 where it has been supposed that the following values for the input variables are available:

$$\overline{n}_{eq} = 5000 \text{ 1/h}$$

$$\overline{h} = 15.0 \text{ m}$$

$$\overline{w} = 30.0 \text{ m}$$

The output value of the whole algorithm depends on the outputs of the four rules, reported in the 7th column of Table 4.2, weighted by the values reported in the 6th column:

$$y = \frac{\sum_{i=1}^{4} L^i_{eq} * \left| L_{AeqT} = L^i_{AeqT} \right|}{\sum_{i=1}^{4} \left| L_{AeqT} = L^i_{AeqT} \right|} = 77.76 \text{ dB}$$

Table 4.2 A schematic example of how the fuzzy model works

	(n_{eq} is small)	(n_{eq} is large)	(h is small)	(h is large)	$\left\|L_{AeqT} = L_{AeqT^i}\right\|$	$L_{AeqT} = L_{AeqT^i}$
Rule 1	0.57		0.88		0.57	–211
Rule 2	0.57			0.11	0.11	–64
Rule 3		0.5	0.88		0.50	400
Rule 4		0.5		0.11	0.50	252

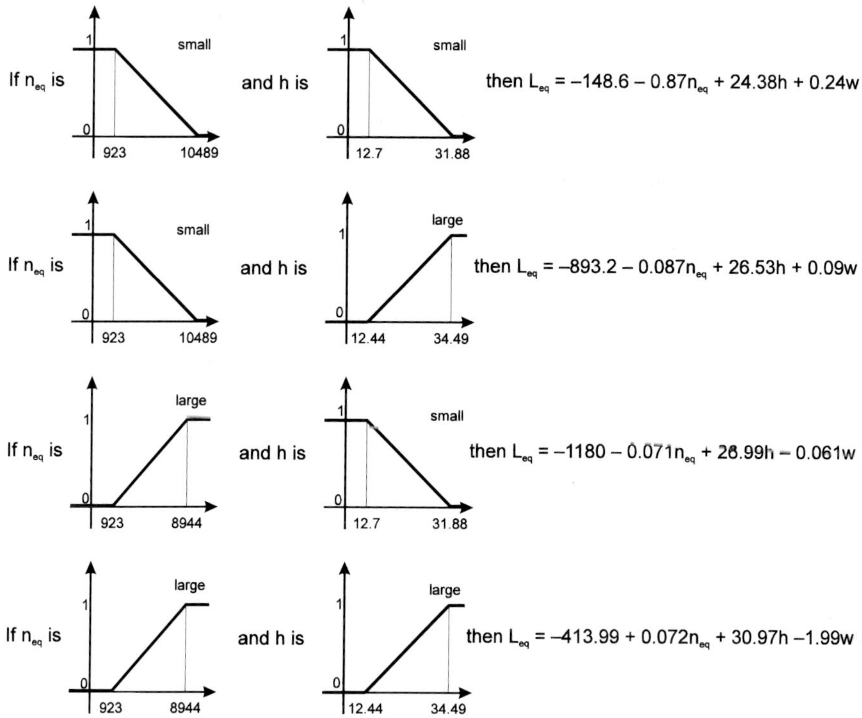

Figure 4.4. The fuzzy model for acoustic pollution

It should be observed that the L_{AeqT^i} values obtained, though expressed in decibels, do not make any physical sense and negative values occurred.

The performance of the proposed model was tested by using both data used for the estimation of the fuzzy models and data not considered for the model identification. The results obtained with the fuzzy model described above for the latter set of data were compared with both those obtained with the model represented by Equation (4-5) and the measured data. This comparison is reported in Figure 4.5.

A visual inspection of Figure 4.5 shows that the fuzzy model has a better capacity to fit the measured data, with respect to the classical empirical model. Even though the fuzzy approach revealed a good performance, the optmization phase, needed to derive the model, required a great computational effort. To avoid this drawback a neural network approach was considered, as described in the following paragraphs.

4.5 THE NEURAL MODEL

On account of the function approximation capabilities offered by certain classes of neural networks, and in particular by Multi-Layer Perceptron (MLP) with sigmoidal activation function for the hidden units [18,19], a neural approach has been also adopted to identify the traffic noise model. The MLP is, in fact, a mapping network which can

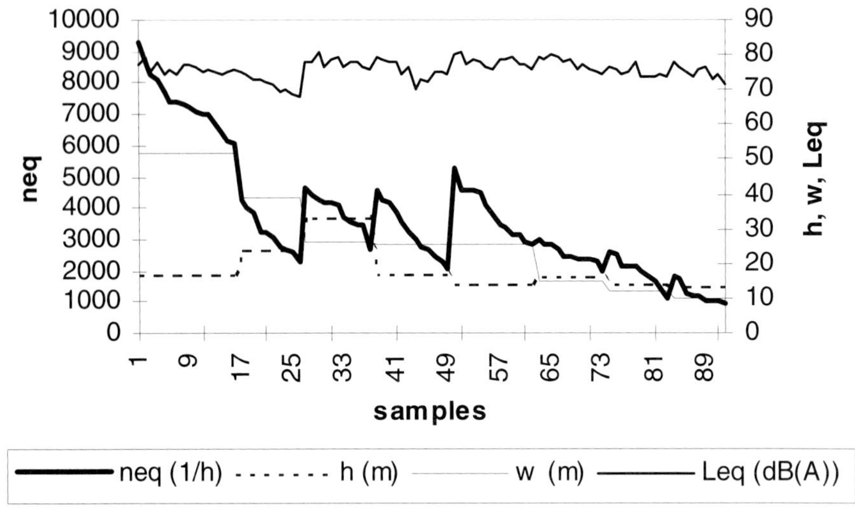

Figure 4.5 Comparison between the measured data, the classical model and the fuzzy model.

approximate, with the desired accuracy, a nonlinear continuous function $f(\cdot)$ from a finite subset of n-dimensional Euclidean space $A \subset \Re^n$ to a finite subset $f[A] \subset A \subset \Re^m$ of m-dimensional Euclidean space, using a set of examples of correct mapping $((x_1, y_1),..,(x_k, y_k),...,($with $x_k \in A$ and $y_k = f(x_k)))$ [18].

The MLP network has been used to determine the functional relationship between the equivalent sound pressure level and the three urban parameters—the equivalent number of vehicles, the average height of the buildings and the width of the road as previously done with the fuzzy approach.

The MLP therefore has three inputs, i.e., n_{eq}, h, w, and a single output represented by the L_{AeqT} value. Among the whole set of measurements described above, a subset of 250 data has been used to train the MLP with the back-propagation algorithm while a different set of 92 data, not used during the learning phase, was considered to test the model.

A key issue in the use of neural networks like the MLP is the determination of the most suitable architecture for the particular problem to be solved. Defining the architecture means to fix particular parameters which characterize it, the most critical of which are the number of hidden slabs and the number of neurons in each hidden slab. In fact while too low a number of hidden neurons causes a bad approximation, too large a number of hidden units determines a network with poor generalization capabilities. Without loss of generality, networks with one hidden layer were considered.

In order to determine the suitable number of hidden neurons, several structures were trained with the same set of data. The best performance was obtained with 9 hidden neurons. In Figure 4.6 the comparison between the neural model, the classical model, and the measured data used to test the model is reported.

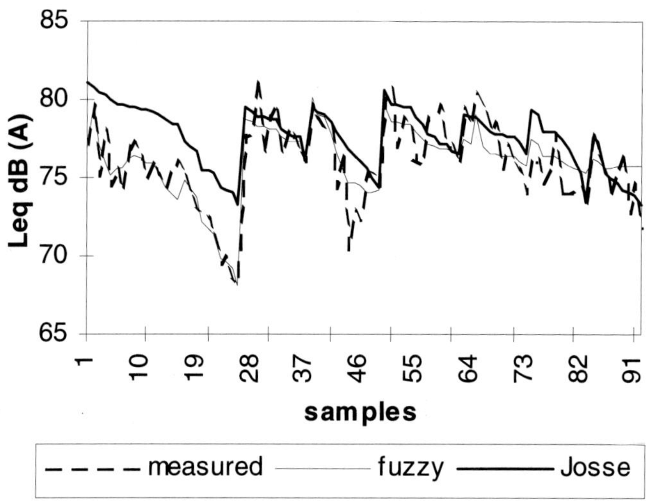

Figure 4.6 The comparison between of the classical model, the neural approach and the measured one.

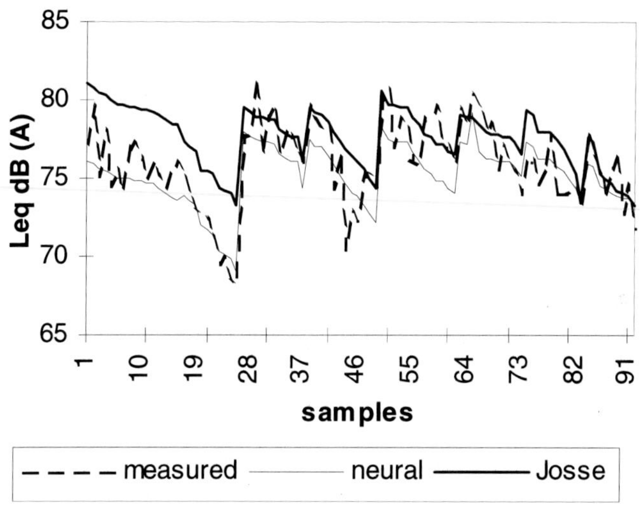

Figure 4.7 The comparison between of the fuzzy model, the neural approach, and the measured one.

A comparison between the performances of the neural and the fuzzy model is reported in Figure 4.7, where the output of the two models and the measured data are depicted.

As can be observed from Figure 4.7 the fuzzy model behaves better than the neural one, even if the complexity of the computation is greater. This suggested using the neuro-fuzzy approach in order to merge the advantages of both the strategies.

4.6 NEURO-FUZZY NETWORKS

Neuro-Fuzzy Networks (FNNs) have been proposed in order to join the advantages of both neural network theory and fuzzy logic in dynamical system control and identification problems [9,15].

The basic idea of the composition method of the FNN is to realize a fuzzy reasoning by a neural network whose weights represent the parameters associated with a set of fuzzy rules. FNNs are designed in order to automatically identify the fuzzy rules, tuning both the shape of the membership functions and their positions in the universe of discourse. Such a goal is obtained by modifying the weights of the connections of a suitable defined feedforward neural network with a learning procedure based on the back-propagation algorithm.

In the following the fundamental characteristic of FNNs, in comparison with neural networks and fuzzy systems, are outlined:

- FNNs realize an automatic procedure for deriving, in the same time, both the consequents and the antecedents of a set of fuzzy rules starting from a set of I/O data of a system; moreover they also tune the shape of the membership functions.
- FNNs require a number of parameters which is very small with respect to the number of connections of a MLP; moreover the number of neurons of a FNN is fully determined from the number of membership functions on each input variable;
- FNNs require a short training phase due to the possibility of fixing a "meaningful" initial configuration for the parameters; moreover FNNs allow including the knowledge of an expert in choosing the topology; this is not possible for MLP;
- FNNs leads to a model of the system which is easily understandable in comparison to the model obtained from a MLP;

In order to derive a learning algorithm for a FNN with a gradient descent technique, the inference rule must lead to a crisp output which is a differentiable function of its parameters. The "product" operator is therefore adopted instead of the "min" operator in order to evaluate the activation degree of each rule and the membership functions are built as a composition of differentiable functions, as for example, sigmoidal functions, instead of triangular or trapezoidal ones.

Several types of FNN structures have been proposed in literature [9,15], which differ for the consequent part of the fuzzy rules, for the type of inference rule and the shape of the membership functions. One of the possible structures of FNN is described in [9,15,19] and implements a set of ifthen rules with sigmoidal-type membership functions and crisp consequents. In Figure 4.8 the described FNN is reported in the case of two inputs, one output and three membership functions in each premise. The structure of the rules realized with this network is therefore:

$$R_i\text{: if } x_1 \text{ is } A_{j,1} \text{ and } x_2 \text{ is } A_{k,2} \text{ then } y \text{ is } y_i \tag{4-11}$$

where j, k = 1..3 indicates the membership function for each variable; i = 1..9 indicates the ith rule; y_i is the ith output; and $A_{j,h}$ is the jth membership function associated with the hth input (h = 1, 2).

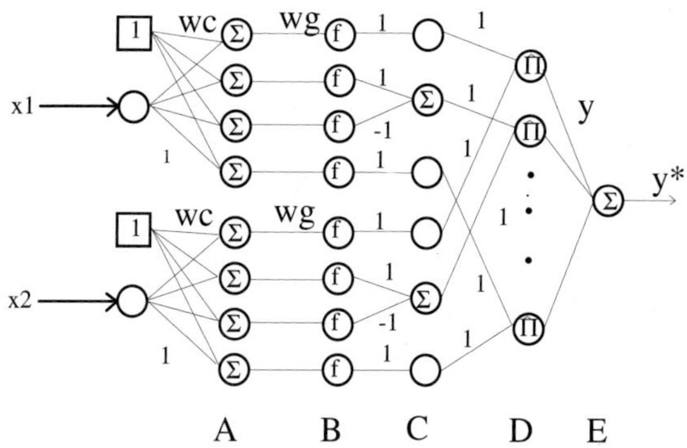

Figure 4.8 The structure of a FNN in which the consequents are constants.

The output of the network is realized as a combination of the outputs of each rule and is denoted with y^*.

As can be observed from Figure 4.8, the topology of the network is fully determined from the number of inputs and the number of membership functions for each input and therefore a "trial and error" phase is not required in order to determine the number of neurons. However the number of membership functions for each variable must be chosen.

The network layers can be viewed as pipeline stages corresponding to each step of the implemented fuzzy reasoning. In particular, the antecedent part is processed in the layers from A to C. The output of each neuron of layer D represents the activation degree of the corresponding fuzzy rule. The remaining layer processes the consequent part of the rules and computes the crisp output.

The relations which describe the forward phase of the network are reported in the following:

$$O_{h,s}^{A} = wc_{h,s} + x_h \qquad h=1,\ 2;\quad s=1..4;\quad p=1..3;\ q=1..3 \tag{4-12}$$

$$O_{h,s}^{B} = \frac{1}{1+\exp(-wg_{h,s}O_{h,s}^{A})} \tag{4-13}$$

$$O_{h,1}^{C} = O_{h,1}^{B} \tag{4-14}$$

$$O_{h,2}^{C} = O_{h,2}^{B} - O_{h,3}^{B} \tag{4-15}$$

$$O_{h,3}^{C} = O_{h,4}^{B} \tag{4-16}$$

$$O_{i}^{D} = \frac{O_{1,p}^{C}O_{2,q}^{C}}{\sum_{i=1}^{9} O_{i}^{D}} \tag{4-17}$$

$$O^E = y^* = \sum_{i=1}^{9} y_i O_i^D \tag{4-18}$$

where O^L denotes the output of the Lth layer

As can be observed from the reported relations, the weights "wc" and "wg" determine the shape of the membership functions and their allocation in the universe of discourse, while the weights y_i represent the crisp output of each rule.

By suitably choosing the connection weights "wc" and "wg", the membership functions can be initialized as shown in Figure 4.9, where the membership functions "medium" is composed of the sum of two sigmoidal function, as can be observed from Equation (4-15).

However the considered structure can be easily extended to a generic number of inputs and membership functions.

Even if more complex structures which realize different forms of the consequent parts might seem more adequate to approach a complex problem or to obtain better performances, some considerations should be kept in mind.

At first, it can be proven [19] that, under certain conditions, the simplified topology of Figure 4.8 produces fuzzy inference systems which are able to arbitrarily approximate quite accurately any nonlinear function on a given compact set.

Nevertheless this is true also for other topologies, they often requires a time-consuming training phase, due to the complexity of the defuzzification process, which does not result in a corresponding improvement of the network performance.

A further modification can include Gaussian membership functions instead of sigmoids. In this case the FNN structure can be represented as in Figure 4.10.

The relations which describe the forward phase of the considered FNN are reported in the following.

$$O_{h,s}^A = x_h - wc_{h,s} \qquad h = 1,2; \quad s = 1,2,3; \quad i = 1..9; \; p = 1,2,3 \tag{4-19}$$

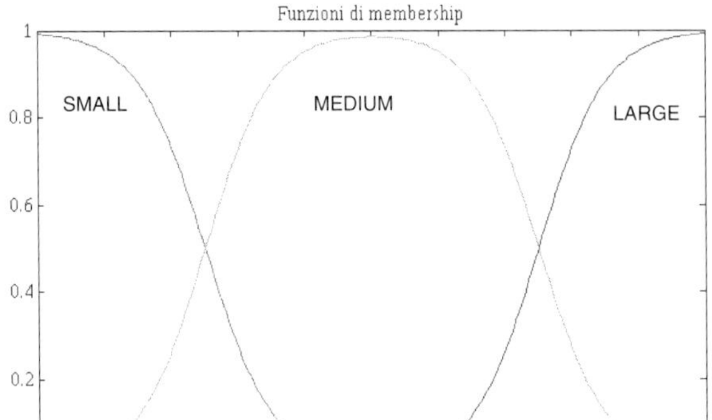

Figure 4.9 The initial configuration of the membership functions.

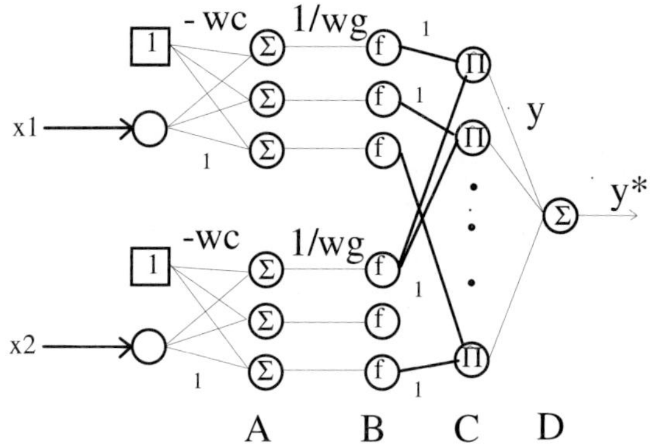

Fig 4.10 The structure of a FNN with Gaussian membership functions.

$$O_{h,s}^{B} = \exp\left(-\left(\frac{O_{h,s}^{A}}{wg_{h,s}}\right)^{2}\right) \tag{4-20}$$

$$O_{i}^{C} = \frac{O_{1,s}^{B} O_{2,p}^{B}}{\sum_{i=1}^{9} O_{i}^{C}} \tag{4-21}$$

$$O^{D} = y^{*} = \sum_{i=1}^{9} y_{i} O_{i}^{C} \tag{4-22}$$

Also in this case, O^L denotes the output of the Lth layer

In the following application, the FNN structure with Gaussian membership functions has been adopted.

The learning algorithm for the proposed FNN can be derived by suitably modifying the classical back-propagation algorithm developed for multi layer perceptrons [8].

The quadratic performance index that will be minimized is defined as:

$$E = \sum_{i}^{N} \left(d_{i} - o_{i}\right)^{2} \tag{4-23}$$

where d_i is the desired value of the ith output of layer F, o_i.

The weights are updated in order to perform a gradient descent on the error surface, as in the classical back-propagation algorithm and the derivatives are computed by applying the chain rule.

The general weights updating formula can be written as:

$$w(t+1) = w(t) + \varepsilon \frac{\partial E}{\partial w} \tag{4-24}$$

where ε is the learning rate.

In order to perform the sound equivalent pressure prediction with FNNs, the structure with Gaussian activation functions was adopted. A set of 250 patterns was used during the learning phase, while the remaining part of measurements was considered to evaluate the approximation capability of the network. In order to determine the topology of the FNN, two fuzzy sets for each input variable were chosen, obtaining a set of 8 rules. Referring to Figure 4.10, the following number of neurons are fixed:

- input layer: 3 inputs;
- layer A: 6 neurons;
- layer B: 6 neurons;
- layer C: 8 neurons;
- layer D: 1 neuron.

The learning phase required about 300 iterations.

The membership functions obtained for each variable are reported in Figure 4.11(a), (b), (c) where "input 1" is the equivalent number of vehicles, "input 2" is the height of the buildings and "input 3" is the width of the roads .The values of the consequents are reported in Figure 4.11(d).

The obtained model has 8 rules which are described in the Tables 4.3, 4.4, and 4.5 in more detail.

Table 4.3 The description of the input variables fuzzy sets

Input Variable Names	Fuzzy Set Names	Centroids
Input Variable 1	FUZZYSET1_1	–0.525801
	FUZZYSET1_2	0.583036
Input Variable 2	FUZZYSET2_1	–0.766005
	FUZZYSET2_2	0.558134
Input Variable 3	FUZZYSET3_1	–0.460455
	FUZZYSET3_2	0.565475

Table 4.4 The description of the consequences

Consequences Names	Values
CONS1_1	0.248745
CONS1_2	0.777778
CONS1_3	0.508815
CONS1_4	–0.582857
CONS1_5	–0.171717
CONS1_6	0.350870
CONS1_7	0.212121
CONS1_8	–0.030303

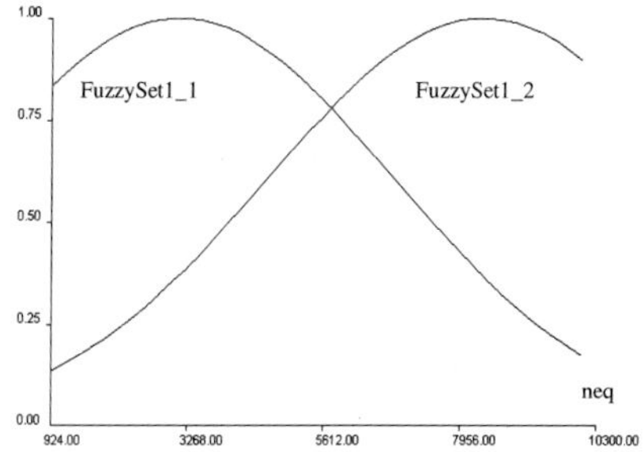

(a) Fuzzy sets for input 1

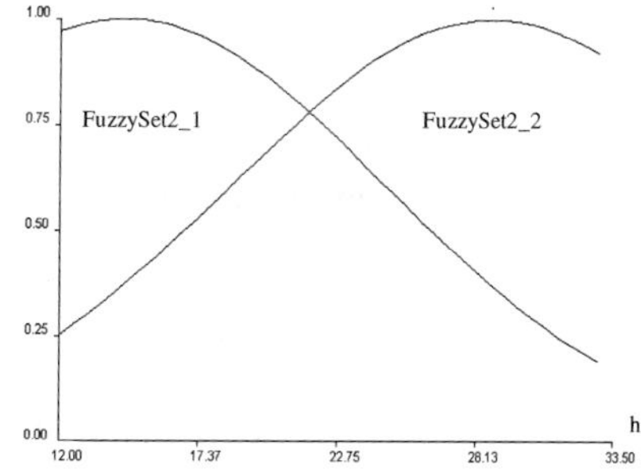

(b) Fuzzy sets for input 2

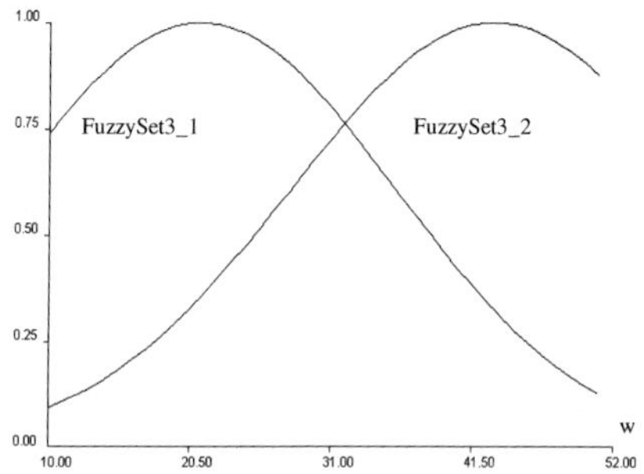

(c) Fuzzy sets for input 3

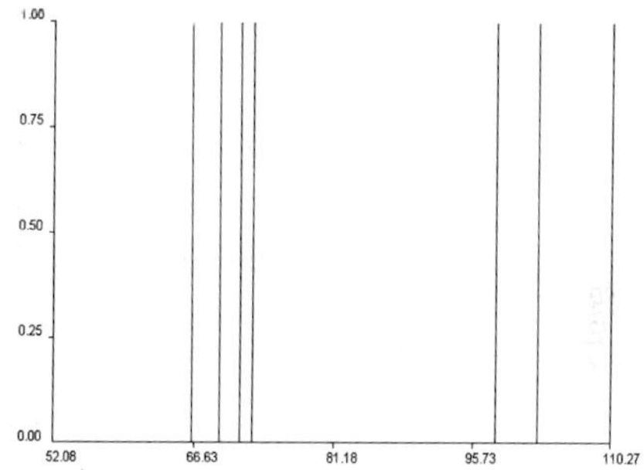

(d) Consequents of the rules set

Figure 4.11

Table 4.5 The obtained set of rules

R1	IF neq is FUZZYSET1_1 AND h is FUZZYSET2_1 AND w is FUZZYSET3_1 THEN Leq is CONS1_1
R2	IF neq is FUZZYSET1_1 AND h is FUZZYSET2_1 AND w is FUZZYSET3_2 THEN Leq is CONS1_2
R3	IF neq is FUZZYSET1_1 AND h is FUZZYSET2_2 AND w is FUZZYSET3_1 THEN Leq is CONS1_3
R4	IF neq is FUZZYSET1_1 AND h is FUZZYSET2_2 AND w is FUZZYSET3_2 THEN Leq is CONS1_4
R5	IF neq is FUZZYSET1_2 AND h is FUZZYSET2_1 AND w is FUZZYSET3_1 THEN Leq is CONS1_5
R6	IF neq is FUZZYSET1_2 AND h is FUZZYSET2_1 AND w is FUZZYSET3_2 THEN Leq is CONS1_6
R7	IF neq is FUZZYSET1_2 AND h is FUZZYSET2_2 AND w is FUZZYSET3_1 THEN Leq is CONS1_7
R8	IF neq is FUZZYSET1_2 AND h is FUZZYSET2_2 AND w is FUZZYSET3_2 THEN Leq is CONS1_8

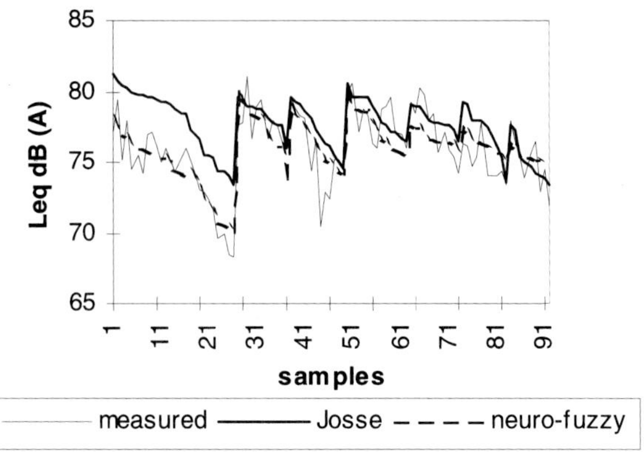

Figure 4.12 The comparison between the classical model, the neuro-fuzzy model and the measured data.

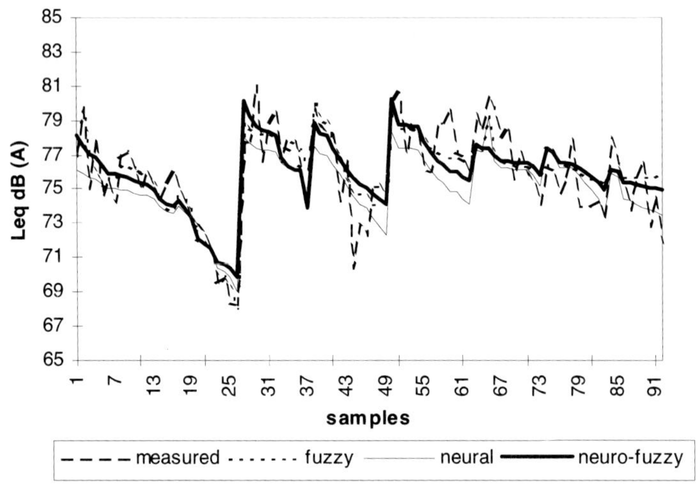

Figure 4.13 The comparison between the neuro-fuzzy model, the fuzzy model, the neural model and the measured data.

Figure 4.12 reports the comparison between the obtained neuro-fuzzy model, the Josse model and the measured data.

In order to verify the performance of all the obtained models, Figure 4.13 is also presented. As can be observed from the figure, the neuro-fuzzy model has intermediary performance with respect to the fuzzy and the neural models. Taken into account the low computational approach of the neuro-fuzzy modeling strategy in comparison with the classical fuzzy approach, its performance can be considered quite satisfactory.

4.7 CONCLUSIONS

In this chapter the problem of predicting traffic noise pollution has been approached with different soft-computing techniques of modeling. The classical approaches to noise pollution modeling consist of deriving linear empirical relationships between the equivalent sound pressure level and some parameters which characterize the road section, i.e., the width of the road, the median height of the buildings and the vehicle flow. These linear models do not fit the measured data with sufficient accuracy. New techniques were adopted to derive more accurate models. At first an optimized fuzzy model was considered. Even if the performance of such a model was quite a bit better than the classical one, the great computational effort of the optimization strategy induced us to try a neural network approach. As expected this model involved a reduced computational complexity but the performance obtained was not satisfactory. To join the advantages of both techniques, a neuro-fuzzy model was also derived.

A comparison of the described techniques has shown that the neuro-fuzzy model represents a good compromise between the neural and fuzzy performances and moreover it is a fully automatic procedure which does not require the knowledge of an expert or complex optimization strategies.

4.8 REFERENCES

1. "Public Health and Welfare Criteria for Noise." Rep. N. 550/9-73-002, 1973.

2. "Information on Levels of Environmental Noise Requisite to Protect the Public Health and Welfare with an Adequate Margin of Safety." Rep. N. 550/9-74-004, 1974.

3. Josse. *Notions d"Acoustique*, Ed. Eyrolles, Paris, 1972.

4. Lamure C. and Auzou D. S., "Le Niveaux de Bruit au Voisinage des Autoroutes Dégagées," Cahiers du Centre Scientifique et Tecnique du Batiment, Dec. 1984.

5. Burgess M. A. "Noise Prediction for Urban Traffic Condition. Related to Measurements in the Sydney Metropolitan Area." *Appl. Acoust.*, Vol. 10, No. 1, 1977.

6. Don C. G. and Rees I. G., "Road Traffic Sound Level Distributions" *Journ. of Sound Vib.* Vol. 100, No. 1, 1985, pp. 41-53.

7. Lofti, A. Zadeh "Outline of a New Approach to the Analysis of Complex Systems and Decision Processes." *IEEE Trans. on Syst. Man and Cyb.* Vol. 3, No. 1, 1973.

8. Rumelhart D. E., Hinton G. E. and Williams R. J., "Learning Internal Representation by Error Propagation." *Parallel Distributed Processing: Exploration in the*

Microstructure of Cognition (Rumelhart, D.E., Mc Clelland, J.L., eds.). MIT Press, pp. 318-362.

9. Shing, R. Jang. "ANFIS: Adaptative_Network_Based Fuzzy Inference System." *IEEE Trans. on Sistem Man and Cybernetics*, Vol. 23, Vo. 3, May 1993.

10. DPCM 1.3.1991, All. B. "Limiti massimi di esposizione al rumore negli ambienti abitativi e nell"ambiente estrerno."

11. Beranek L. and Ver, I. L., *Noise and Vibration Control Engineering*, Wiley Interscience, 1992.

12. Harris, C.M. *Handbook of Noise Control*, McGraw-Hill, 1979.

13. Cammarata G., Fichera, A., Graziani, S. and Marletta, L. "Fuzzy Logic for Urban Traffic Noise Prediction." *J. Acoustic. Soc. Am.* October 1995

14. Cammarata G., Cavalieri S. and Fichera, A. "A Neural Network Architecture for Noise Prediction." *Neural Network,* Vol 14, 1995.

15. Baglio, L., Fortuna, M.G., Xibilia, P., Zuccarini, "Neuro-Fuzzy to Predict Urban Traffic." Proc. EUFIT94, Aachen, Germany, September 1994.

16. Takagi, Sugeno. "Fuzzy Identification of Systems and Its Applications to Modeling and Control." *IEEE Trans. on Systems, Man and Cybernetics*, Vol 15, No. 1, 1985.

17. Press, B. I. Flannery, S. A. Teukolsky, W. T. Vetterling. *Numerical Recipes: The Art of Scientific Computing*. Cambridge, University Press, 1986.

18. Cybenko, "Approximation by Superposition of Sigmoidal Functions." *Math. Control Systems*. Vol. 2, 1989, pp. 303-314.

19. Horikawa, T. Furuhushi, Y. Uchikawa, "On Fuzzy-Modeling Using Fuzzy Neural Networks with the Back-Propagation Algorithm." *IEEE Transaction on Neural Networks*. Vol. 3, No. 5, Sept. 1992.

5

SYNERGISTIC FUZZY CONTROL

R. E. King
University of Patras
GREECE

In production chains commonly found in the process industry, the control decisions at each stage of the production chain invariably affect the state of other stages both downstream as well as upstream and it becomes essential therefore that *synergism*, i.e. cooperation, between the various controllers be established in order to avoid interactions which could lead to unsatisfactory performance and even instability of the production chain. The process industry was one of the first to realize the importance of intelligent control for the control of large-scale industrial processes involving many interacting stages which had hitherto defied automatic control by conventional methods. Operating in a typical process environment characterized by vagueness and uncertainty, *soft computing* methods have been developed over the last two decades to resolve this problem. Autonomous fuzzy and neural controllers are two examples of *soft computing* which have recently been applied in the process industry with excellent results. It is obvious that synergism of intelligent controllers is essential in order to maximize production flow, product quality and energy conservation. With the diffusion of these new software techniques for the intelligent control of a broad class of industrial processes, the problem of synergism is becoming acute and techniques for the analysis and design of this class of controllers are called for since much of what is known today relies on empirical expert knowledge.

This chapter describes in some detail a synergistic fuzzy control system. The system, comprising a cluster of fuzzy controllers arranged in a multilayer hierarchical architecture, controls every stage of the cement production process at one of Europe"s largest cement manufacturing plants.

5.1 INTRODUCTION

Intelligent control [1,5,6,27,28] is the discipline that pertains to machines that perform human-like control tasks under uncertain or unfamiliar environments with minimal interaction with human operators. The structure of an intelligent control system is modular: facts and other knowledge about a specific domain are separate from the inference procedure for applying those facts. Intelligent control systems differ from conventional control systems in yet one other aspect: in conventional control systems, knowledge pertinent to the process under control and the methods for using this knowledge are interrelated while in intelligent control systems there is a clear demarcation of knowledge of the process and the information about the process data as well as the inference mechanism for applying this knowledge. Intelligent control systems are especially attractive when the expertise available to control a process is a consequence of empirical associations that have been acquired from operational experience, a situation that is typical in the process industry.

Fuzzy logic [8,12,21,32,33] and neural networks [8,11,19,24] are examples of a new class of soft computing methods being used in control. Both are vehicles for implementing intelligent control systems which are being applied in the process industry with great success. Indeed in the process industry there are numerous examples of conventional control systems being replaced by intelligent control systems. Fuzzy control [2,4,9,16,17,18,23,31], chronologically the first example of soft computing, has led to significant improvements to product quality, productivity and fuel economy in a variety of process industries and is now firmly established as one of the leading *advanced process control* techniques [3,4,7,29] in use today.

Fuzzy control possesses the essential property of working with the nonnumerical, inexact, and vague data which human operators comprehend. Indeed the ability of fuzzy controllers to deal with incomplete and ill-defined information permits implementation of human-like control strategies which have hitherto been impossible using any of the established control methodologies, both classical and modern. Fuzzy control, though by no means a panacea, has offered rational solutions to many production and quality control problems that have plagued the process industry [20,25].

Over the last ten years or so there has been an increasing trend to apply fuzzy control to the individual stages of large production chains. Usually these controllers operate autonomously with no regard for preceding and succeeding stages in the production chain. The overall behavior can, however, be disappointing as a result of inevitable interactions. Just as classical interacting control led to the development of multivariable control, here also it is essential that the effects of interactions of the various autonomous intelligent controllers be taken into account. Antsaklis et al. [1] have presented an extensive discussion on the issue.

In practice the cluster of fuzzy controllers controlling the various stages in a production chain must cooperate so as to satisfy the overall quantitative and qualitative demands of the production process. These new developments are posing anew problems of analysis of the interactive effects of cooperating or *Synergistic Fuzzy Control Systems* (SFCS) [15,16,17,18,31] to control the various stages of large-scale production chains. The enormous potential and importance of such systems and the need for cooperation of their components are apparent. From a practical control viewpoint, it is essential to understand the causal and temporal effects of this synergism and to be able to systematically correct for unwanted interactions which may adversely affect the overall performance of the production process so as to assure operational reliability, integrity and stability. These matters now are the subject of ongoing research.

5.2 SYNERGISTIC RULE-BASED CONTROL

Knowledge in controlling a specific process in a production chain is usually represented as a set of conditional *if-then* rules which reflect human operator understanding on *how* to control the process. These rules are elicited from domain experts such as plant operators, production managers and process engineers through questionnaires and repeated interviews which are essential to minimize rule uncertainty and conflict. Indeed knowledge elicitation has often been quoted as the most critical and difficult stage in the development of intelligent control systems and one which has often limited the wider adoption of this technique. It is evident that inaccurate or incomplete knowledge on how to control a process inevitably leads to unsatisfactory control and even failure. In the past there have indeed been numerous examples of fuzzy control systems that have failed to come up to expectations because of a lack of the necessary knowledge and expertise about the process and how to control it.

Synergism is a desirable feature in the control of any large-scale manufacturing process. Where the manufacturing process is a cascade of production stages, it is inevitable that each stage affects preceding and succeeding stages even if there are intermediate buffer storage areas. It is obvious that proper operation of all stages in the production chain is essential to maintain the desired production flow levels and quality. Such factors as delays and stoppages of equipment either due to equipment malfunction or materials flow problems upstream or downstream in the production chain can and do have adverse effects on the overall production flow. Just as human operators of the various subprocesses in the production chain must communicate and cooperate with their fellow operators elsewhere in a plant when their subprocesses are directly subjected to such adverse condition, so too synergistic intelligent controllers must possess an analogous facility to communicate and exchange information on the state of all the other stages in the chain in order that they may take the necessary corrective or preventive control actions. Thus information, and perhaps knowledge, must be exchanged between the various intelligent controllers in a plant.

An architecture wherein the cluster of cooperating fuzzy controllers are embedded in different layers offers an effective structure for synergism. The architecture proposed by Saridis [1,27,28], based on the principle of "decreasing precision with increasing intelligence" as one moves up the hierarchy, comprises three basic layers: *coordination, organization* and *execution.*

The uppermost *organization* layer of a synergistic fuzzy control system is the production management layer and is responsible for defining the desired product quality assurance parameters as well as the production demands of the overall plant. It is evident that production management skills are emulated by the intelligent controller at this layer. Under normal operating conditions the *organizer* layer operates aperiodically and at very low frequency and is at the command of production management.

Each stage of the production process is supervised by a cluster of interconnected *coordinators* at the intermediate *coordination* layer which lies immediately below the *organization* layer from which it receives commands. In turn the *coordination* layer supplies the upper layer in the hierarchy with the production information essential to maintain product specifications and production quotas. Human skills in controlling the various stages in the plant are emulated at this layer.

The *coordinators* normally operate periodically and are responsible for maintaining supervisory control of the production process by providing the corrective actions necessary to compensate for any deviations in process behavior from that desired. The repetition period for each fuzzy controller is determined from a knowledge of the dynamics of each subprocess and after extensive experimentation. It should be evident that the overall SFCS is therefore a multirate control system.

Control actions on the various subprocesses in the production chain are effected by a cluster of distributed remote units which form the un-intelligent *execution* layer at the lowest layer of the hierarchy of the SFCS. This layer involves conventional Programmable Logic Controllers (PLC) which perform data acquisition and communication, the necessary interlocking and safety aspects of the equipment and classical three term control on the subprocess. The *execution* layer is thus responsible for the low-level control of the production process and the communication of information to and from the real-time data base or a *system blackboard* of the SFCS. The system blackboard is essentially a depository of information pertaining to the state of the process that can be posted and accessed by all the cooperating fuzzy controllers.

The use of distributed cooperating fuzzy controllers to control the various stages of a large production plant necessitates the interchange of information between the various controllers. In a distributed architecture the various components of an SFCS may be viewed as nodes of a Local Area Network (LAN) [as in a classical industrial Digital Control System (DCS)] as shown in Figure 5.1. Information interchange among the fuzzy controllers is achieved through communications protocols attuned to the manufacturing environment.

When centralized control of a production plant is employed, as in the paradigm described in Section 5.3, the fuzzy controllers are normally co-resident, as shown in the

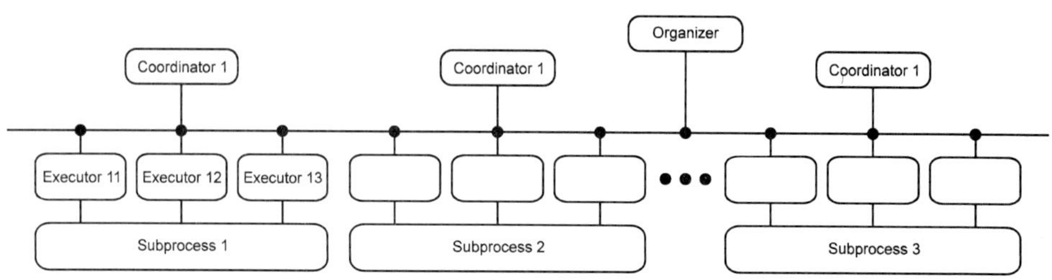

Figure 5.1 The architecture of a distributed network of fuzzy controllers.

schematic of Figure 5.2. With the ready availability of fault tolerant process computers this architecture is a viable alternative for SFCS through its simpler star-like topology. At each layer, information which is pertinent to the functional domain of that layer is exchanged with adjacent layers through a common real-time data base on which all relevant process information is posted. The functional task of each layer is supported by information contributed from adjacent layers. Upon completion of these tasks, decisions are generated that are transmitted to the process actuators via the lowest *execution* layer. In this topology the hierarchical nature of the overall system is explicit.

5.3 THE GENERIC FUZZY CONTROLLER

The implementation of any industrial fuzzy controller [11,31] requires an effective inference mechanism that permits the handling of the linguistic statements used by human operators to control a process. Today there exist numerous software tools in the form of *expert system shells* which complement logic programming languages to ease the task of their implementation [11]. The disadvantage of some of these shells is that they do not always offer the facilities necessary for real-time industrial control of complex large-scale processes.

Fuzzy logic forms the theoretical foundation of fuzzy controllers used as the basic components in any SFCS. The generic fuzzy controller has the general form shown in Figure 5.3. Here each subprocess output vector is first conditioned (implying some degree of signal processing to smooth out random instrumentation noise) prior to being

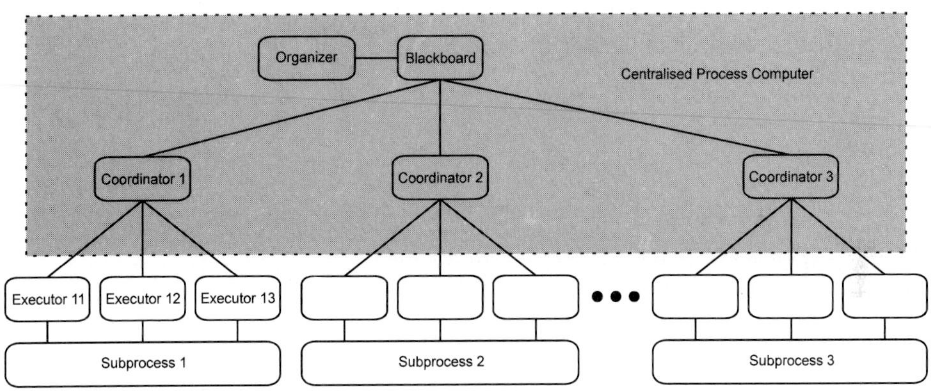

Figure 5.2 The architecture of a centralized SFCS.

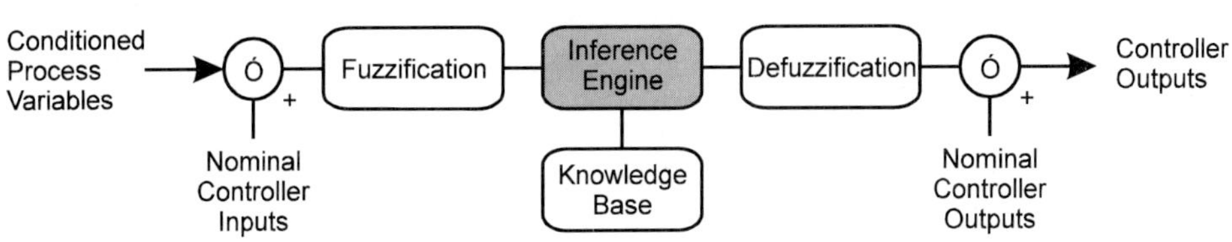

Figure 5.3. Schematic of the generic fuzzy controller elements.

subtracted from the nominal (or desired set-point) operating vector to form the input deviation, or input offset, to an incremental fuzzy controller. The fuzzy controller in turn yields a crisp output deviation which is transmitted to the appropriate low-level controller (typically a PLC) which executes the command. Both input and output nominal set points must be specified by the plant operator interactively and may be changed by him at any time in response to perceived changes in the operating conditions. An experienced plant operator will normally specify the desired subprocess state with considerable accuracy. This assists the fuzzy controller to track the equilibrium conditions rapidly since the controller can only approach the steady-state conditions in a sequence of bounded increments.

The generic fuzzy controller input offsets, which are suitably conditioned, are quantized into a finite number of quanta over the *universe of discourse*. The number of quanta directly defines the *degree of fineness of control* of the controller. Five distinct fuzzy sets, one for each linguistic descriptor (**VL** for very low, **LO** for low, **OK** for acceptable, **HI** for high, and **VH** for high, respectively) were used in specifying the generic fuzzy control rules. Experience has shown that in an industrial environment this number usually suffices to define the rules by which the fuzzy controller infers the desired control action. Rarely, in any case, do human operators describe their control actions in more terms.

The set of fuzzy membership functions associated with the controller input variables are discretized into 13 *singletons* symmetrically ordered about zero over the *universe of discourse* as shown in Figure 5.4. Each fuzzy set may be specified independently. Naturally what each descriptor represents in terms of real subprocess variables is defined by the plant operator who must specify initially the minima and maxima of the process output and control variables through interaction with the intelligent system.

An example of a linguistic rule used in the kiln controller is:

if

the Kiln Power Demand is **HI**

and

the Outlet Oxygen Content is **LO**

and

the Primary Fuel Feed is **OK**

then

make a **LO** change to the Primary Fuel Feed

and

make **OK** (i.e., no) change in the Induced Draught Fan Speed

This is precisely the manner in which plant operators are trained to control the process. Such expressions specify fuzzy relationships between the subprocess input linguistic variables and the corresponding changes that have to be made to the subprocess output variables. Rules may be conveniently displayed as chromatic pixels [20] as shown in Figure 5.5 (shown here monochromatically for lack of color). Each color of the tile pattern in the bounded control space specifies its linguistic value. With a capacity of 3 inputs and 3 outputs, the generic fuzzy controller has a rule base which can contain a

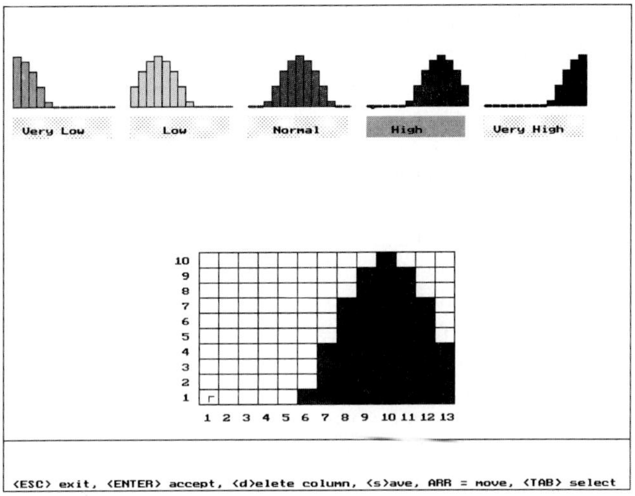

Figure 5.4 Fuzzy membership functions.

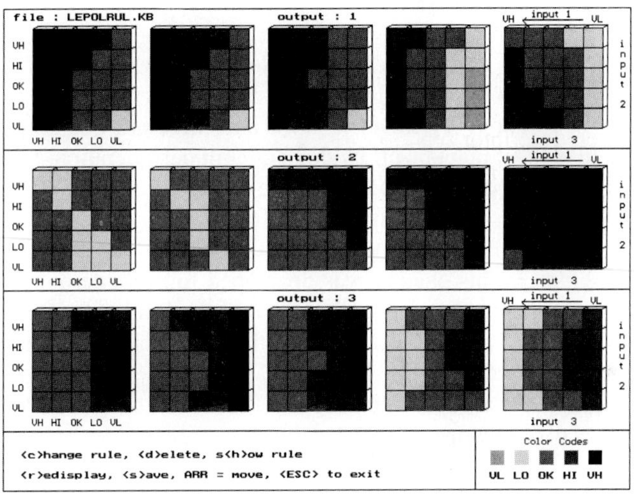

Figure 5.5 Rulebase depicted in pixel form.

maximum of 5×5×5 = 125 rules. In practice not all of these rules need be specified and consequently rule bases are generally sparse.

At specified time instants the process variables are sampled and fuzzified. Every rule in the rulebase is then examined to ascertain to what degree it participates (i.e. its "degree of fulfillment") in the final control decision [11]. In practice only a few rules contribute towards the final decision. Depending on whether the Mamdani or the Larsen inference [10,11,23,24] is selected, the results are combined to yield the composite output membership function.

To derive the crisp incremental control action which must be transmitted to the executive layer and then applied to the process actuators, the composite output membership function must accordingly be defuzzified. Though the Center of Area (COA) defuzzification method is normally used, the Mean of Maxima (MOM) method may alternatively be selected.

Finally, in implementing the generic fuzzy controller, it proved convenient to compute the outputs sequentially, i.e., to partition the controller into a cascade of multi-input-single-output controllers. This minimizes the workload considerably and the controller can thereby have any number of outputs. In all cases the principal control variables never exceed three; thus the generic controller was limited to this number of inputs.

The software was written in FORTRAN and the platform used is a DEC PDP11/44 operating under RSX11M. Overall computing time for each multivariable controller is of the order of 10 seconds. This includes the communications time.

5.4 APPLICATION PARADIGM: SYNERGISTIC CONTROL IN A CEMENT PLANT

5.4.1 The Cement Production Chain

The cement industry has difficult control problems which hitherto have defied solution using any of the conventional analytic control techniques. Indeed this process industry was one of the first to adopt intelligent control and there are hundreds of examples of fuzzy controllers in cement plants worldwide in continuous operation. Initially introduced to control the kilning process [8,9,12,13,14], the most critical and difficult stage in the cement production chain, these intelligent controllers are now being used to control virtually all stages of the cement production chain.

The SFCS presented in this chapter was developed for a modern cement plant with a production capacity of 1.8 m t/a, comprising one raw meal mill, one coal-fired rotary kiln with a 5-stage precalciner, grate cooler, and ancillary coal mill, and two finish ball mills. The plant is situated on the island of Euboea in Greece. A schematic of the production chain and materials flow is shown in Figure 5.6 and a brief outline of the various production stages is given below.

The basic components of clinker (the principal component of the end product cement), limestone (quarried on-site), shale, iron pyrites, and low alkaline siliceous additives, are blended together and ground to a fine meal in a roller *raw mill*. Samples of the raw meal are taken periodically via pneumatic sampler and analyzed spectrographically. From this analysis raw meal indices are computed and compared to those desired. Should deviations occur from the desired specifications, then minor corrections must be made to the component feed rates to the mill. At all times the internal conditions in the mill must be controlled to maintain the desired temperature, humidity, etc., of the meal.

The next stage in the cement production chain is the *rotary kiln*, its precalciners and ancillary grate coolers, all of which constitute the principal production stage. In parallel there is a *coal mill* supplying granulated coal to the fuel burners of the kiln. In the kiln, the raw meal undergoes a gradual chemical transition, culminating in the production of

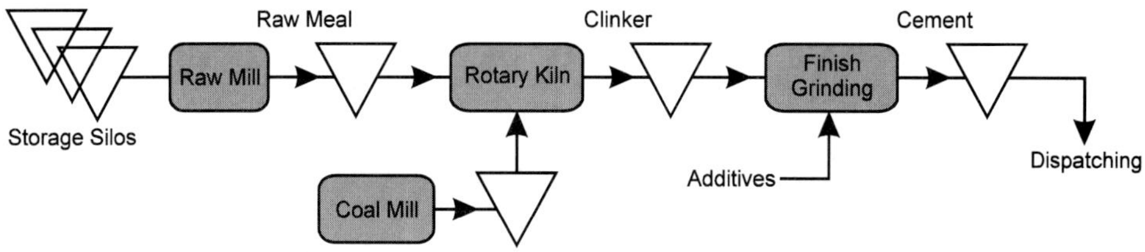

Figure 5.6 Simplified materials flow in the cement production chain.

clinker. Clinker is the primary component of the end product cement, accounting for up to 95% of cement in the case of high quality Portland cement.

Shown also in Figure 5.6 are the intermediate storage *silos*, or buffers, between the various stages in the production chain. These are essential in order to compensate for the various production rates of the intermediate stages and to maintain an even materials flow. Thus limited stoppages due to equipment malfunctions or materials interruptions can be accommodated without disrupting the normal materials flow.

The final stage in the cement production chain is the *finish grinding mill* in which gypsum (a retardant) and other additives are added to the clinker from the preceding stage and ground to a fine powder. The end product of this chain is *cement.* In order to conform to international standards samples of the cement produced must be continuously taken and analyzed as to their chemical properties. Should deviations from these standards be found, then appropriate corrections must be made to the materials components.

5.4.2 The HERACLES Synergistic Fuzzy Control System [18]

A cluster of cooperating fuzzy control systems structured in a multilayer SFCS architecture was designed, implemented and commissioned in 1986 at the HERACLES II plant on the island of Euboea, the newest and most modern plant of the General Cement Company in Greece. The SFCS, whose components are shown schematically in Figure 5.7 involves one fuzzy controller in the role of *organizer* in which the desired quality specifications on clinker CaO and $CaCO_3$ are specified. In turn the *organizer* computes the nominal kiln power demand and the kiln precalciner feed rates and temperatures in order to yield the desired product quality.

Eight fuzzy controllers in the *coordinator* layer control every stage in the cement production chain. Two fuzzy controllers support the raw mill, three support the rotary kiln and its grate cooler, one supports the coal mill and one each support the two finish grinding mills. The SFCS was developed in stages over a period of three years and was completed in 1986. When commissioned, the HERACLES SFCS had one of the highest *Machine Intelligence Quotients* (MIQ) of any cement production plant in Europe.

The *Raw Mill* (with a maximum capacity of 400 t/h) is controlled by two fuzzy controllers: the first, identified as *Mill Control* in Figure 5.7, controls the raw feed rate, the induction fan speed and the temperature of the water spray feed to the mill. The inputs to this controller are the mill underpressure, the meal exit temperature, and the mill differential pressure. The sampling period is set at 10 minutes. The second fuzzy controller, identified as *Blend Control* in Figure 5.7, performs raw materials blending in

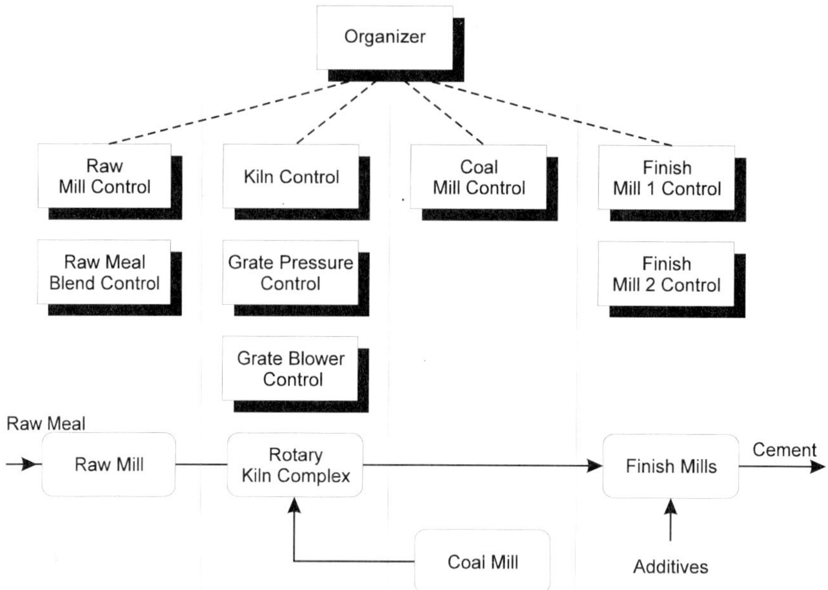

Figure 5.7 The components of the HERACLES SFCS.

response to directives from periodic on-line measurements or laboratory analyses. This controller proved superior to the conventional analytical blending system which had been in use previously.

The *Rotary Kiln* system is by far the most complex and synergistic. No less than three cooperating fuzzy controllers are required to control the kiln/precalciner complex and its ancillary grate cooler. The function of the primary *Kiln Controller* is to maintain the required burning conditions within the kiln. This is achieved by measuring the kiln motor power, the fuel feed rate, and the oxygen content at the top of the precalciner. The corresponding controller outputs are the speed of the precalciner induction fan and the kiln fuel feed rate. Two *Grate Controllers* are required to maintain the pressure and temperature in the cooler at the required levels. Thus in the first, the fuel feed to the kiln, the secondary air temperature from the kiln, and the tertiary air temperatures in the grate cooler are used to control the grate speed. In the second fuzzy controller, the kiln power, the total raw feed to the kiln, and the kiln fuel rate in conjunction with the grate tertiary air temperature control a series of grate blowers in parallel.

The *Coal Mill* fuzzy control system controls the materials feed rate, the recirculation rate and the entry damper. The controller inputs are the coal entry temperature, the mill underpressure and differential pressure. Unlike the other fuzzy controllers, this controller operates periodically, i.e., in batch mode for a time sufficient to produce an adequate supply of granulated coal to the rotary kiln and precalciner. Fast data sampling and computing are therefore essential in this case and a sampling period of 60 seconds is used.

Finally the two *Finish Mills* (each with a capacity of 120 t/h) fuzzy controllers maintain cement fineness control. Each controller controls the total feed rate to each mill, their separator speed, and the returns damper. Inputs to the controller are cement Blaine measurements obtained automatically or from conventional analytical lab analysis, the mill power, and the mill underpressure. The fuzzy controllers maintain Blaine deviations to within ±3% of the desired specifications as against over 10% under manual control.

5.5 CONCLUSIONS

The primary objectives of the SFCS are productivity improvement, quality assurance and energy reduction, tasks which the system has performed extremely well. The system has yielded figures of the order of 5% for productivity and energy reduction since its introduction, paying off its development in a matter of a few months. These are typical figures reported for the improvements brought about by the introduction of fuzzy controllers.

The process industry is now reaping the benefits of many years of theory and research in fuzzy logic. Fuzzy control, an early contender in the field of *soft computing* for *advanced process control*, has amply demonstrated its usefulness in hundreds of applications in a variety of applications over the past two decades or so. The further diffusion of synergism in the process industry will offer new opportunities for enhanced product quality and productivity as well as energy saving.

5.6 ACKNOWLEDGMENT

The work described in this chapter was funded by the HERACLES General Cement Co. of Greece as part of an extensive project to introduce modern advanced process control methods to the company's cement production plants. The author acknowledges the support of the management in this endeavor. The author is thankful to E. Anagnostou, now Plant Manager at the HERACLES II plant whose expertise and knowledge were elicited to establish the rule bases for the SFCS. Special thanks are due to F. C. Karonis, formerly with the HERACLES Group and now Principal Engineer with CAP SESA Tertiaire in Paris, France, for his invaluable contributions to the project.

5.7 REFERENCES

1. Antsaklis, P. J. and Passino, K. M. "Towards Intelligent Autonomous Control Systems: Architecture and Fundamental Issues", *J. of Intelligent and Robotic Systems*, Vol. 1, 1989, pp. 315-342.

2. Åström, K. J., Anton, J. J. and Arzen, K.-E. "Fuzzy Control." *Automatica*. Vol. 22, No. 3, 1986.

3. Bernard, J. A., "Use of a Rulebased System for Process Control", *IEEE Control Systems Magazine*. Vol. 8, No. 5, 1988.

4. Boverie, S., Demaya, B. and Titli, A. "Fuzzy Logic Control Compared with Other Automatic Control Approaches", *Proc. 30th CDC Conference*, Brighton, England, 1991.

5. Francis, J. C. and Leitch, R. R. "Intelligent Knowledge Based Process Control", *Proc IEE Conference on Control*, London, 1985.

6. Gupta, M. M., Saridis, G. N. and Gaines, B. R. (eds.). *Fuzzy Automata and Decision Processes*, Elsevier/North Holland, 1977.

7. Harmon, Ray W. *Advanced Process Control*. Butterworths, 1989.

8. Harris, C. J., Moore, C. G. and Brown, M. *Intelligent Control—Aspects of Fuzzy Logic and Neural Nets*. World Scientific, 1993.

9. Holmblad, L. P. "Experience with Automatic Kiln Control Applying Computers and Fuzzy Logic." *Proc VDZ Congress*, Dusseldorf, Germany, 1985.

10. Holmblad, L. P. and Ostergaard, J. J. "Fuzzy logic control: Operator Experience Applied to Automatic Process Control." *Zement, Kalk, Gips*. Vol. 34, No. 3, 1988.

11. Hunt, K. J., "Neural networks for control—a survey." *Automatica*. Vol. 28, No. 6, 1992.

12. Jamshidi, M., Vadiee N. and Ross T. J. (eds.) *Fuzzy Logic and Control—Software and Hardware Applications*. Prentice-Hall, 1993.

13. Karonis, F. C. and R. E. King "Knowledge-Based Expert Systems in Process Automation." *Proc. First Intl. Fuzzy Sets Association Conference,* Majorca, 1985.

14. King, R. E., "Fuzzy Logic Control of a Kiln Precalciner Flash Furnace", *Proc. IEEE Conference on Adaptive and Multivariable Control*, Hull, 1982.

15. King, R. E. and Karonis, F. C. "Rule-Based Systems in the Process Industry." *Proc. 25th IEEE CDC.* Athens, 1986.

16. King, R. E. and Karonis, F. C. "Multi-Layer Expert Control of a Large-Scale Industrial Process." In *Fuzzy Computing*. (Gupta, M. M. and Yamakawa, T. eds.). Elsevier Science Publishers, 1988.

17. King, R. E., "Distributed Intelligent Industrial Control—a Process Industry Paradigm." *Proc. IEEE Conference on Control ICCON"89*, Jerusalem, 1989.

18. King, R. E., "Expert Supervision and Control of a Large-Scale Plant." *J. Intelligent and Robotic Systems,* Vol. 2, No. 3, 1992.

19. King, R. E., Magoulas, G.D. and Stathaki A. "Multivariable Fuzzy Controller Design." *IFAC Control Engineering Practice*. Vol. 2, No. 3, 1993.

20. King, R. E. and A. Stathaki, "Rule-Based Control of Mechatronic Systems." *Proc. Intl. Conf. on Recent Advances in Mechatronics—ICRAM"95*, Istanbul, Aug. 1995.

21. Kobayashi, S., "The Present and Future in Expert Systems Technology", *Control Theory and Advanced Technology*, Vol. 2, No. 3, 1986.

22. Kosko, B. *Neural Networks and Fuzzy Systems*. Prentice-Hall, 1992.

23. Mamdani, E. H. "An Application of Fuzzy Algorithms for the Control of a Dynamic Plant." *Proc. IEE*. Vol. 121, No. 12, 1974.

24. Negoita, C. V. *Expert Systems And Fuzzy Systems*. Benjamin Cummings Publishing Co., 1985.

25. Olsson, G. and Piani, G. *Computer Systems for Automation and Control*. Prentice Hall, 1992.

26. Perray, K. E. and Waddel, J. J. *The Rotary Cement Kiln*. Chemical Publ. Co., 1972.

27. Saridis, G. N. "Toward the Realization of Intelligent Control." *Proc. IEEE*, Vol. 87, No. 8, 1979.

28. Saridis, G. N. and Valavanis, K. P. "Analytical Design of Intelligent Machines." *Automatica*, Vol. 24, pp. 123-133, 1988.

29. Shirley, R. S. "Some Lessons Learned Using Fuzzy Systems for Process Control." *IEEE Trans Control Systems*. Vol. 7, No. 6, 1987.

30. Steelman, D. M. "Distributed Cement Plant Control—the Intelligent Approach." *IEEE Trans on Industry Applications*. Vol. IA-18, No. 2, 1982.

31. Sugeno, M. *Industrial Applications of Fuzzy Control*. Elsevier Science Publishers, 1985.

32. Zadeh, L. A. "Fuzzy Sets." *Information and Control*. Vol. 8, 1985.

33. Zadeh, L. A. "Outline of a New Approach to the Analysis of Complex Systems and Decision Processes." *IEEE Trans Systems, Man & Cybernetics*. Vol. SMC-3, No. 1, 1973

6

AUTOMOTIVE POWERTRAIN CONTROL WITH FUZZY LOGIC

F. Graf
Siemens AG - Automotive Systems
GERMANY

W. Hauptmann
Siemens AG - Corporate R & D
GERMANY

With the rapid development of electronics and the increasing demand for higher performance with respect to safety, comfort, fuel efficiency, and emissions, modern automotive systems are supposed to perform more and more sophisticated tasks. In powertrain control, which integrates engine and transmission control, the application of knowledge processing techniques will play an essential role in the management of available information. This chapter presents two examples of automotive applications realized by fuzzy logic and neuro-fuzzy to approach the trend for new "software" technologies of the future. The first example proposes a fuzzy logic system to determine the optimal shift strategy for the control of an automatic transmission which adapts to driver and some environmental conditions. An extended approach is presented in the second part of this chapter. Here, the incentive for using knowledge processing methods again is the need to integrate human expert knowledge and, in addition, to learn from available sensor data. To take the next step towards intelligent automotive systems on a broader scale, the problem of environment recognition is addressed and a neuro-fuzzy approach for the identification of the driving situation is introduced.

6.1 INTRODUCTION

The progress of car electronics provides countless technological improvements in vehicular performance and functions. They are mainly supported by the rapid development of microcomputers which are currently used for the electronic control of engines, transmissions, chassis, suspensions, brakes, etc. In addition, there is still a strong growth of functionality and system complexity driven by the continuing demand for the improvement of comfort and safety but also for the reduction of fuel efficiency and exhaust emissions. To face these increasing challenges, future automotive control systems will be composed of enhanced and more powerful hardware. Software technology will progress from the pure implementation of numerical algorithms to a higher degree of knowledge based and self-learning solutions.

This contribution presents two examples of automotive applications realized by fuzzy logic and neuro-fuzzy to approach this trend for new "software" technologies of the future. The first example proposes a fuzzy logic system to determine the optimal shift strategy for the control of an automatic transmission which adapts to driver and some environmental conditions. This knowledge-based system is derived from the human expert knowledge of application engineers. Thus, the implementation of the corresponding functionality was facilitated by the use of the fuzzy logic methodology and the fact that a corresponding mathematical model did not have to be defined. Test results demonstrate the beneficial use of this technology in an application of man-machine interaction and environmental recognition.

As a continuation to the first example of applying fuzzy logic to automotive systems an extended approach is presented in the second part of this chapter. As more and more sensors are used in the vehicle delivering an increasing amount of measurement data to be analyzed there is an urgent demand for an efficient processing of the provided information. Here, the incentive for using knowledge processing methods again is the need to integrate human expert knowledge and, in addition, to also use available measurement data. To take the next step toward intelligent automotive systems on a broader scale, the problem of environment recognition is addressed and a neuro-fuzzy approach for the identification of the driving situation is introduced. The presented recognition system will classify the type of driving situation (urban area, stop and go, highway, etc.) and the road characteristics (flat, hilly, straight, winding road, etc.). It uses fuzzy logic for the classification task, is generated and optimized by means of a neural network, and allows the bi-directional conversion between the fuzzy and the neuro domain. This new method leads to superior classification results and reduced development time compared to "manual" system development.

6.2 FUZZY LOGIC ASSISTED AUTOMATIC TRANSMISSION CONTROL SYSTEM

Over the last few years the number of vehicles equipped with an automatic transmission has increased substantially in North America, Japan and Europe. Besides the mechanical system needed for the shift operation, there is the trend to have an "intelligent" control module, which changes gears based on the observations of the environment. This trend to electronic Transmission Control Units (TCU) is also driven by the replacement of mechanical components through Electric Wire (SW)-based functions. The advantage of such a system is that an intelligent environment-dependent behavior can be implemented cost-effectively. It can deal with varying conditions such as a changing environment due to the road gradient, changing drivers' intentions, road type, and special driving maneuvers. This kind of adaptivity plays a growing role in the whole powertrain to ensure driveability, efficiency, less pollution, and improved performance. For applications of transmission control, adaptivity in this sense will be standard in 3 to 5 years.

6.2.1 Drawbacks of Conventional Control of Automatic Transmissions

The majority of automatic transmission controllers marketed today are designed to control the planetary gear type transmission with a torque converter. Conventional transmission control systems determine the required transmission gear ratio with a manually selected predetermined shift pattern. The shift pattern consists of characteristic curves, which describe when a shift operation takes place according to the measured

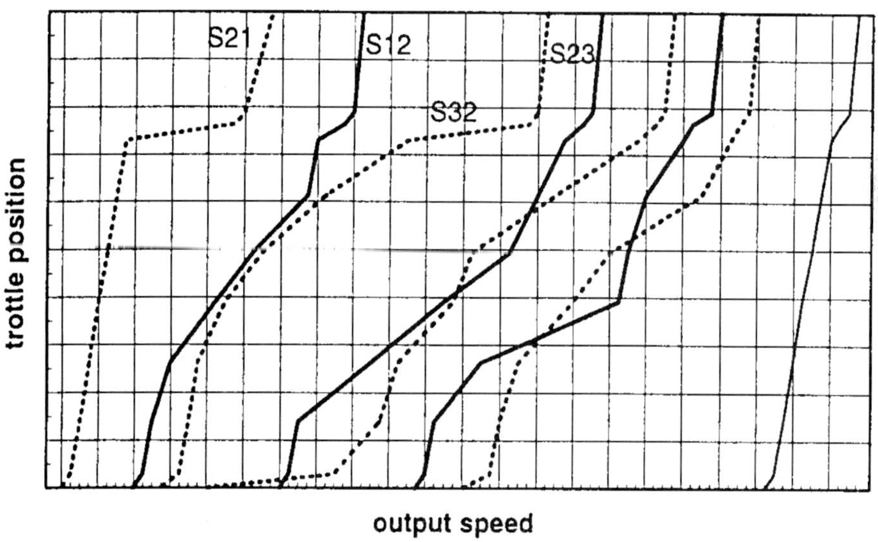

Figure 6.1 Conventional shift pattern.

vehicle speed (or transmission output speed) and throttle opening (Figure 6.1). There are at most three different shift patterns available in today's automatic transmissions (e.g., sporty, defensive, winter). With these shift patterns it is impossible to optimally adapt the transmission gear ratio to all possible driving conditions. In addition the shift patterns are not adapted to the driving conditions automatically but must be selected manually.

Only the transmission output speed n_{out} and the throttle opening, tv, are considered when determining a suitable gear position (Figure 6.1). That is not enough to deal with a changing environment (e.g., hillside), changing vehicle load or detecting a driver's intention (sporty, defensive). Therefore automatic adaptation to the driving conditions can hardly be implemented. A typical case of the weakness of today's automatic transmission is shown in the following example. When driving down a "steep incline," the driver reduces the throttle opening and possibly actuates the brakes. The intention is to drive in a gear as low as possible to utilize the increased braking torque of the combustion engine. The transmission control system does quite the opposite. The shift pattern in Figure 6.1 shows that the reduction of the throttle opening leads to a higher gear position. That means the braking torque of the engine is reduced and therefore the mechanical brakes have to compensate by absorbing more energy. This behavior is not desirable for vehicles.

An intelligent control module should take these effects into consideration. It therefore needs additional information about the environment, which means additional sensors. Modern cars are equipped with a lot of sensors for different control systems, e.g., cruise control, engine management systems, electronic anti-lock brakes, etc. In the fuzzy logic TCU we only use sensors already available on these vehicles. To improve the above mentioned behavior, several solutions have been realized and implemented in products [1,2]. All of these use methods with conventional control strategies and are based on mathematical models, which, e. g., describe and classify the behavior of the human driver. Though it is necessary to set up such a description model, that means corresponding efforts to reach the targeted functionality. One example could be:

$$\begin{aligned} \text{class driver} = {} & \text{k1} \times \text{throttle opening} \\ & +\text{k2} \times \text{speed of throttle opening} \\ & +\text{k3} \times \text{vehicle acceleration} \end{aligned}$$

Another well-known method to classify is the use of 3-dimensional maps to combine two inputs to assign one result. Such a map has to be implemented by a complete dot matrix. Those methods also have poor capability to implement rule-based knowledge in the form of "if ... then" rules. Therefore, the goal is to overcome this methodology and introduce methods which do not force implementation by mathematical description, especially if there is no obvious physical background or description available.

6.2.2 Description of fuzzy logic Based Solution

Structure of control system. Figure 6.2 gives an overview of all components interacting and the location of the transmission controller.

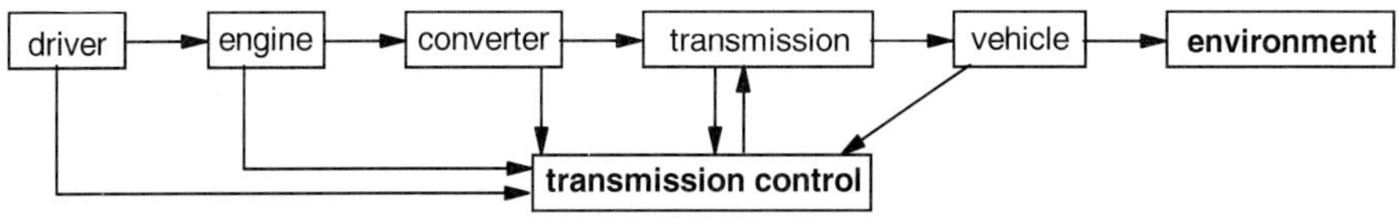

Figure 6.2 Integration of TCU in the driving environment.

Figure 6.3 shows the overall structure of the fuzzy logic TCU. Generally, the fuzzy logic system plays the role of an observer which selects the well-adapted shift strategy. Shift behavior itself is still fixed in shift patterns which are precalculated and represent crisp relations between the active gear (ratio) and transmission output speed/ throttle position. The fuzzy controller receives seven input signals and calculates three output signals. One output signal indicates load detection, one classifies the driver, and one shift signal inhibits or releases shifts dependent on the driving situation.

In the *fuzzification module* input parameters (numerical) are transformed to degrees of membership of linguistic states. An *inference system* working with predetermined rules calculates each rule relevance and conclusion for the linguistic output variables. These rules determine the behavior of the controller in the system. The transformation from linguistic states to crisp output signals for process control is done by the *defuzzification module* [3]. The *conventional control portion* includes the choice and correction of conventional shiftmaps. In [4] detailed information on methods and components of fuzzy control are given. The fuzzy controller calculates three output signals. Therefore three rule bases must be set up which describe the desired classification task in linguistic terms.

- *Load detection*:

There are three different load states used in the fuzzy controller: *mountain, plain, and valley*. The rules for load detection are based on the differential torque. The measured acceleration and output torque together with the mathematical model allow us to calculate a reference torque. This reference torque is compared to the actual torque minus the known parts of the driving resistance. The resulting differential torque, T_{diff}, is a direct

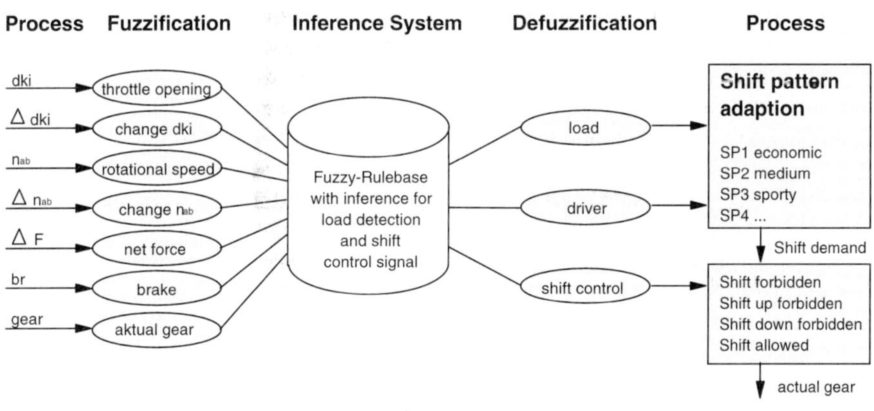

Figure 6.3 Overview of the control system.

indicator for vehicle load and/or additional climbing resistance. One simple rule for detecting adapted shift behavior due to inclinations is:

$$\text{If } T_{diff} = \text{positive AND dki} = \text{not small THEN load} = \text{mountain.}$$

- *Driver classification*:

The driver is classified as being either *defensive, medium,* or *sporty*. The classification depends, e.g., on the moving average of the absolute changes Δ dtv. If the moving average rises, it means that the driver is demanding a more sporty shift behavior and vice versa. Another significant input for driver classification is also the average value of throttle opening, dki_avg. Two rules using these values for driver classification are:

$$\text{If } \Delta \text{ dtv} = \text{medium THEN driver} = \text{medium.}$$

$$\text{If } \Delta \text{ dtv} = \text{high AND dki\_avg} = \text{grt\_med THEN driver} = \text{sporty.}$$

- *Shift inhibition signal/dynamic correction*:

This signal has four discrete states: *shift allowed, shift down forbidden, shift up forbidden, shift forbidden*. System security is affected by the shift operation and so it is important to give priorities to special rules or states.

The output of each rule reflects the relevance for at least one linguistic output variable. To generate an output signal these different variables assigned to one output have to be condensed. In the fuzzy transmission controller this is done by a modified MAX-DOT inference model [5]. Each Membership Function (MSF) belonging to a conclusion is weighted with the related rule relevance. The defuzzification uses all weighted MSFs of one output to calculate the crisp output signal by computation of the center of gravity. The chosen defuzzification method is well suited for the described transmission control system. Depending on the shape and the area of the MSFs, the influence of special rules on the output signal can be reduced or amplified. Therefore the effect of rules can be adapted to their purpose. On the other hand the influence of highly prioritized rules can be maximized.

The presented system also consists of a conventional control portion: Both the load and driver signals are used to automatically select situation-dependent shift patterns. The shift pattern adaptation to the driver and the load is shown in the following example. With increasing load or a sporty driver there is an automatic selection of shift patterns which are optimized to vehicle acceleration (output torque) and increased shift hysteresis to avoid shift oscillation when in uphill mode. The signal "shift control" helps to overcome the drawbacks of shift maps under dynamic driving conditions. Generally, all shift demands resulting from crossing a shift line are inhibited or delayed if they do not fit into the driving situation. This may happen, e.g., if the accelerator is released because a curve is entered (avoidance of unnecessary upshift) or slip of driven wheels is detected (no downshift so as not to destabilize vehicle). For a Continuously Variable Transmission (CVT) it makes sense not to use shiftmaps but a second fuzzy logic system to make gear ratio the accordingly "fuzzy."

Methodology for design and implementation. The presented system works as a "knowledge-based expert," not like a closed-loop controller. Existing human knowledge was translated into if-then constructs within heuristic rules. This knowledge originates from human experience, systematic analysis, and observation of the behavior of different types of drivers and driving conditions. To set up membership functions, it is necessary to assign fuzzy domains of input signals. Preprocessing (filtering, calculation of derivatives) to the input signals is more significant than the use of raw input signals. To design, test and implement the fuzzy controller, the Computer Aided Engineering (CAE)-tool SIEFUZZY was used. This tool is PC-based and offers a Windows-oriented graphical user interface to define fuzzy logic systems (membership functions, rule bases, structure) and simulate or test their functionality. This increases transparency of the which defined fuzzy logic system, is very important because of its complexity.

A further feature of SIEFUZZY is the capability to generate standard C code or even assembly codes for some microcontroller-CPUs. In terms of code efficiency (size, run time) results are comparable to those of conventional solutions. In the example, the whole transmission controller software (including fuzzy logic) was implemented on an 8-bit microcontroller. This capability to generate codes means that the expert is not engaged in writing these in a programming language but is concentrating on the efforts to design and test his fuzzy logic system on a much higher description level. The consequence is a distinct reduction of development time.

6.2.3 Results of Road Test

A. Passing slow vehicle on a country road (see Figure 6.4) The driver follows a vehicle with low vehicle velocity, represented by transmission output speed n_{out}. The passing maneuver starts at $t \geq 56$ s that is recognized by activities of the driver (fast opening of throttle). Corresponding rules to identify a sporty driver (or the need to ensure best acceleration) are rules called Fuzzy 1 - 3 shown with their degree of relevance in Figure 6.4. Rule "Fuzzy 1" is mainly based on throttle position tv itself, whereas rule "Fuzzy 2" uses driver's activity on throttle Δdtv to conclude a sporty driver. Rule "Fuzzy 3" becomes relevant, if throttle opening mean value tv_avg arises in an according domain. After inference and defuzzification the crisp output "driver" reflects the final classification of the driver's intention, which could also have its origin not only in the individual style to drive but also in the driving situation. A "driver" output immediately triggers immediately the sporty shift pattern ("2") at the beginning of the passing procedure, a down-shift is started, and upshifts are delayed. This leads to optimized acceleration, which is also a contribution for safety. Rules "Fuzzy 1 – 3" are tuned in such a way, that they complement each other. After the passing maneuver, the "economy" shift pattern is reactivated again (not shown in Figure 6.4).

B. Covering two successive road inclines (see Figure 6-5) Road gradient is represented by differential torque T_{diff} ($0 \hat{=}$ flat, positive $\hat{=}$ uphill, negative $\hat{=}$ downhill); after the first road incline (about 12%) follows a down grade. Rule "Fuzzy 1" effects at the start of the first incline (98 s > t > 60 s) a significant rise of crisp output "load." Subsequently, load adaptive shift pattern ("3") is activated which leads to a 43-downshift. This downshift is

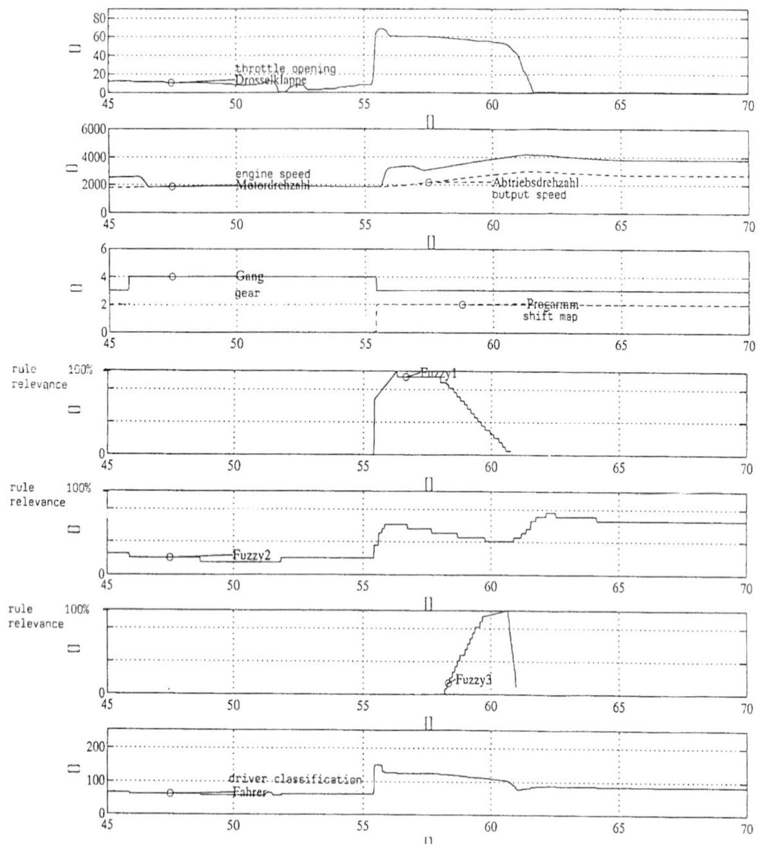

Figure 6.4 Driving maneuver "passing slow vehicle."

not initialized by a driver's action on the accelerator but by the ability of the control system to identify the need of a changed gear ratio adapted to road gradient. The third gear is not only engaged during driving uphill but also riding downhill between the two upgrades that is effected by rule "Fuzzy 2." This rule weights continuous presence of inclines. The consequence is a better engine brake effect when driving downhill what is visible by reduced application of the brake (see "brake" signal). Another effect of keeping the lower gear downhill is that at the following second road incline (t > 115 s) the appropriate gear is activated. Signal "throttle" also demonstrates capabilities of the system: there is no tendency to shift oscillations or demands for shift that would be expressed by large changes of throttle, caused by the driver.

6.3 DRIVING SITUATION IDENTIFICATION

During the ongoing research and development concerning the fuzzy logic advanced transmission control, described in the first part of this chapter, two major aspects turned

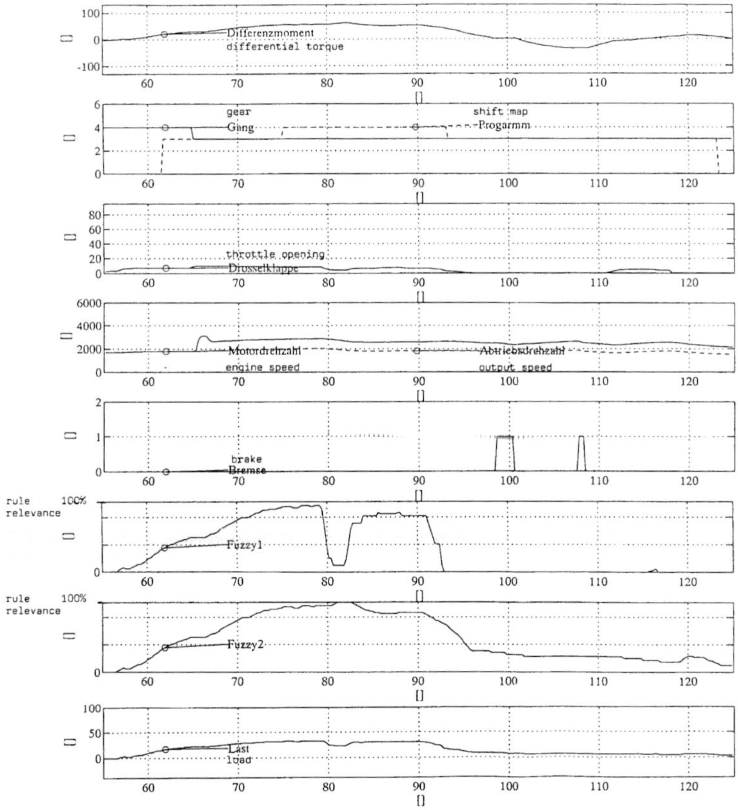

Figure 6.5 Driving maneuver "two successive road inclines."

out to become increasingly important for future intelligent vehicle control systems like the powertrain control. First, there is a need to perceive the driving environment by using different sensors and taking into account more sensor data. Secondly, to facilitate an adaptive system and to be able to handle the increasing number of design and application parameters of a complex system, a learning component is needed. Consequently, as the next step toward intelligent automotive systems on a broader scale, the problem of environment recognition is being investigated and a prototype of a driving situation identification system is introduced.

6.3.1 Intelligent Vehicle Control Systems

The progress of car electronics provides countless technological improvements in vehicular performance and functions. They are mainly supported by the rapid development of microcomputers which are currently used for the electronic control of engines, transmissions, chassis, suspensions, brakes, etc. Such controls have been applied individually in each of these components but will tend to be integrated to provide total control of a vehicle. An essential part of such an overall control system will be an intelligent module, termed driving situation identification system here, that is able to recognize the current driving situation based on the observation of the surrounding environment and the processing of the incoming information.

Figure 6.6 illustrates how the major system components will complement each other in the intelligent automobile of the future. When driving, different signals are picked up by various sensors around the car. These sensors are already installed as a component of an existing electronic system, like an Anti-Lock Braking System (ABS) or an engine control; they are also installed for the recognition system only. From these sensor measurement data the current driving conditions will be derived and classified into certain types of situations, like steep hill, curvy road, town, highway, etc. Depending on the result of the recognition, an overall controller can change the parameters of the different adaptive systems in the car accordingly. In this way the vehicle characteristics are adjusted to optimize the vehicle performance. A driving situation recognition stage can be independent from the specific subsystem to be controlled and possible systems to be adapted to the driving situation are automatic transmission, electric power steering, shock absorbers, level control, hybrid motor, etc.

6.3.2 Combining Expert Knowledge and Data

The proposed recognition system is based on the fusion of ideas from the areas of fuzzy classification and artificial neural networks within an integrated neuro-fuzzy system. By applying a neuro-fuzzy approach the advantages of fuzzy systems and neural networks can be used both in the design and in the development phase. On the one hand existing experience and expert knowledge can be processed by linguistic variables, fuzzy membership functions, and if-then rules (fuzzy rule-base). On the other hand, the power

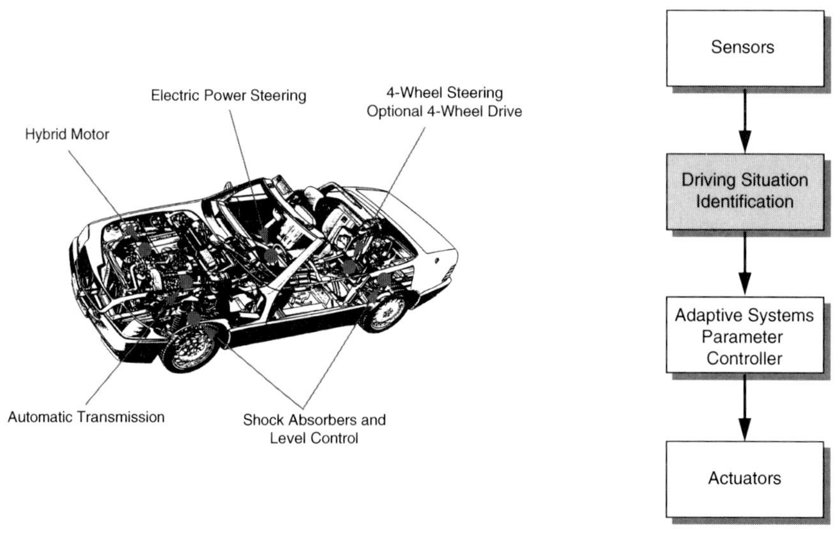

Figure 6.6 Intelligent vehicle control.

of neural networks to learn nonlinear functions from a given set of sensor data can be used as well. The combination of expert knowledge and process data allows the creation of a more accurate and complete model of the real world. In the neuro-fuzzy recognition system, continuously recorded measurement data from up to twelve different sensors are used as an input. Important sensor signals are, for example, current vehicle speed, throttle opening degree, current gear, lateral acceleration, and motor speed. After a preprocessing and sensor fusion step these primary sensor signals are fed into a fuzzy system which will characterize the momentary driving situation, e.g., into the categories "stop," "city traffic," "road (straight, curvy, hilly, etc.)," highway, etc.

6.3.3 System Overview

Figure 6.7 gives an overview of the complete driving situation recognition system as used for the design and development phases. The overall system can be divided into two major parts, A and B, as indicated by the dotted line. Tied together by the integrated neuro-fuzzy system, the key processing element, these subsystems basically operate independently from one another. Part A includes a signal buffer, a preprocessing and sensor fusion block, and the actual fuzzy system for classification. The full sequence of operations can be best shown by starting out with the measured sensor data and then following the data flow.

In the vehicle the different sensor and control signals are continuously monitored, measured, recorded, and stored in a signal buffer. These raw sensor data are then fed into a preprocessing stage where the signals are filtered and processed with respect to their history in time. In addition, a time compression is performed and by fusion of different sensor signals a reduction of the input dimension is achieved. As indicated in the figure, eight sensor signals are combined to be represented by only four fusion signals by suitably joining different sensor information. By this kind of preprocessing both a

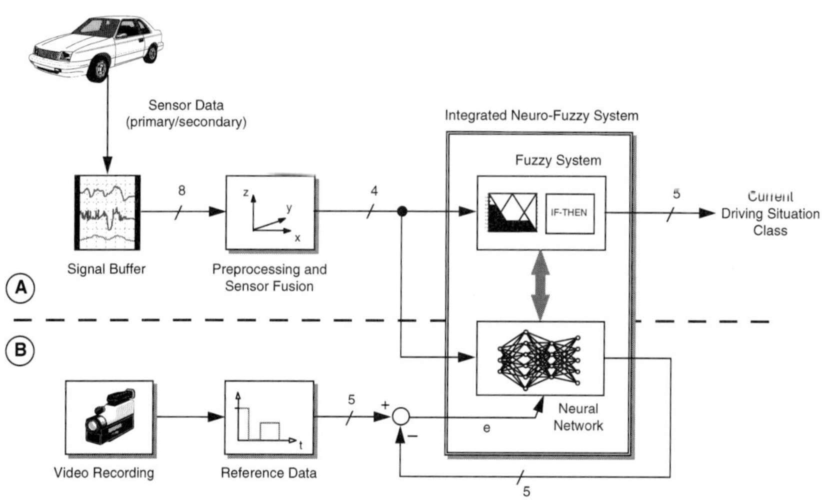

Figure 6.7 System overview.

reduction of the sampling rate as well as a reduction of the dimension of the classification problem are achieved.

The preprocessed sensor signals are then used as inputs for the fuzzy classification system which will identify the current driving situation based on the incoming input data and classify them using a given number of situation classes. As subsystem A performs the actual recognition task during the on-line mode, the described sequence of processing steps exactly corresponds to the final operation in the vehicle while driving.

In contrast, subsystem B is needed for generation and optimization of the fuzzy classification system in off-line mode by the use of an artificial neural network. This processing block is only used in the design, development, and optimization phase and is not part of the final on-line identification system in the vehicle. Part B includes a data base with video recordings of performed test drives, a reference data set, and a neural network. Parts A and B are connected with each other via the integrated neuro-fuzzy system. The complete mode of operation and the interaction of the two subsystems can be best explained by going through a typical development cycle.

The basis for the identification system is formed by the fuzzy classification system which is set up in the first step of the development phase based on the available information and the existing expert knowledge. Here, the fuzzy system is already given its initial structure to perform the classification task, like the number of input and output variables, the number of membership functions, and an initial rule base. This fuzzy system will eventually perform the actual classification of the driving environment during the on-line operation. Within the integrated neuro-fuzzy system this initial fuzzy system will then be completely converted into a neural net structure without any loss of information Thus a neural net is available in block B that identical reproduces the fuzzy system from block A with exactly the same input/output behavior. This Fuzzy-neuro transformation carries out a translation of the classifier from its representation in the fuzzy domain into a representation in the neuro domain. Hereby, a bi-directional transformation is applied which preserves the correspondence, i.e., the neural net topology can be retranslated into the fuzzy representation at any time.

Continuing with the development in part B, the driving situation classifier is trained by means of the neural network to perform a data-driven optimization. In this optimization process several components of the original fuzzy classifier are either modified, discarded or generated. In off-line mode the training of the neural network is done with training data containing a suitable set of input/output pairs of the classifier. As input, the sensor data which have been recorded in the vehicle during the performed test drives and subsequently preprocessed are used. During the training phase the classification result at the net output is compared with the ideal reference output. By using the error resulting from the difference between the classifier and reference output, the neural net is trained applying the well known backpropagation algorithm [6,7]. The supervised training of the classifier requires a reference data set be provided giving for every point in time the actual driving situation class in correspondence to the recorded sensor measurement data. For this reason, in addition to measuring the sensor data, the driver's view is recorded by a front-mounted video camera during the different test drives. This video recording is then analyzed and, depending on the desired number of output classes, the time sequence of the driving situation classes is extracted and stored as the target data set. During training a sensor data combination can be presented to the neural net as an input vector, e.g., four sensor fusion signals, and the corresponding situation classes as an output vector, e.g., five classes.

After completion of the training and optimization phase of the classifier the resulting neural net is completely retranslated into the fuzzy representation of part A. Having performed the data-driven optimization, part B is detached and is not part of the final driving situation recognition system implemented in the vehicle. The modifications of the fuzzy classifier components caused by the training (shape and position of input and output membership functions, fuzzy rule-base, etc.) can then be visualized and assessed. After subsequent evaluation in a suitable fuzzy simulation tool the development phase is finished, having utilized and integrated both expert knowledge and measured sensor data.

The full cycle of the development phase within the frame of the neuro-fuzzy system can be summarized as following:

Step		**Subsystem**
1	Design of the initial fuzzy system (expert knowledge)	A
2	Translation of the fuzzy system into a neural network	A-B
3	Training of the neural network (sensor data and video information)	B
4	Retranslation of the neural network into the fuzzy system (neuro-fuzzy transformation)	B-A
5	Final driving situation classifier in the vehicle	A

6.3.4 Preprocessing and Sensor Fusion

The raw sensor signals picked up in the vehicle are first fed into a preprocessing block where the inputs of the fuzzy classification system are derived from the recorded measurement data. This step performs a filtering and a processing of the signals with respect to their time history and also a fusion of the different measured parameters. While driving the following sensor signals are being monitored in the vehicle:

nab:	wheel revolution speed (proportional to vehicle speed)
bl:	brake (on/off)
nmot:	motor revolution speed
a_lat:	lateral acceleration
m_diff:	torque total at gear box output
gear_act:	active gear
driver:	classification of driving style

In order to reduce the number of necessary input variables for the fuzzy classification system and at the same time use all available measurement data, the sensor signals are combined to give four new fusion signals. The objective of this fusion is a representation of the vehicle's current dynamic driving behavior in the three spatial dimensions x, y, z (see Figure 6.8) and by a purely speed dependent component w. By the fusion of the raw signals and the resulting reduction of the input space of the problem, the model set up and the semantic formulation of a comprehensible and clear rule base for the classification task is made feasible. In this case the four fuzzy input variables are being derived from the seven original sensor signals via the following relations:

$$x(t) = f_x\,\{gear\_act\,(t, t-1, t-2, t-3), nmot\,(t, t-1, t-2, t-3), bl\,(t)\}$$
$$y(t) = f_y\,\{a\_lat\,(t, t-1, t-2, t-3), nab\,(t, t-1, t-2, t-3), driver\,(t)\}$$
$$z(t) = f_z\,\{m\_diff\,(t, t-1, t-2, t-3), nab\,(t, t-1, t-2, t-3), driver\,(t)\}$$
$$w(t) = f_w\,\{nab\,(t, t-1, t-2, t-3)\}$$

The fuzzy input variable x, for example, is computed from the sensor signals *gear_act*, *nmot*, and *bl*. Here the notation *gear_act* $(t, t-1, t-2, t-3)$ describes that the present time window t and, additionally, the last three time windows $t-1$, $t-2$, $t-3$ are used in the computation. The basic principle of the processing of the signals' histories is depicted on the right side of Figure 6.8. To achieve a data compression a jumping window $w(t)$ of length L is moved along the signal and each window is represented by the average of the signal values lying inside its section. For the sensor signal *gear_act*, this mean value would be denoted *gear_act*(t). To be able also to take into account part of the past signal history into the computation, the mean values of the three most recent overlapping windows $w(t-1)$, $w(t-2)$, $w(t-3)$ are included, however with a different weight factor. Therefore, the present value for variable x results from four time windows of sensor signal *gear_act*, four windows of signal *nmot*, and the current window of signal *bl*. Then the time windows are moved forward in time by the window length L. By using only the average values and a jumping window, considerable data compression is achieved. For instance, using a window length L of 250 samples and an overlap of 50% on a raw signal that has been sampled at a rate of T_{ab} = 40 ms a new result of $x(t)$ is computed every 10 seconds. By modifying the window length the system's reaction time and the lag between two successive classifier outputs can be adjusted.

6.3.5 Structure of the Fuzzy Classification System

A possible structure for the fuzzy system which performs the actual classification task is shown in Figures 6.9 and 6.10. The inputs of the fuzzy system are the four derived sensor fusion parameters. The left side of Figure 6.9 shows a possible distribution of the input membership functions. The domain of every input variable is divided into four

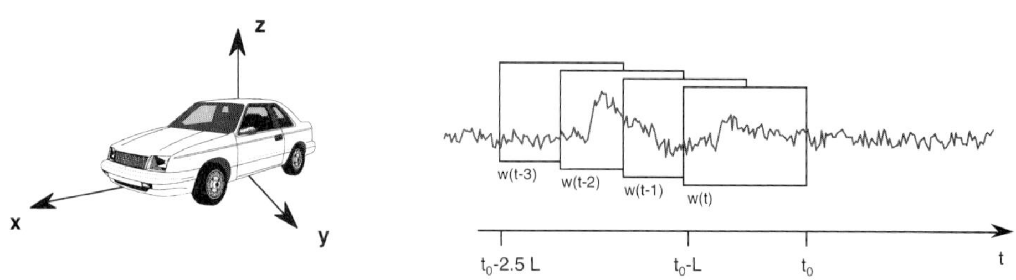

Figure 6.8 Sensor fusion and data compression.

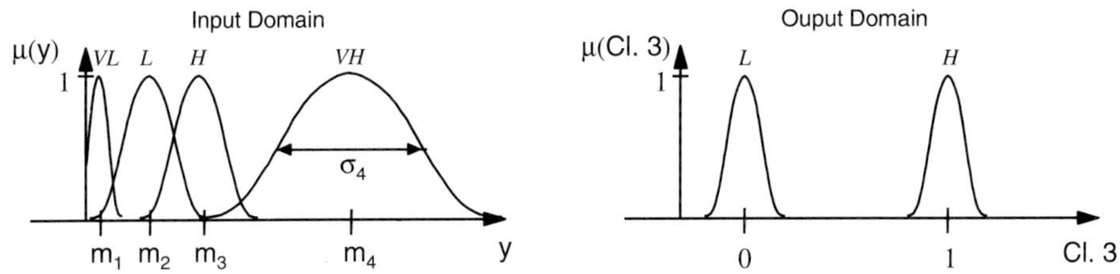

Figure 6.9 Configuration of input and output membership functions.

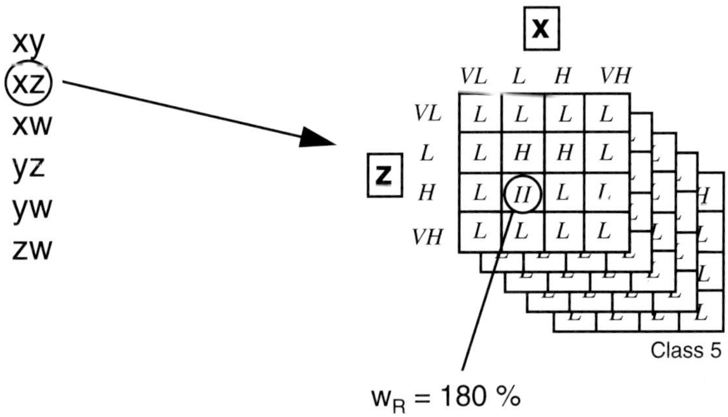

Figure 6.10 Fuzzy rule-base.

membership functions VL (very low), L (low), H (high), VH (very high), and the incoming crisp input variables are fuzzified accordingly and labeled with the corresponding linguistic terms. To enable the later transformation of the fuzzy system into corresponding structures of a neural network topology, Gaussian membership functions are used. The adjustable parameters center m_i and σ_i will then be represented by corresponding weights in the neural net.

From the parameters that are presented to its inputs, the classification system has to identify the current driving environment. As output, this situation is to be characterized as one of the following five possible situation classes:

Class 1: Very high speed, continuous driving situation, straight section, low side acceleration, no high gradients or inclines

Class 2: High speed, continuous driving situation, straight section, low side accelerations, no high gradients or inclines

Class 3: High/medium speed, continuous or discontinuous driving situation, possible high side accelerations and/or high gradients or inclines (curvy, hilly roads, transitions between class 1 and 2, etc.)

Class 4: Low speed, often non-continuous driving situation, high traffic density, medium/no side accelerations, medium/low gradients or inclines (downtown, stop and go, outer city area)

Class 5: Very low speed or stop (traffic light, intersection, parking, etc.)

Figure 6.10 shows the structure of the fuzzy rule-base. There is exactly one output for each of the five driving situation classes, each having only two Gaussian output membership functions (see right side of Figure 6.9), one centered at output value 0 (membership function "Low") and one at value 1 (membership function "High"). For every single time step the fuzzy system will give as defuzzified output a vector with five components [Cl_1, Cl_2, Cl_3, Cl_4, Cl_5] describing the estimates for the five classes by a value between 0 and 1. If the classification result were the output vector [0, 1, 0, 0, 0] it would mean that in this case the driving situation would be characterized as a class 2 situation with a degree of 100% and with 0% in all other classes. In contrast, with a resulting output vector of [0.0, 0.0, 0.1, 0.2, 0.7] the current situation would be classified all of the five classes with different weights, however with a dominating weight in class 4 (20%) and class 5 (70%).

Input and output variables are connected with each other via the fuzzy rule base where existing expert knowledge is expressed in linguistic format. Shown in Figure 6.10 is a rule base that is set up in a rather systematical way. Using the four input variables of the fuzzy system, denoted as x, y, z, w, all six possible pairwise permutations are taken, i.e., xy, xz, xw, yz, yw, zw, and for each of the pairs a rule matrix is filled out. To illustrate this, an example matrix for the input pair xz is shown. Since both of the inputs x and z are each fuzzified with the four membership functions VL, L, H, VH, the resulting rule base is a 4×4 matrix. As the sixteen matrix elements, the linguistic conclusions for the linguistic combinations of the inputs according to the corresponding fuzzy rule are filled in. Thus the marked element in the upper matrix implements the rule:

$$R_{xz,23}\text{: IF } x \text{ is low AND } z \text{ is High THEN Cl\_1 is High}$$

The matrix entry H for "High" denotes the fuzzy conclusion for the output Cl_1 (Class 1) which, according to the rule, results from the AND combination of the partial conclusions x = Low and z = High. The remaining elements of this matrix are given by the rule base for all AND combinations of x and z in a similar fashion. As indicated the total rule base consists out of 5 matrices, one for each output variable Cl_1, Cl_2, Cl_3, Cl_4, and Cl_5. Thus each of the matrices describes the conclusions of the different rules with respect to one specific output, the uppermost describing output Cl_1 and the lowest output Cl_5. Thus the complete fuzzy rule for the example is:

$$\begin{aligned}R_{xz,23}\text{: IF } x \text{ is low AND } z \text{ is High THEN Cl\_1 is High and Cl\_2 is Low}\\ \text{and Cl\_3 is Low and Cl\_4 is Low and Cl\_5 is Low.}\end{aligned}$$

Each of the rules gives a statement for all of the existing output variables. Due to the systematic setup of the rules, there are $4 \times 4 = 16$ rules for each of the six possible pairwise permutations resulting in a total number of 96 rules for the complete rule base.

6.3.6 The Integrated neuro-fuzzy System

After having finished the design of the initial fuzzy classification system, it will be converted completely and without any loss of information into the structure of a neural network to perform the data-driven optimization. Doing so a neural network is generated which exactly represents the same classifier as the fuzzy system with the identical input/output behavior. The Fuzzy-neuro transformation facilitates the translation of the classification system from its representation as a fuzzy system to its representation as a neural net. We call this specific process a bi-directional transformation since it allows the retranslation of the neural network topology into the fuzzy domain after adaptation during the training phase has been performed.

For the functional equivalent realization of a fuzzy model within the framework of a neural topology, the single processing steps, fuzzification, rule base, and defuzzification, are reproduced by corresponding substructures of a neural network. Although an important restriction is the use of only standard network elements and the applicability of established learn and adaptation algorithms [6], a special network architecture facilitates the complete translation of all fuzzy model operations.

6.3.7 neuro-fuzzy Topology

A schematic overview of the complete neural net topology for the neuro-fuzzy system is given in Figure 6.11. The structure shown is based on the fuzzy classification system with the four input variables x, y, z, w, and the five output variables Class 1–Class 5, and a rule base consisting of 96 rules. Clearly visible are the three main substructures for the fuzzification, rule base, and defuzzification. The resulting topology is a feed-forward type neural net with eight layers, three of them being responsible for the functional equivalent reproduction of the defuzzification operation.

The neuro-fuzzy architecture represents a complete reproduction of the underlying fuzzy model. The four input signals I_X, I_Y, I_Z, and I_W are being fuzzified in the second and third layers. The number, shapes, and positions of the input membership functions as well as the correct assignment to each variable are determined in the neurons of the second layer. Here the inputs are each fuzzified by the four membership functions *VS, S, B, VB* (very small, small, big, very big), resulting in a layer of sixteen neurons. Subsequently in the third layer the single rules R_1 to R_{96} of the rule base are reproduced. For each rule there exists a corresponding rule neuron with the appropriate premises at its input. These premises are combined with each other as described above and a rule activation degree α is computed and multiplied by the rule weight w_{Rule}. Via the partial conclusions these rule activations are then assigned to the membership functions of the output variables Class 1–Class 5. The partial conclusions for one output variable resulting in layer 5 are combined in the following defuzzification part. Between layers 5 and 6 a weighting of the conclusions with the precomputed areas A_k and moments M_k belonging to the specific

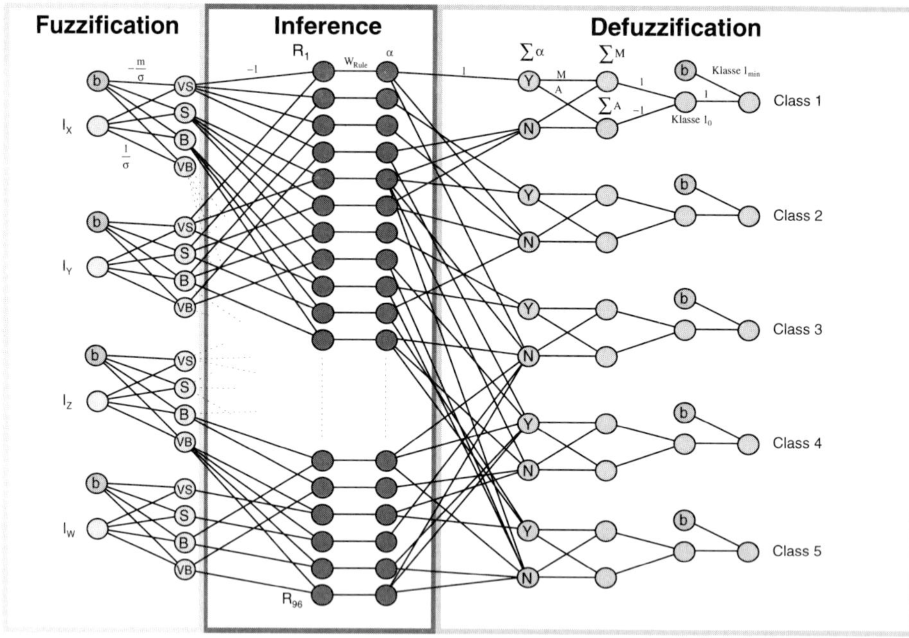

Figure 6.11 Neuro-fuzzy topology.

output membership function is performed. The computation of the moment/area division is finally performed in layers 6 and 7 using a logarithmic function and a simple subtraction of the arguments.

6.3.8 Results

To evaluate the performance of the fuzzy driving situation identification system and the neuro-fuzzy approach for the optimization phase, several test drives were carried out and driving data for different environments were recorded for off-line processing. Table 6-1 gives the relevant parameters of different fuzzy systems for environment recognition and their respective classification results. Shown are the results for a representative drive which comprises equal amounts of all the five different driving situations. In this case sensor measurement data for a driving time of about 32 minutes were processed, corresponding to roughly 48000 samples at a sampling rate of 40 ms.

The use of a time window length of L = 250 samples in the preprocessing stage results in a total number of 192 classifier outputs for the whole duration of the drive, delivering one output every 10 seconds. From left to right in the table columns the following parameters are given: The description of the specific classification system, details about the fuzzy system parameters trained (IM: input membership functions, OM: output membership functions, RW: rule weights, C: conclusions), the size of the fuzzy rule base, the number of input membership functions, the mean square error J as the

Table 6.1 Classification results

	Trained Parameters								
System	IM	OM	RW	C	# Rules	# IM	*J*	R/W/N	R/W/N %
initial	–	–	–	–	96	16	0.058	106/11/75	55/6/39
man. opt.	–	–	–	–	96	16	0.057	142/11/39	74/6/20
lrn_mbf	+	–	–	–	96	16	0.033	159/9/24	82/5/13
lrn_rw_mbf	+	–	+	–	44	16	0.016	179/6/7	93/3/4
lrn_all	+	–	+	+	85	16	0.013	181/4/7	94/2/4
rule_red	+	–	+	–	20	13	0.021	179/12/1	93/6/1

distance to the target data set, and the classification results as both absolute value and in percentage (R/W/N: right/wrong/not classified).

The initial fuzzy system, being set up by using expert knowledge alone, will classify the driving situation for this test drive correctly in 106 out of 192 cases (55%), wrong in 11 (6%), and will not make a statement in 75 cases (39%). The manually optimized system achieves a result of 74% for correctly identified situations. For the trained recognition systems the results depend on the parameters that were free for adaptation during the training in the neural network. Here the available sensor data were divided into a training set and generalization set applying a ratio of 7:3. The results show that a considerable improvement of the recognition rate is achieved, and development time and cost are reduced dramatically at the same time by using the data driven optimization with a neuro-fuzzy approach. With a reduced rule base of 85 rules, for example, a classification result of 94% "right," 2% "wrong," and 4% "not classified" is achieved.

6.4 CONCLUSION AND OUTLOOK

With respect to the example of the transmission controller, we can summarize the following benefits as a result from the application of fuzzy logic:

- Fast transfer from human expert knowledge to a technical control system (especially when a mathematical model is not available).
- Variable classification of characteristic parameters (e.g., driver) is set up easily based on fuzzy conclusions.
- Fuzzy-CASE-tool accelerates process of functional design and implementation (easy definition of fuzzy system implementation is executed by automatically generated software).
- No fuzzy-specific hardware, only software implementation.
- Higher transparency of functions and rules (important for complex systems).

The proposed neuro-fuzzy approach is a considerable improvement in the direction of a self-learning component:

- Initial human expert knowledge may be used.
- Training optimizes system in terms of quality of classification result and size of rule base.

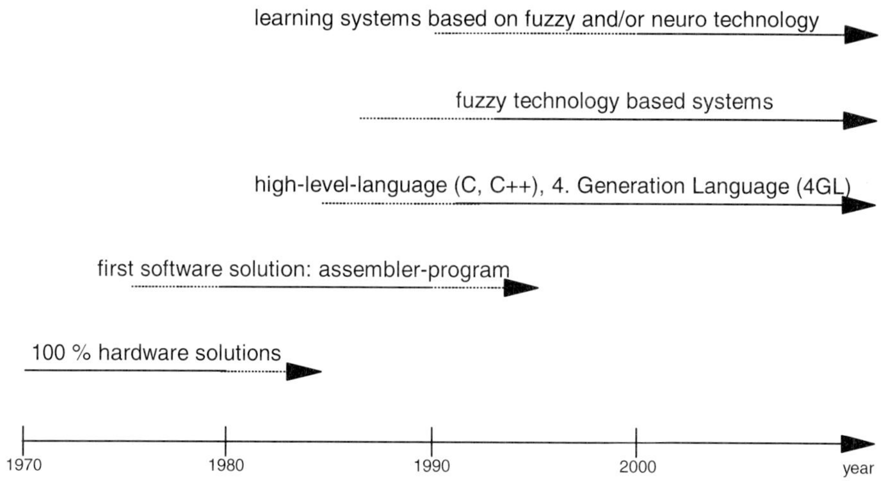

Figure 6.12 Method overview for implementation of automotive electronics.

- Setup of knowledge-based system by trained neural network leads to dramatic reduction of development time and cost compared to "manual" system design => examples for "correct behavior/conclusion" of system needed.
- No need for a human expert to analyze data and define rules.

Figure 6.12 sets the fuzzy logic methodology into chronological context with other means of establishing functionality. In the past, assembler programming has been replaced by high-level language programming. Reasons for this trend are the growing complexity of functions, portability, better software quality, and structures. But the use of high-level language is far and away an appropriate means to define functionality on a description level directly done by the application expert. Implementation of software is still a matter for software experts. A solution is the use of specification languages (4GL) which have the capability to produce software code.

fuzzy logic technology is one further step ahead. It offers the possibility to implement functionality in terms of "if-then"-rule based knowledge which means a direct link to expert experience. Fuzzy reasoning is an interesting and important conclusion to crisp decision making and numerical methods. Fuzzy logic technology is continued by the application of neural networks. These make it unnecessary to define knowledge for control or classification tasks, but are trained before or simultaneously to the process by correct examples or other reference data.

6.5 REFERENCES

1. Welter, A. and Renoth, F. "Adaptive Getriebesteuerung für BMW-Automobile." *ATZ.* Vol. 95, 1993, p. 420.

2. Spira, J.C. "Die Porsche-Tiptronic im Detail." *Krafthand*, Heft 23, 1989.

3. Mamdani, E. H. "Application of fuzzy logic to Approximate Reasoning." *IEEE Trans. Comput.* Vol. 26, 1977, pp. 1182-1191.

4. Driankov, D., Hellendoorn, H. and Reinfrank, M.. *An Introduction to Fuzzy Control.* Springer Verlag, 1993

5. Zimmermann, H. J. *Fuzzy Set Theory and Its Application.* Kluwer Academic Publishers, 1991

6. Kung, S. Y. *Digital Neural Networks.* Prentice Hall, 1993

7. Rojas, R. *Theorie der neuronalen Netze.* Springer Verlag, 1993

7

FUZZY CONTROL OF HEAVY-DUTY HYDRAULIC MACHINES

N. Sepehri
Department of Mechanical and Industrial Engineering
The University of Manitoba
CANADA

In nonautomated, unstructured environments, a machine such as an excavator has no model of either itself or the world, and the operator has to control it at a joint-mode level. It has recently been shown [1] that the operator can be more productive by controlling the machine in task coordinates, whereby the implement direction and speed are decided by the deflection of a 3D joystick handle, and the appropriate control signals to the hydraulic actuators are determined by a computer control system. In this chapter the application of the rapidly growing field of fuzzy logic to the development of a real-time hydraulic control system for such machines is demonstrated. Hydraulic systems are more complex, nonlinear, and difficult to control than electric-driven ones, and therefore demand different requirements. A demonstration of the benefits and a performance evaluation of a fuzzy control application to a Caterpillar 215B-based log loader is presented.

Figure 7.1 Caterpillar 215B-based log loader.

7.1 INTRODUCTION

The human interface for controlling heavy-duty equipment, for example the excavator-based machine shown in Figure 7.1, has not changed significantly over many years. Traditionally two joysticks, each with two degrees of freedom, have been used by operators to control the joint rates of the excavator arms independently. Machine operators have to implicitly understand the kinematics of the machine and have to adapt to the mapping between joysticks axes and arm joint axes. In certain tasks such as digging or leveling, the operator is required to rely on secondary information such as engine noise or cab level for a better control.

Motivated by potential benefits such as reduction in learning/adaptation time, productivity gains, reduction in fatigue and enhanced safety, a project has been initiated to examine new control interfaces for such hydraulic machines [1]. A resolved-endpoint-velocity control system was found to be suitable for implementation on a Caterpillar 215B-based log loader. With reference to Figure 7.2, the new system uses a 3D single coordinated hand controller to directly correspond to the linear implement velocities. The operator can command the endpoint to move in whatever direction the joystick is deflected from the spring-centered position. This concept which is also referred to as a coordinated control [2], follows the intuitive nature of the human operator, since it combines more than one axis in an attempt to come closer to a rational environment.

The coordinated control, however, requires a computer-based system to first translate the joystick movements into joint commands— it then decides on the control inputs that are required for each link, and at each moment in time, in order to obtain the desired direction, speed and/or position of the implement. Each link in a heavy-duty machine, moves through the application of a torque, based on a pressure differential in the hydraulic actuator. The pressure differential is achieved by means of a spool displacement in a main valve that controls the fluid flow into and out of the corresponding actuator.

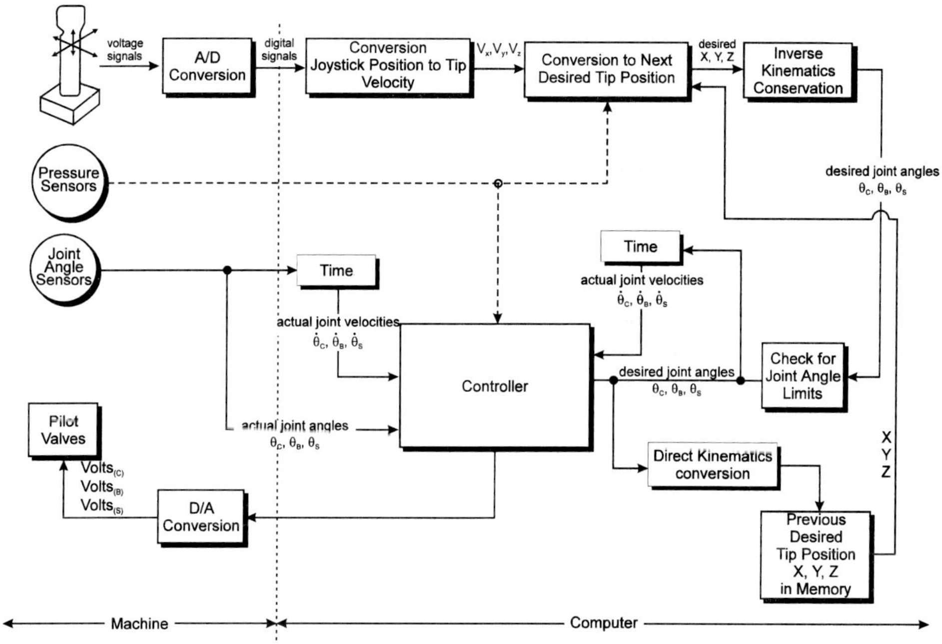

Figure 7.2 Coordinated control of log loader with a single hand joystick.
***Legend:* V = tip velocity; θ = joint angle; C = cabin;**
B = boom; S = stick; X,Y,Z = Cartesian coordinates.

The main valve spools are pushed by special electrohydraulic pilot valves [3]. A PID controller was initially applied on the prototype machine shown in Figure 7.1, to control each actuator. In order to improve the response and to compensate for nonlinearities in the hydraulic functions, a feedforward velocity control scheme was later added which improved the tracking accuracy [4]. The scheme uses the measured line pressures and requires a knowledge of the hydraulic system including the interconnection between the valves and their deadbands. Niemela, Zhao, and Virvalo from the Institute of Hydraulics and Automation in Finland have suggested using a fuzzy rule-based look-up table to assist the operator in coordinating movements of a three-degree-of-freedom crane. The initial study was encouraging although it was remarked that the accuracy of the tracking was highly dependent on geometrical dimensions, and the control signals were not optimal and required smoothening.

One characteristic of a fuzzy controller is its applicability to systems with model uncertainty or even with unknown models. This chapter demonstrates, for the first time, how fuzzy logic can be directly applied to control the nonideal hydraulic functions in heavy-duty machines. The emphasis has been placed on the development of methodology and the investigation of the promise of the technique. The approach taken was to first study the controller performance in a simulation model. The simulation study led to construction of suitable control plans and provided valuable insight before it was validated experimentally. The experiments were then performed on an instrumented

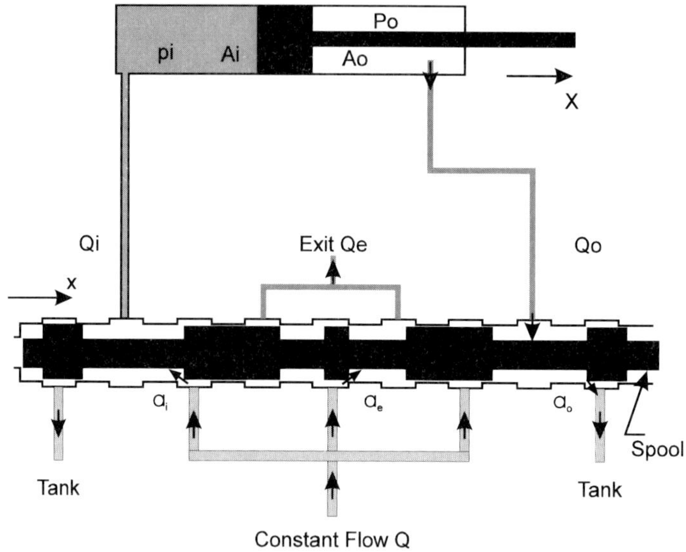

Figure 7.3 Heavy-duty hydraulic open-center valve.

Caterpillar 215B excavator machine located at the University of British Columbia, Vancouver.

7.2 HEAVY-DUTY HYDRAULIC FUNCTIONS

Figure 7.3 shows the schematic of a hydraulic drive unit used for the class of mobile machines such as the one shown in Figure 7.1. The linear piston actuator is connected to an open-center spool valve through flexible hoses. These valves are normally used along with a torque-limited pump system, i.e., a pump that sees the highest load and changes its output, Q, to meet the engine power requirement. The three orifices in the main valve (a_i, a_o and a_e) are controlled by a single spool displacement, x, and thus are interdependent. The flow is controlled by changing these orifice areas. When the spool is in the neutral position, orifices a_i and a_o are closed and the pump flow, Q, returns to the tank through the open orifice a_e. As the spool is displaced, for example to the right (as shown in Figure 7.3), orifices a_i and a_o open (after the spool travels its deadzone at the null position) and the orifice a_e closes. This results in a rise in the pump pressure, P_s, due to a_e restricting the constant flow from the pump, until it exceeds the cylinder pressure, P_i, allowing fluid flow into (Q_i) and out of (Q_o) the actuator. The governing nonlinear equations that describe the fluid flow distribution in the valve can be written in their simplest forms as follows [5]:

$$Q_i = kwx\sqrt{P_s - P_i}$$

$$Q_o = kwx\sqrt{P_o - P_e}$$

$$Q_e = Q - Q_i = k(1-x)\sqrt{P_s - P_e}$$

where Q_e is the remaining pump flow back to the tank having an exit pressure P_e, P_i and P_o are the input and output line pressures, respectively; k is the orifice coefficient; and w is the area gradient. In the above equations, the orifice areas a_i, a_o and a_e were assumed to be linearly proportional to the normalized spool displacement, x. This was done to simplify the derivation of the transfer function. The model used in the simulation study, however, applies the actual nonlinear relations.

The above equations are now linearized using a Taylor series expansion about an operating point and neglecting the higher-order terms;

$$Q_i = K_x x - K_p P_i$$

$$Q_o = K_x x + K_p P_o$$

where

$$K_x = kw\sqrt{\frac{P_s - P_l}{2}} = \textit{flow gain}$$

$$K_p = \frac{kwx}{2\sqrt{2(P_s - P_l)}} = \textit{flow pressure gain}$$

$$P_l = P_i - P_o = \textit{load pressure}$$

Note that in the above analysis, the pump pressure, P_s, is variable depending on the load or the spool displacement.

The line pressures (P_i and P_o) and the cylinder linear acceleration ($\ddot{X}$) are related as below:

$$F = \left(P_i A_i - P_o A_o\right) = J\ddot{X} + f_d \dot{X}$$

where f_d is the equivalent viscous damping coefficient and J is the inertia driven by the actuator. A_i and A_o are the piston areas. Note that for the case of a rotary actuator, such as the excavator cab swing motor, $\ddot{X}$ is taken as the angular acceleration and $A_i = A_o$. Using lumped parameter theory[5], the first order transmission line model is

$$\dot{P}_i \frac{V_i}{\beta} = Q_i - A_i \dot{X}$$

$$\dot{P}_o \frac{V_o}{\beta} = A_o \dot{X} - Q_o$$

V_i and V_o are the volumes of fluid trapped at the sides of the actuator and β is the effective bulk modulus of the hydraulic fluid.

Using a Laplace transformation, the input-output relationship between the actuator displacement, X, and the main valve spool displacement, x, in an explicit form is written as follows:

$$\frac{X(s)}{x(s)} = \frac{K_x(A_i + A_o)}{(JC)s^3 + (JK_p + f_d C)s^2 + (f_d K_p + A_i^2 + A_o^2)s}$$

where $C = \frac{V_i}{\beta} \approx \frac{V_o}{\beta}$ is the hydraulic compliance [5]. The above equation is a third-order system and has the following form:

$$\frac{X(s)}{x(s)} = \frac{\kappa}{s\left[\frac{s^2}{\omega_n^2} + \left(\frac{2\varsigma}{\omega_n}\right)s + 1\right]}$$

where

$$\kappa = \frac{K_x(A_i + A_o)}{K_p f_d + (A_i^2 + A_o^2)}$$

$$\omega_n = \sqrt{\frac{K_p f_d + (A_i^2 + A_o^2)}{JC}}$$

$$\zeta = \frac{K_p J + f_d C}{2\sqrt{JC(K_p f_d + (A_i^2 + A_o^2))}}$$

The inclusion of 1/s in the above equation basically means that the actuator is pushed by fluid flow. The oscillatory roots are lying in the second-order part and they are functions of ζ (damping) and ω_n (hydraulic natural frequency). The increase in compliance also decreases damping which together with low natural frequency may cause undesirable response. The effect of damping in stability is quite important. The sources of damping excluding J, C, A_i, and A_o are load damping which appears in the form of K_p in the equations, structural damping included in f_d, dry friction, and leakage. Leakage effect can be included in the linearized model by changing K_p to $(K_p + K_l)$, where K_l is the leakage coefficient.

For the excavator-based machines, the fluid flow is controlled through a two-stage valve arrangement. The first stage is the pilot electrohydraulic servovalve which, given a control signal, provides enough pressure to move the main valve spool [3]. For such an arrangement, a first-order relation between the input (i.e., control signal to the pilot valve) and the output (i.e., main valve spool displacement) proved to be accurate enough [3]. Therefore, the response of the whole system can be described as a fourth-order system;

$$\frac{X(s)}{x(s)} = \frac{\kappa}{s(\tau s+1)\left[\frac{s^2}{\omega_n^2} + \left(\frac{2\varsigma}{\omega_n}\right)s + 1\right]}$$

where τ is the response time of the main valve dynamics.

7.3 FUZZY CONTROL OF HYDRAULIC ACTUATORS

7.3.1 General

In spite of many articles on the successful application of fuzzy control to electric-driven actuators [6,7], the application of fuzzy control to hydraulic manipulators is sparse and can only be found in a few research papers. Zhao and Virvalo [8] combined a linear state controller with a fuzzy rule evaluation method. The method indirectly detects the presence of a load based on a fuzzy rule evaluation of actuator velocity and adjusts the position, velocity, and acceleration gains of a state controller accordingly, in order to make it insensitive to load variations. Chou and Lu [9] developed a more direct fuzzy control for a class of indoor hydraulic robots. The robustness and tracking ability of the controller were demonstrated through simulations and limited experimental studies.

The basic fuzzy controller used here is a simple two-input controller. The inputs are the scaled angular position error, G_a*(Desired Angle – Actual Angle)*, and scaled angular velocity error, G_v*(Desired Velocity – Actual Velocity)*. G_a and G_v are the scaling factors that put the error values within the controller universe of discourse. The fuzzy controller then fuzzifies these input quantities and applies a series of rules to these variables which are now converted to linguistic notions. It then outputs a control action based on a weighted result of the rule evaluation. Rules are normally written based on experience, common sense, observations, and understanding how the system responds and the attributes it must contain. The reasoning method used in this work is based on Mamdani's Minimum Operation Rule (MOR) and the Center Of Area (COA) defuzzification technique [10,11]. In the following section the membership functions and rules proposed for heavy-duty hydraulic functions are described.

7.3.2 Membership Functions and Rules

When using fuzzy logic, one should not expect to implement similar control strategies in both electric and hydraulic manipulators. Each class has different control requirements and characteristics which must be accounted for when designing the rules. For example, in a hydraulic manipulator, given a zero spool displacement (i.e., valve closed) the arm can be kept in place due to the oil trapped on both sides of the cylinder. This means that the arm can come to rest quite fast when the valve spool returns to its neutral position. In electric robots on the other hand, the link may pass the target point with a zero motor voltage. The amount of the overshoot depends on the inertia and the velocity. Negative motor voltage is often applied before the arm reaches the desired position to prevent the excessive overshoot. Also, a constant control input (motor voltage) is required in order to keep the arm in a desired set position in the presence of gravity.

In this work, five degrees of position errors and three degrees of velocity errors are the inputs to the fuzzy controller having three degrees of control output (see Figure 7.4). The membership functions that represent the values of the input and output degree of truth for each set of linguistic variables are simple symmetric triangular functions. They have sufficient overlap to produce smooth controls. The 5-output regions of control signals are also shown in Figure 7.4, MIN being the centroid of the negative large control action and MAX the centroid of the positive large control action. Referring to Figure 7.4, the width

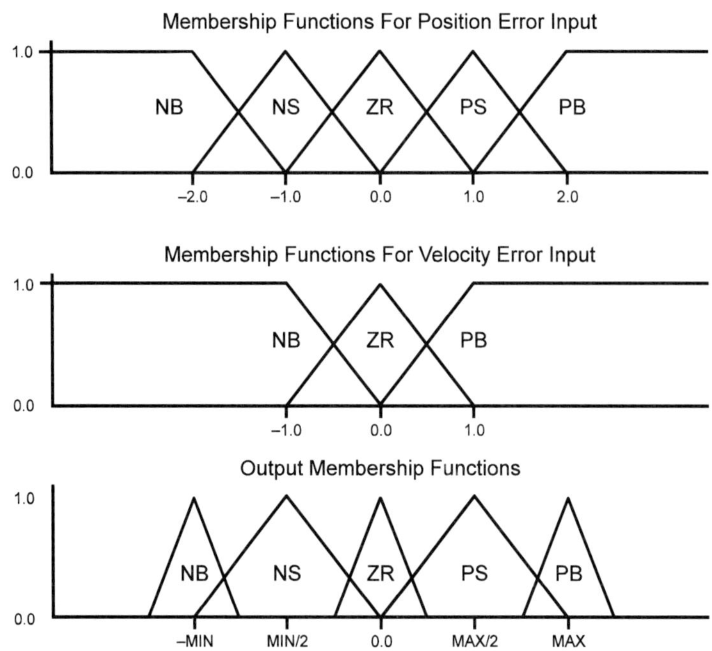

Figure 7.4 Membership functions.
***Legend*: NB = Negative Big; NS = Negative Small; ZR = Zero; PS = Positive Small; PB = Positive Big.**

of the zero output membership function is chosen to be narrow to produce large control values for the position errors close to zero. This is done in view of the existence of nonidealities inherent to heavy-duty manipulators such as friction, leakage, slow valve response, pump pressure changes and valve deadbands. Also from Figure 7.4, the output membership functions contain narrow large (positive or negative) regions. This allows large control signals (i.e., full spool travel) to be mostly effective in the presence of absolutely large errors; they are less active when the position error has membership in both small and large error zones.

For any set of scaled position and velocity errors, an output is produced based on the rules summarized in a matrix notation shown in Figure 7.5(a). The membership functions and rules have been designed and adjusted based on simulations, testing and a knowledge of the characteristics and response for the class of hydraulic robots under investigation. These are common approaches to the design of fuzzy logic rules [10]. Referring to the rule table of Figure 7.5(a), in the presence of large-scaled position errors, large control actions are ruled. The velocity error is not effective in this region. This was required in order to have a response with a fast risetime. The rules associated with small-scaled position errors indicate positive (negative) control signals for positive (negative) errors, regardless of the sign of the velocity error. Note that for a zero velocity error or a velocity error that decreases the position error, a zero spool displacement may stop the link spontaneously, preventing it from reaching the required position. This leaves us with only a small control input option in those regions. A large control action is only applied when the velocity error does not appear to decrease the position error, e.g., a case where a change in the direction of travel is desired. When the scaled position error is within the fuzzy zero region and in the presence of a velocity error outside the zero region, the control action works in the direction of the velocity error.

The rule table described here is to some extent different from the one [refer to Figure 7.5(b)] that is normally applied to electric-driven servo systems [12]. This is due to

		Position Error				
		NB	NS	ZR	PS	PB
Velocity Error	NB	NB	NB	NS	PS	PB
	ZR	NB	NS	ZR	PS	PB
	PN	NB	NS	PS	PB	PB

(a)

		Position Error				
		NB	NS	ZR	PS	PB
Velocity Error	NB	NB	NB	NS	ZR	PS
	ZR	NB	NS	ZR	PS	PB
	PN	NS	ZR	PS	PB	PB

(b)

Figure 7.5 Fuzzy associative memories: a) proposed set; b) conventional set.

different dynamics of the hydraulic actuators as compared to the electric actuators. The fuzzy control surfaces of both sets are shown in Figure 7.6, where the dashed line depicts the conventional rule surface for the fuzzy associative memories shown in Figure 7.5(b). These fuzzy surfaces show how the control signal (having threshold of ± 2.5 volt) varies with scaled position and velocity errors. A close-up of these surfaces is shown in Figure 7.7; comparing operating points **a** and **b** it is seen that for a given scaled velocity error, the hydraulic control rules require smaller scaled position error in order to fire a zero spool displacement command, which is expected.

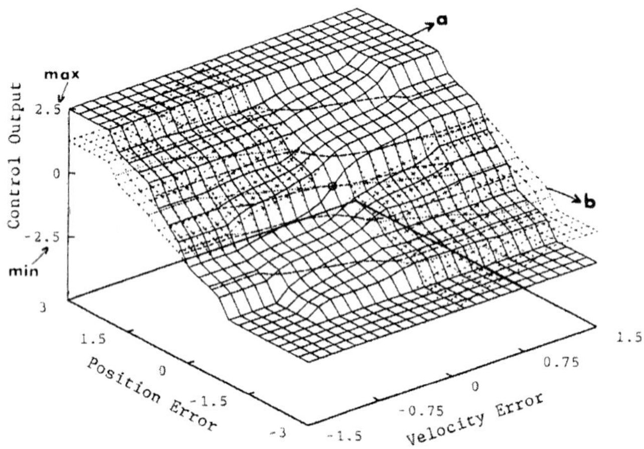

Figure 7.6 Fuzzy control surfaces: a) proposed set; b) conventional set.

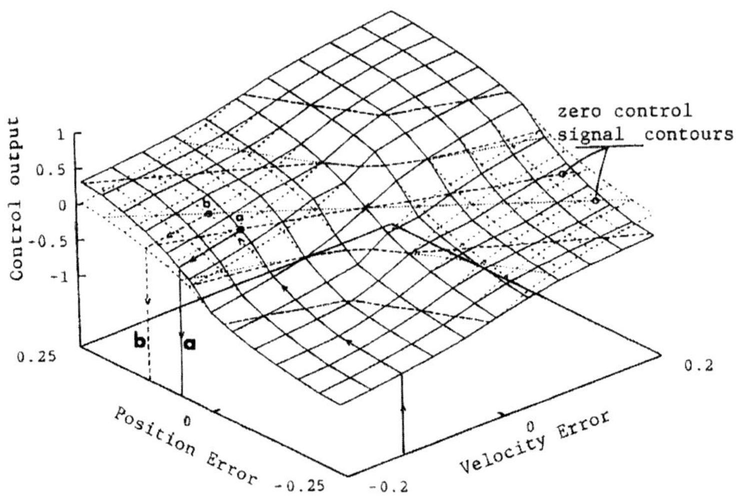

Figure 7.7 Close-up of fuzzy control surfaces.

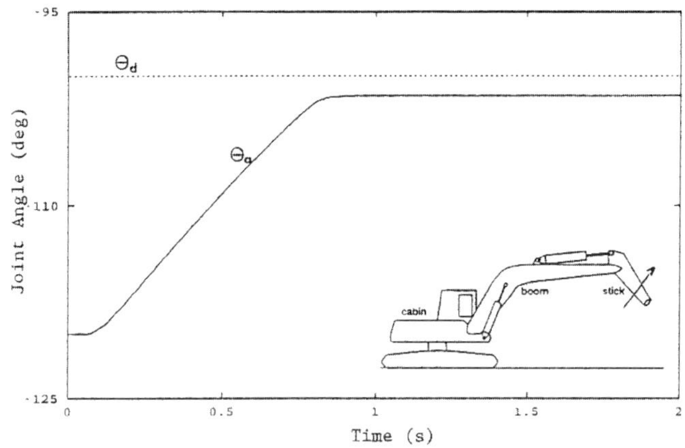

Figure 7.8 Step input response of fuzzy PD controller (simulation):

$$G_a = 0.2\frac{1}{\text{deg}};\ G_v = 0.02\frac{1}{\text{deg/s}}$$

7.3.3 Simulation Study

The setting of the position and velocity error gains can be a tedious and time consuming procedure [13]. There exist, however some practical guidelines for adjusting these parameters for particular industrial processes [9,14,15,16]. In this section a simulation study is conducted to observe how the changes in the fuzzy input scaling factors affect the response of the control system having membership functions and rules proposed in the previous section. The Caterpillar 215B-based log loader simulator developed previously is used here. The machine dynamics is complex and nonlinear in both the structure and the actuation. The simulation model combines analytical, numerical, and steady-state solutions for a fast simulation of the machine without losing any significant state information [17]. The accuracy of the simulations has been verified previously and has been reported elsewhere [3].

Figure 7.8 demonstrates the step response of the out motion of link three (stick) using a relatively low position error gain, G_a. This link was commanded to move from −120° to −100°. Referring to Figure 7.8, there is a large steady-state error. Changing the velocity error gain, G_v, did not improve the response; it even slightly increased the steady-state error. The value of the steady-state error was, however, limited to the flow deadband which was a function of machine configuration, loading and valve spool overlap [17]. Increasing the position error gain produced a response as in Figure 7.9 (*curve* 2) with a smaller steady-state error as compared to the response in Figure 7.8 [also shown in Figure 7.9(a) (*curve* 1)]. There was, however, an undesirable overshoot. Increasing the velocity error gain, G_v, reduced the amount of overshoot as seen in Figure 7.9(a) (*curve* 3). Further increase in velocity error gain produced a response plotted as (*curve* 4).

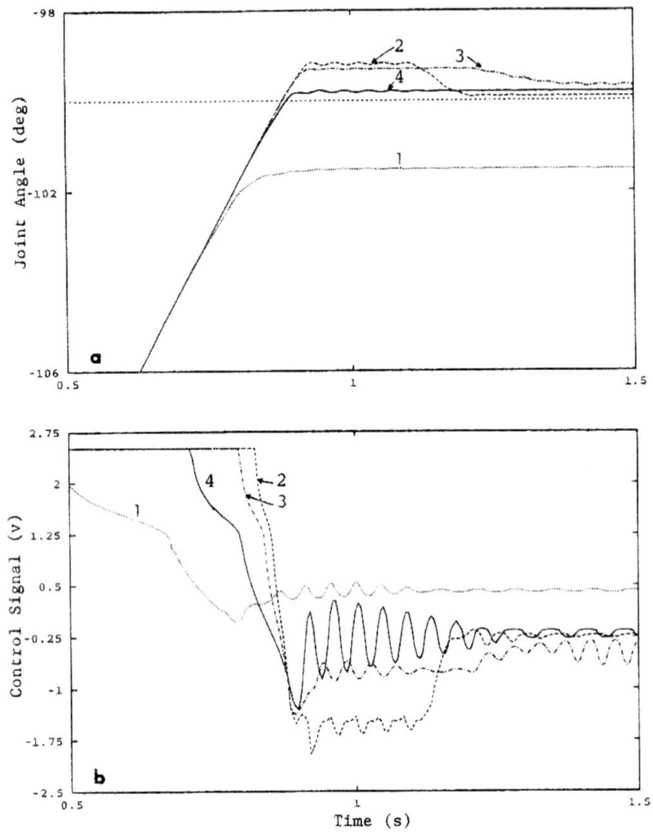

Figure 7.9 Responses with different input gains (simulation): ***curve 1*** **–** $G_a = 0.2$; $G_v = 0.02$; ***curve 2*** **–** $G_a = 1.0$, $G_v = 0.02$; ***curve 3*** **–** $G_a = 1.0$; $G_v = 0.05$; ***curve 4*** **–** $G_a = 1.0$, $G_v = 0.15$.

The control signal trajectories are shown in Figure 7.9(b)—the maximum control signal in the simulation—was 2.5 volt which allowed the full travel of the main valve spool. Referring to Figure 7.9(b), increasing the velocity error gain produced large control signal oscillations about the set point. Due to the existence of flow deadband, these oscillations did not affect the response much but added strain to the main valve system. In order to reduce the control signal oscillations, the width of the zero control output membership function was doubled. The result is shown in Figure 7.10 (*curve* 1) along with (*curve* 2) from the previous test. It is seen that the control pattern changed although the response was practically unaffected. No further improvement was achieved in the response by increasing the position or velocity error gains.

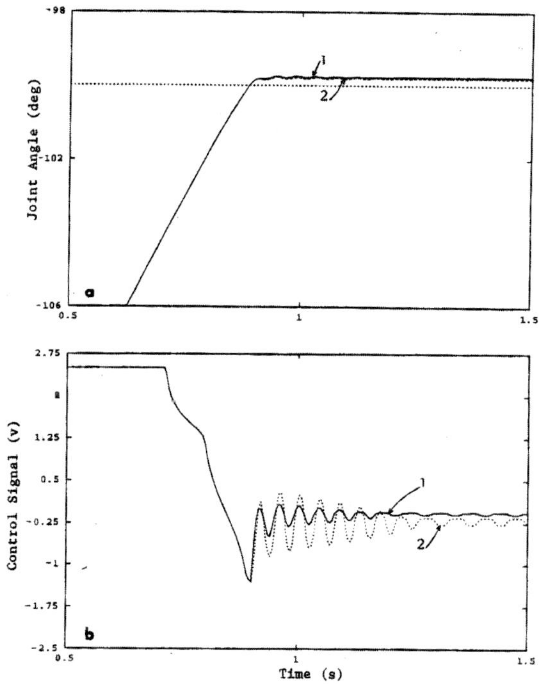

Figure 7.10 Response to increased fuzzy zero output width (simulation):
***curve 1-* G_a = 1.0; G_v = 1.5; increased width;**
***curve 2-* G_a = 1.0; G_v = 1.5; original width.**

7.4 IMPLEMENTATION

7.4.1 Equipment

Figure 7.1 shows the Caterpillar 215B excavator located at the University of British Columbia. The machine has been configured as a log loader and is instrumented with a flexible VME-bus-based computer system for machine control system development and human factors study. The computer interface system has been designed to provide a real-time control using a 20 MHz Inmos-T800 transputer-based processor board and a UNIX-based SPARC-1E VME bus board. The machine hydraulic system has been modified to include electrohydraulic pilot valves. The maximum control input signal to each pilot valve is 6 volt. Resolvers have been mounted on each of the excavator joints to sense the joint angles. An on-line low-pass 5-point differentiator software package is used to calculate the joint velocities from the joint angle measurements. The description of the computer system interfacing with the associated components and software packages can be found elsewhere [2]. The experiments described in the following were performed at the control frequency of ≈ 40 Hz.

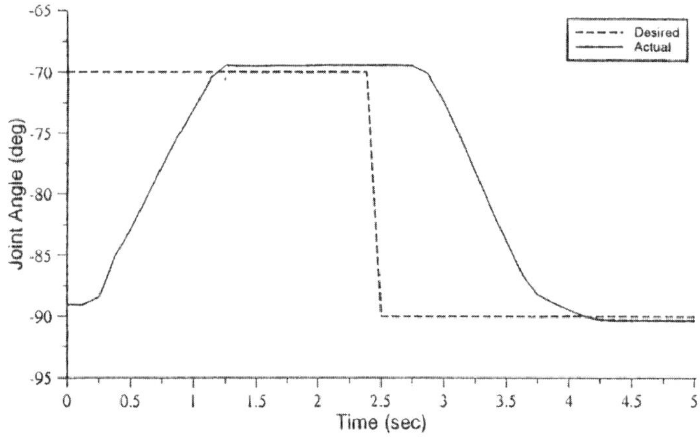

Figure 7.11 Step input response of fuzzy PD controller (experiment):

$$G_{\alpha} = 0.3\frac{1}{\text{deg}};\ G_{v} = 0.05\frac{1}{\text{deg/s}}$$

7.4.2 Results

Two sets of experiments were performed. The first set was to test the positioning ability of the controller. The second test was to examine the tracking ability. Figure 7.11 shows the step input response of the fuzzy controller to link three (stick) of the log loader. The response for well-tuned gains was fast with an acceptable steady-state error and with no overshoot or oscillations about the set point. The steady-state error is due to the main valve deadband, dry friction as well as the open-center characteristic of the main valve system. The same control gains as in the step input were then used and kept fixed in the trajectory tracking experiment. The trajectory profile [see Figure 7.12(a)] included a gradual change in the required velocity at the start and at the end. A constant error is seen when following the ramp. This is expected and is due primarily to the nature of the fuzzy PD controller as applied to hydraulics. Zero position and velocity errors would result in the controller closing the valve and stopping the motion; thus, some error must be present in the response. The response of the fuzzy controller for a similar trajectory but at a faster speed is shown in Figure 7.12(b).

Similar tracking tests were performed this time with a well-tuned nonfuzzy PD controller. The results are summarized in Figure 7.13. The PD controller responded well for slow tracking; with the same gains it failed to produce a satisfactory response at a higher speed. The nonfuzzy PD controller showed substantial oscillations during the stick-out motion. Reducing the gains to improve the response produced a very large steady-state error. The fuzzy controller displayed characteristics which were difficult to achieve with a conventional linear PD controller. This is due to the fact that the effect of the velocity error on the fuzzy control action is considered in union with the position error. This nonlinear characteristic of the controller is of great advantage. The minor

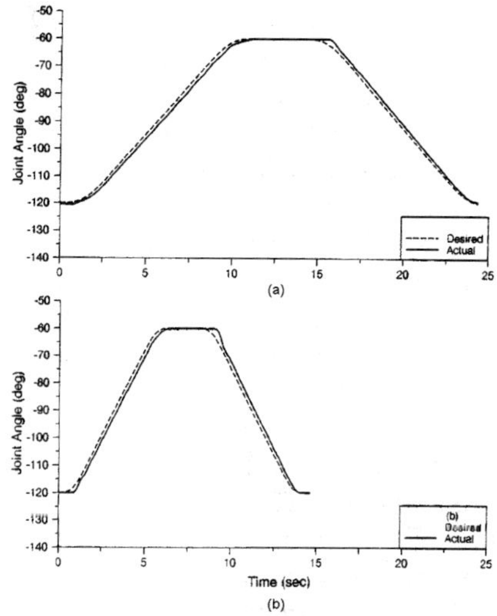

Figure 7.12 Tracking response of fuzzy PD controller (experiment): a) low speed; b) increased speed.

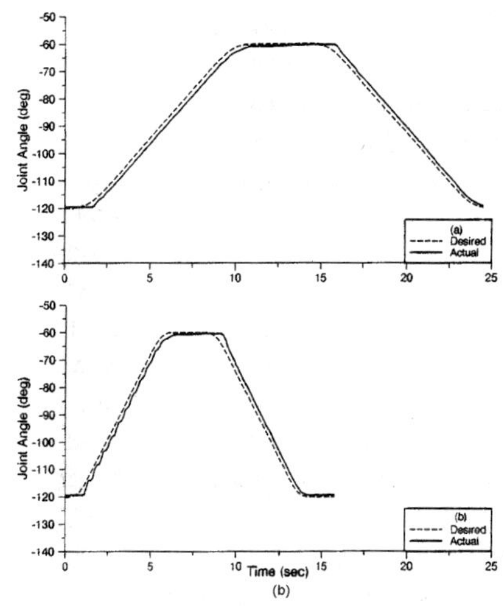

Figure 7.13 Tracking response of nonfuzzy PD controller (experiment): a) low speed; b) increased speed.

increase in the computational complexity when using a fuzzy control instead of a non-fuzzy PD control was not a problem for the application reported here.

7.5 STEADY-STATE ERROR ELIMINATION

7.5.1 General

Eliminating the steady-state error is normally done by adding an integral control action. Zheng [18] utilized a summation of control actions from a PD-type fuzzy controller to produce a fuzzy PI controller. PI controllers are, however, known to give poor performance in transient response due to the internal integrating operation. In order to improve the performance of PI controllers, Lee [19] proposed two methods which take out appropriate amounts of the accumulated control inputs according to a set of rules based on fuzzily described states on error, error rate and control input. Chen and Yen [20] suggested the use of PID-type fuzzy controllers to avoid some of the shortcomings when only PD or PI controllers are used. Mizumoto [21] constructed a PID controller based on the product-sum-gravity method and a simplified fuzzy reasoning method. To improve the transient and steady-state performance of control systems containing deadzones, Kim et al. [22] used a precompensator which magnified according to a fuzzy rule table, the position error input to a linear conventional PD controller. Recently, two methods of steady-state error elimination were developed and were examined on a Unimate hydraulic manipulator [23]; the most promising one namely, reset offset compensator will be described in the following section with relation to its applicability to the excavator machines.

7.5.2 Reset Offset Compensator

The reset offset compensator utilizes the control measure of the steady-state error obtained from the fuzzy PD control part as an offset to be added to the current control action. The algorithm is simple and can easily work in parallel with the main controller. Referring to Figure 7.14, the compensator is only active when the absolute velocity error becomes smaller than a certain value, V_r. A zero velocity error is the indication of a system which has come to a steady-state and that the fuzzy PD controller has done all it could to reduce the error. The controller output is then essentially a measure of the steady-state error; it therefore can be utilized as a compensation control term namely, an offset. Referring to Figure 7.14, upon the activation of the compensator, two loops become active: the integrator loop and the timer loop. The timer loop delays incrementing the compensation control signal, $U_o(t)$, until a certain time period, T_r, is elapsed. It then forms a new signal, $U_o(t) = [U_o(t - T_r) + U_{PD}]$. As long as the velocity error is within a range $\pm V_r$, the compensator block remains active and at each time period, the signal is augmented. When the velocity error exceeds a zero neighborhood, defined by V_r, the offset is set to zero and stays inactive. This is done by multiplying the forward path of the integrator loop (see signal *c* in Figure 7.14) by a zero.

Alternatively, the compensator signal, $U_o(t)$, can be updated by adding a portion of the fuzzy PD control, i.e., μU_{PD}, at each time period, T_r $0 < \mu \leq 1$ is the degree of membership of zero velocity error defined for the compensator block. When the velocity error is greater than V_r, then $\mu = 0$; when the velocity error is precisely zero, $\mu = 1$. μ scales the integration and contributes to governing the rate of increase on the compensator

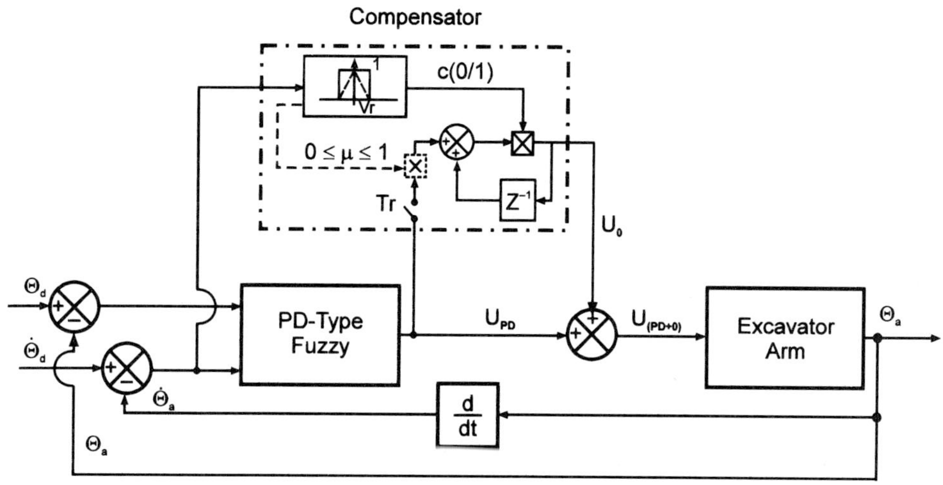

Figure 7.14 PD-type fuzzy plus compensator control block.

signal. This addition to the compensator which is shown as dashed lines in Figure 7.14, was not implemented in the experiments described below.

7.5.3 Experimental Results

The application of the above compensation scheme to link 3 of the excavator produced a result which is shown in Figure 7.15. The ramp input is followed almost exactly [(Figure 7.15(a)]. The resulting velocity profile is shown in Figure 7.15(b). Referring to Figure 7.15(c), as soon as the velocity error falls within a predetermined zero region, a value equal to the current fuzzy control is assigned as an offset. Thus, the time between the method activation until a rough measure of the steady-state error control is achieved is short. It is also seen that as the error is eliminated, the fuzzy PD control contribution goes to zero while the control contribution from the compensation block keeps the hydraulic valve open.

In the experiments reported here, the zero velocity error region defined for the compensator was chosen as 1/10 of the one used in our fuzzy PD controller. The velocity error input to the compensator was first scaled using the same velocity error gain used for the fuzzy PD controller. Depending on the designer's choice and application, narrower or wider zero velocity error regions may be used for the compensator block. Note that by keeping the velocity error threshold low enough, the compensator would not be active during the initial response to a step, thereby eliminating the integral windup problem seen under conventional integral actions.

In this experiment, the rate at which the compensator updated its signal was ≈ 1.5 Hz. It is important to allow enough time between the updates so that the system can respond to the control change and that the fuzzy PD controller can realize the changes and accommodate them in its future control outputs. A fast updating rate should not be used since it does not allow enough time to observe the full effect of the offset on the response

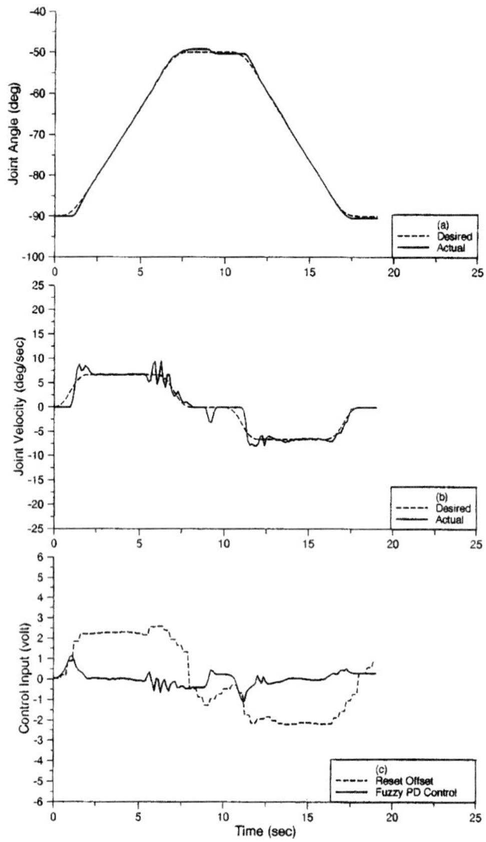

Figure 7.15 Tracking response of compensated fuzzy controller (experiment).

and therefore, the system may tend to become unstable. A slow updating rate may result in a slow elimination of the steady-state error. The response time of the system being controlled should be considered when choosing the offset updating rate. For the log loader under investigation, updating rates between 1 Hz to 3 Hz produced results with no appreciable differences. Figures 7.16 shows another result for the case of slow speed.

7.6 CONCLUSIONS

A fuzzy PD controller was implemented, for the first time, to a link of an instrumented Caterpillar 215B-based log loader. The controller utilized a specialized set of membership functions and rules to meet the basic control requirements of hydraulic functions. In the proposed fuzzy control scheme, the universe of discourse of position and velocity errors were partitioned into only 5 and 3 simple sets, respectively. The position and velocity-related scaling factors were tuned off-line, and were kept fixed during the experiments. A simulation study was first performed and the effects of variation of controller gains were

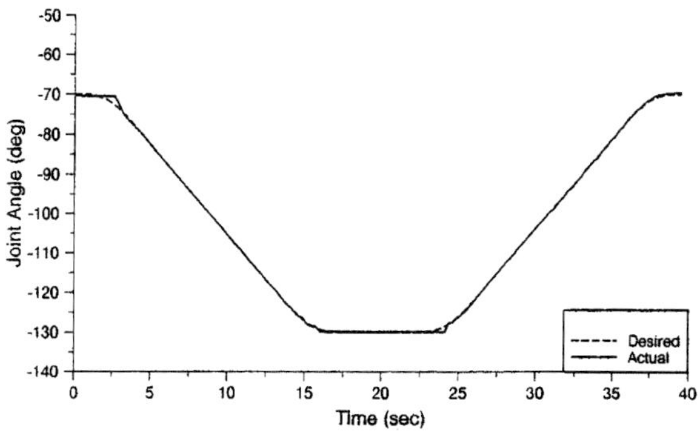

Figure 7.16 Slow tracking response of compensated fuzzy controller (experiment).

examined before the actual implementation was carried out. A novel and conceptually simple compensation scheme was further employed in parallel with this controller, in order to eliminate the steady-state errors observed in both regulating and tracking responses. The compensator used velocity error as a criterion to boost the control signal obtained from the fuzzy controller. The compensation signal is augmented at each time period for as long as the velocity error stays within a set threshold. By keeping the velocity error threshold low enough, the compensator would not be active during the initial response to a step, thereby eliminating the integral windup problem seen under simple integral action.

The results of this work, which is believed to represent a further contribution to the published work on fuzzy logic applications, have investigated the promise of the technique and have shown certain benefits over a classical PD controller when applied for controlling heavy-duty hydraulic functions.

7.7 ACKNOWLEDGMENTS

The author is sincerely grateful to Professor Peter Lawrence of the Department of Electrical Engineering, University of British Columbia, for allowing the conduct of the experiments on the Caterpillar 215B excavator machine and for helpful discussions. The author would like to also acknowledge the contributions of Todd Corbet in implementing the control software, and of Dan Chan and Simon Bachmann in implementing the hardware and sensor packages.

7.8 REFERENCES

1. Lawrence, P. D., Sassani, F., Sauder, B., Sepehri N., Wallersteiner, U. and Wilson J. "Computer-Assisted Control of Excavator-Based Machines." *Int. Off-Highway Powerplant Congress & Exposition*, Milwaukee, WI, 1993. SAE Technical Paper No. 932486.

2. Lawrence, P. D., Sepehri, N. and Sassani, F. "Coordinated Hydraulic Actuation Systems." *Proc. 2nd Int. Conf. on Machines Automation* (ICMA '94), pp. 355-367, Tampere, Finland, 1994.

3. Sepehri, N., Lawrence, P. D., Sassani, F. and Frenette, R. "Resolved-Mode Teleoperated Control of Heavy-Duty Hydraulic Machines." *ASME J. of Dynamic Systems, Measurements, and Control* Vol. 116, 1994, pp. 232-240.

4. Sepehri, N. and Lawrence, P. D. "Model-Based Sensor-Based Velocity Control of Teleoperated Heavy-Duty Hydraulic Machines." *Proc. IEEE/DSJ Int. Conf. on Intelligent Robots and Systems*, Raleigh, pp. 859-862, NC, 1992.

5. Merritt, H.E. "*Hydraulic Control Systems.*" John Wiley & Sons, 1967.

6. Kumbla, K. , Moya, J., Baird, R. , Rajagopalan, S. and Jamshidi M. "Fuzzy Control of Three Links of a Robotic Manipulator." *Robotics and Manufacturing*; *Proc. ISRAM'92* (Jamshidi, M. et al., eds.). ASME Press, 1992, pp. 687-694.

7. Ken, C., Jinn-Ya, L. and Xiang, L.Y. "Fuzzy Control of Robot Manipulator." *Proc. IEEE Int. Conf. on Systems, Man, and Cybernetics*, Vol. 2, pp. 1210-1212, Beijing, China, 1988.

8. Zhao, T. and Virvalo, T. "Fuzzy Control of a Hydraulic Position Servo with Unknown Load." *Proc. IEEE Int. Conf. on Fuzzy Systems*, pp. 785-788, San Francisco, CA, 1993.

9. Chou, C.H. and Lu, H. C. "Design of a Real-Time Fuzzy Controller for Hydraulic Servo Systems." *Computers in Industry*. Vol. 15, 1993, pp. 129-142.

10. Lee, C.C. "Fuzzy Logic in Control Systems: Fuzzy Logic Controller—Parts I & II." *IEEE Trans. on Systems, Man and Cybernetics*. Vol. 20, 1990, pp. 404-433.

11. Kickert, W. J. M. and Mamdani, E.H. "Analysis of a Fuzzy Logic Controller." *Fuzzy Sets and Systems*. Vol. 1, 1978, pp. 29-44.

12. Zhou, J. and Coiffet, P. "Fuzzy Control of Robots." *Proc. IEEE Int. Conf. on Fuzzy Systems*, pp. 1357-1364, San Diego, CA, 1992.

13. Boscolo, A. "Computer Aided Tuning Validation of Fuzzy Systems." *Proc. IEEE Int. Conf. on Fuzzy Systems*, pp. 605-614, San Diego, CA, 1992.

14. Daugherity, W. C. "Performance Evaluation of a Self-Tuning Fuzzy Controller." *Proc. IEEE Int. Conf. on Fuzzy Systems*, pp. 389-397, San Diego, CA, 1992.

15. Smith, S. M. and Comer, D. J. "An Algorithm for Automated Fuzzy Logic Controller Tuning." *Proc. IEEE Int. Conf. on Fuzzy Systems*, pp. 615-622, San Diego, CA, 1992.

16. Maeda, M., Sato, T. and Mirakami, S. "Design of the Self-Tuning Fuzzy Controller." *Proc. Int. Conf. on Fuzzy Logic & Neural Networks*, pp. 393-396, Lizuka, Japan, 1990.

17. Sepehri, N. "Dynamic Simulation and Control of Teleoperated Heavy-Duty Hydraulic Manipulators." Ph.D. thesis, Department of Mechanical Engineering, University of British Columbia, 1990.

18. Zheng, L. "A Practical Guide to Tune Proportional and Integral Like Fuzzy Controllers." *Proc. IEEE Int. Conf. on Fuzzy Systems*, pp. 633-640, San Diego, CA, 1992.

19. Lee, J. "On Methods for Improving Performance of PI-Type Fuzzy Logic Controller." *IEEE Trans. on Systems, Man and Cybernetic*. Vol. 1, 1993, pp. 298-301.

20. Chen, Y. Y. and Yen, C. C. "PD-Type vs PID-Type Fuzzy Controllers." *Proc. IEEE 10th Region Conf.*, pp. 341-345, Melbourne, Australia, 1992.

21. Mizumoto, M. "Realization of PID Controls by Fuzzy Control Methods." *Proc. IEEE Int. Conf. on Fuzzy Systems*, pp. 709-715, San Diego, CA, 1992.

22. Kim, J.H., Park, J.H., Lee, S.M. and Chong, E.K.P. "Fuzzy Precompensation of PD Controllers for Systems with Deadzones." *J. of Intelligent and Fuzzy Systems*. Vol. 1, 1993, pp. 125-133.

23. Sepehri, N., Corbet, T. and Lawrence, P. D. "Fuzzy Position Control of Hydraulic Robots with Valve Deadbands." *Int. J. of Mechatronics*. Vol. 5, 1995, pp. 623-643.

8

TWO PARADIGMS OF AUTOMOTIVE FUZZY LOGIC APPLICATIONS

M. Schröder and R. Petersen
VOLKSWAGEN AG
GERMANY

F. Klawonn and R. Kruse
University of Braunschweig
GERMANY

In the last years many fields of fuzzy logic applications have developed with a high commercial success. However publications in the automotive field are comparatively rare. In this chapter we present two applications of fuzzy logic on the motor vehicle. First we consider the engine idle speed control and the second part of this chapter is devoted to the shift-point determination of an automatic transmission. The idle speed controller is realized by a volumetric control. For this control task we have used a fuzzy controller according to Mamdani, which is based on equality relations to interpret the procedure as an interpolation of a partially specified control function in a vague environment, which reflects the indistinguishability of measurements or control values. The shift-point determination with fuzzy methods leads to a continously adaptive gearshift schedule between an economic and a sporting gearshift diagram. Thereby the filtered accelerator velocity, the number of preceding sign changes, and the accelerator position are used to distinguish different types of drivers or driving situations, respectively. A calculated sport factor carries out an individual adaptation of the gearshift movements at the driver's request.

8.1 IDLE SPEED CONTROL

To control the idle speed of a spark ignition engine, there are two possibilities. The first one is to use a spark advancer, where the ignition point will be displaced up to 25 degrees of crank angle (timing range) depending on the deviation of the current revolutions (i.e., the error) to the set point. If the current number of revolutions is too small the engine needs an advanced spark to increase the torque; otherwise if the current number of revolutions is high, a retarded ignition decreases the number of revolutions. The task of the spark advance is to achieve a fast stabilization of the idle speed.

The second method is the volumetric control, where an auxiliary air regulator varies the cross-section of a by-pass to the throttle (see Figure 8.1).

The auxiliary air regulator is pulse-width modulated by the control device. If the control device recognizes a drop in the number of revolutions, it sends a signal to the auxiliary air regulator to increase the by-pass cross-section. The air flow sensor measures the increased air flow rate and sends a message to the control device. There will be calculated the new quantity for the fuel injection. In combination with the higher air flow rate the engine yields more torque leading to a raised number of revolutions (see also [1]). For an introduction into spark ignition engine, see for instance [2].

Analogously it works to reduce the number of revolutions, at which the by-pass cross-section is to decrease.

The statutory and the individual requirements concerning the performance of engines demand a reduction in fuel consumption and pollutant emissions. A slower idle speed satisfies these demands, but a very flexible controller is necessary, because the intensified application of comfort facilities, like the air-conditioning system, power steering or electrical load, leads to big drops in revolutions.

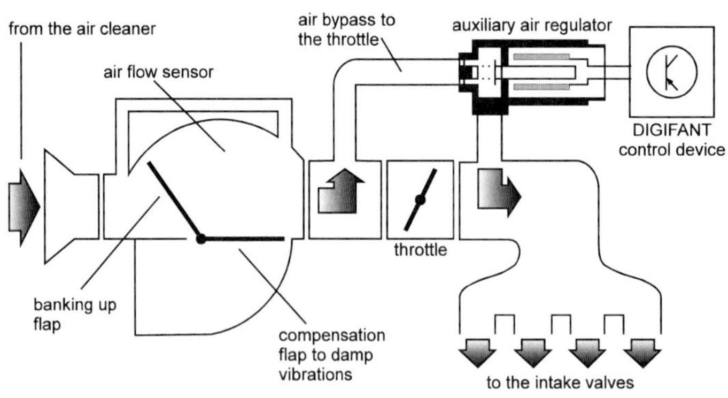

Figure 8.1 Principle of engine idle speed control.

From this point of view the task of a volumetric control is to compensate for different types of loads, which are recognized by the corresponding drop of revolutions. The compensation will be done by means of varying the by-pass cross-section. The current idle speed should reach the correct set point as fast as possible.

In comparison to the spark advance, the volumetric control works quiet slower. From this point of view, these two different types of idle speed control are decoupled in time and may work together.

The delay of the automatic control system coming from the air transfer time (determined by the capacity of the intake manifold) and the inaccuracy of the engine speed information produce very complex solutions with partially unsatisfactory results, if traditional control engineering is used. The inaccuracy of engine speed information can be explained by the imprecision of the available signal of the Hall pick-up in the ignition distributor. One obtains differences of up to ±30 rpm resulting from manufacturing tolerances, gear clearances, and torsional vibrations. Another problem arises from the plurality of stochastic processes in the engine system. As an example, bad combustion or dispersion in the ignition system and in the fuel injection system have to be mentioned, because in these cases a volumetric controller should not react in order to prevent an oscillation of the idle speed. In addition, in classical controller types a local adjustment is often very difficult.

From this motivation there was a cooperation between Volkswagen AG, Wolfsburg, and the University of Braunschweig, where a fuzzy control approach (for theoretical foundations about fuzzy systems [3]) based on the concept of equality relations to interpret the procedure as an interpolation of a partially specified control function in a vague environment reflecting the indistinguishability of measurements or control values, respectively, was investigated for the volumetric control. The spark advance was maintained. The experimental plant was a 2000-cc, 116-hp engine of the current Volkswagen Golf GTI model. An easier design and an improved control behavior become possible by using fuzzy control methods.

8.1.1 Equality Relations

In many real-world applications engineers use fuzzy control with an understanding of the control rules as logical statements or implications. However, fuzzy control may be seen often as an interpolation of a partially specified control function in a vague environment, which reflects the indistinguishability of measurements or control values. The concept of equality relations turns out to be a natural way to represent such vague environments and gives an appropriate mathematical framework in order to be able to develop a knowledge-based model of the control actions of an engineer. From an engineering point of view equality relations are a method of representing indistinguishability or similarity of points with a short distantce in between. By this approach fuzzy sets may be seen as representatives of crisp values in a vague environment described by equality relations.

In physics and engineering science one frequently has to handle inexact measurements. There are two different ways to consider inexactness. In the first view, inexactness may be due to a limited precision of the measurement instruments or to general restrictions enforced by the experiment or process itself and might be categorized as indistinguishability. On the other hand, exaggerated exactness often makes no sense as one may seen for instance in the case of measuring the room temperature, that this kind of inexactness might be understood as similarity. In the following, we do not make a distinction between these two concepts, and call both of them indistinguishability.

In our sense, a vague environment is a set of values for which we take indistinguishability into account. A mathematical approach to consider a vague environment $X \subseteq \Re$ (without loss of generality, we assume $X = \Re$) is to choose $\varepsilon > 0$ as an error or tolerance bound and to identify two values, $x, x' \in X$, if their distance is at most ε, i.e., $|x - x'| \leq \varepsilon$.

Then x and x' may be called as ε-indistinguishable. This ε-indistinguishability can be represented by a relation $R_\varepsilon \subseteq X \times X$ containing all pairs of numbers that are ε-indistinguishable, i.e., $R_\varepsilon = \{(x, x') \in X \times X \mid x - x' \mid \leq \varepsilon\}$. In practical applications it is often reasonable instead of fixing one value for ε to consider a range of values, for example the interval [0, 1], so that we are able to define the mapping

$$E: X \times X \to [0,1] \ , \quad (x,x') \mapsto 1 - \inf\left\{\varepsilon \in [0,1] \middle| (x,x') \in R_\varepsilon\right\}. \tag{8-1}$$

The greater the value $E(x, x')$, the more x and x' are indistinguishable or similar, respectively. In this simple example we obtain $E(x, x') = 1 - \min\{|x - x'|, 1\}$, which may be intuitively read as the degree to which x and x' are equal or to which they can be identified. This motivation leads to the following definition.

Definition 8-1: A mapping $E: X \times X \to [0, 1]$ is an *equality relation* if E satisfies the following axioms:

1. $E(x,x) = 1$
2. $E(x,x') = E(x',x)$ and
3. $E(x,x') + E(x',x'') - 1 \leq E(x,x'')$

at which the third condition is motivated in the view point if $(x, x') \in \Re_\varepsilon$ and $(x', x'') \in \Re_{\varepsilon'}$ then $(x', x'') \in \Re_{e+e'}$.

As an example, if δ is a metric on X (for instance the standard metric $\delta(x, x') = |x - x'|$), then $E_\delta(x, x') = 1 - \min\{|x - x'|, 1\}$ is an equality relation on X. If we consider the equality relation induced by the standard metric on $\Re$, this metric is not an appropriate metric in any case, because it does not take any scaling factors into account. Scaling factors are necessary in order to obtain the same degree of indistinguishability if we consider different units.

Example: Let person A be 29 years 3 months and person B 28 years 6 months. If we consider the measured ages in years and use the standard metric δ, we obtain E_δ(29.25,

28.5) = 1 − min {|29.25 − 28.5|, 1} = 0.25 as the degree of indistinguishability. On the other hand, we have to introduce a scaling factor $1/12 = c \geq 0$ in order to obtain the same indistinguishability if we consider the unit months, and we define the equality relation:

$$E^{(c)}(x, x') = 1 - \min\{|c \cdot x - c \cdot x'|, 1\} \tag{8-2}$$

Depending on the underlying unit, one has to choose the corresponding scaling factor c. In some applications it is useful to generalize the concept of scaling factors in the sense that instead of one scaling factor for all values of a considered real interval $[a, b]$, we determine different scaling factors for different ranges. As an example, we consider a vague environment induced by a measuring device that works with different precision in different ranges. For instance, we can notice a high precision in the range of $[-1, 1] \subseteq [a, b]$ and with less precision outside this area. This might be characterized by choosing a scaling factor 3 for the range $[-1, 1]$ and the scaling factor 0.5 for the area outside $[-1, 1]$. The scaling factor 3 causes a magnification of the interval $[-1, 1]$ and a scaling factor 0.5 leads to reductions of the intervals $[a, -1)$ and $(1, b]$ by the factor 2. Let us assume $a = -3$ and $b = 5$ to "measure" the distance between two elements $x, x' \in [-3, 5]$ (piecewise linearly) to the interval $[0, 9]$. This piecewise transformation is given by the mapping:

$$f: [-3, 5] \to [0, 9]$$
$$x \mapsto \begin{cases} 0.5 \cdot (x+3) & \textit{if } -3 \leq x < -1 \\ 3 \cdot (x+1) + 1 & \textit{if } -1 \leq x \leq 1 \\ 0.5 \cdot (x-1) + 7 & \textit{if } 1 < x \leq 5 \end{cases}$$

at which we are able to define an integrable scaling function c: $[-3, 5] \to [0, \infty)$ by

$$x \mapsto \begin{cases} 0.5 & \textit{if } -3 \leq x < -1 \\ 3 & \textit{if } -1 \leq x \leq 1 \\ 0.5 & \textit{if } \ 1 < x \leq 5 \end{cases}$$

The equality relation on $[-3, 5]$ induced by this transformation is defined by

$$E: [-3, 5] \times [-3, 5] \to [0, 1]$$
$$(x, x') \mapsto 1 - \min\{|f(x) - f(x')|, 1\}$$

This equality relation reflects the modeling of the vague environment induced by a measuring device that does not measure with the same exactness over the whole range. Therefore, this equality relation describes indistinguishability. Analogously, it is also possible and reasonable to define such an equality relation representing similarity (for further details see [4]–[6]).

8.1.2 Points And Sets In A Vague Environment

If we deal with a vague environment with indistinguishability, we have to take into account that in general a crisp value obtained from a measuring device does not correspond exactly to that value according to the indistinguishability. If we specify a crisp value x_0 in a vague environment, generally it is attached to *approximately* x_0 only. Always if we consider crisp points or sets in a vague environment, we have to take into account the indistinguishability which is induced by the environment.

If we identify two values in the sense mentioned in Section 8.1.1, i.e. if their distance is less than a value ε, then a crisp value x_0 induces the interval $[\, x_0 - \varepsilon, x_0 + \varepsilon \,]$. This interval is the set of points that are ε-indistinguishable to x_0. Generally, we do not fix one value for ε, but we consider a whole set of such tolerance bounds. If we approve values between zero and one, x_0 might be associated with the set

$$\{[x_0 - \varepsilon, x_0 + \varepsilon] \mid \varepsilon \in [0, 1]\} \tag{8-3}$$

On the other hand, if we consider the equality relation E characterizing our vague environment X, we can associate the crisp value $x_0 \in X$ by the mapping

$$\mu_{x_0} : X \to [0, 1] \quad , \quad x \mapsto E(x_0, x) \tag{8-4}$$

specifying for each x which x is equal to x_0. The mapping μ_{x_0} is a fuzzy set, which could be characterized as the fuzzy set of all $x \in X$ that are equal to x_0 with respect to the vague environment. From this point of view, certain fuzzy sets are representations of crisp points in a vague environment described by an equality relation.

Definition 8-2: Let E be an equality relation on the set X. For $x_0 \in X$ the fuzzy set

$$\mu_{x_0} : X \to [0, 1],\ x \mapsto E(x_0, x)$$

is called a *singleton* corresponding to x_0.

Remark: In the literature one often uses the term singleton for a fuzzy set that assigns the membership degree 1.0 to one point only and zero to all others (see for example [7]). However, this notion is not contradictory to the given definition (see [6]).

If we consider not a crisp point but a crisp set, we have to associate that with the corresponding fuzzy set of all points, which are "equal" to one of the points in the crisp set. If $M \subseteq X$ is a subset, we associate with that the following set

$$\left\{ \bigcup_{x \in M} [x - \varepsilon, x + \varepsilon] | \varepsilon \in [0, 1] \right\} \tag{8-5}$$

at which $\bigcup_{x \in M} [x - \varepsilon, x + \varepsilon]$ is the set of points that are ε-indistinguishable to at least one of the elements in *M*. In the terms of the corresponding equality relation we obtain the following definition.

Definition 8-3: Let *E* be an equality relation on the set *X*. For a subset $M \subseteq X$ the fuzzy set

$$\mu_M : X \rightarrow [0, 1],\ x \mapsto \sup\{E(m, x) | m \in M\}$$

is called the *extensional hull* of *M*.

Remark: Definition 8-3 is a generalization of Definition 8-2, since $\mu_{x_0} = \mu_{\{x_0\}}$ holds.

As an example, let $X = \Re$ and let $E = E_\delta$ be the equality relation induced by the standard metric δ as mentioned above. The singleton corresponding to $x_0 \in \Re$ is the fuzzy set $\mu_{x_0}(x) = 1 - \min\{|x_0 - x|, 1\}$, at which μ_{x_0} is a triangular membership function. Analogously, if we consider the interval $[a, b]$ instead the single point x_0, the extensional hull $\mu_{[a, b]}$ of $[a, b]$ is given by the trapezoidal membership function

$$\mu_{[a,b]} = \begin{cases} 1 & \textit{if} \quad a \leq x \leq b \\ \max\{1 - a + x, 0\} & \textit{if} \quad x \leq a \\ \max\{1 - x + b, 0\} & \textit{if} \quad b \leq x \end{cases}$$

Steeper or less steep slopes can be obtained by taking scaling factors into account.

If we take more than one vague environment into account, we have to find an adequate equality relation on the product space of the vague environments. One solution is the following suggestion.

Definition 8-4: Let $E_1, \ldots, E_n$ be equality relations on $X_1, \ldots, X_n$, respectively. Define

$$E_{1,\ldots,n} : (X_1 \times \cdots \times X_n)^2 \rightarrow [0, 1]$$

$$\left((x_1, \ldots, x_n), \left(x_1', \ldots, x_n'\right)\right) \mapsto \min\left\{E_1\left(x_1, x_1'\right), \ldots, E_n\left(x_n, x_n'\right)\right\}$$

The mapping $E_{1,\ldots,n}$ is an equality relation on $X_1 \times \cdots \times X_n$ (for proof, see [6]).

8.1.3 Fuzzy Control Based on Vague Environments

In this section, we want to present an approach to the procedure of fuzzy control (according to Mamdani), which is based on the concept of equality relations. A given control problem might be formalized by n input variables $\xi_1,\dots, \xi_n$ taking values in the sets $X_1,\dots, X_n$. For the sake of simplicity we consider control systems concerning one output variable η with values in the set Y only. An adequate control function $\varphi: X_1 \times \cdots \times X_n \to Y$, which assigns to each input tuple $(x_1,\dots,x_n) \in X_1 \times \cdots \times X_n$ an appropriate output value $y = \varphi(x_1,\dots,x_n) \in Y$.

The fuzzy control approach according to Mamdani might be formalized in the following manner. The control function φ is specified by k linguistic control rules R_r in the form

$$R_r : if\ \xi_1\ is\ A^{(1)}_{i_1,r}\quad and \dots and\ \ \xi_n\ is\ A^{(n)}_{i_n,r}\quad then\ \eta\ is\ B_{i_r}\ ,\quad r = 1,\dots,k$$

where each linguistic term $A^{(1)}_{i_1,r}, \dots, A^{(n)}_{i_n,r}, B_{i_r}$ is associated with a fuzzy set $\mu^{(1)}_{i_1,r}, \dots, \mu^{(n)}_{i_n,r}, \mu_{i_r}$ on $X_1,\dots,X_n$, Y, respectively. Concerning the input tuple $(x_1,\dots,x_n) \in X_1 \times \cdots \times X_n$, the output of Mamdani's fuzzy controller is the fuzzy set

$$\mu^{output}_{x_1,\dots,x_n} : Y \to [0, 1]$$

$$y \mapsto \max_{r\in\{1,\dots,k\}} \left\{\min\left\{\mu^{(1)}_{i_1,r}(x_1),\dots,\mu^{(n)}_{i_n,r}(x_n),\mu_{i_r}(y)\right\}\right\}$$

on Y. In order to obtain a crisp output value, the fuzzy set $\mu^{output}_{x_1,\dots,x_n}$ has to be defuzzified, by which one often uses the center of area method.

Next, we will consider the concept of knowledge-based control based on vague environments. We will see, that although we start from a quite different approach, we obtain the same calculations as in Mamdani's model. Therefore, we are able to see a fuzzy controller according to Mamdani from a view of reasonable semantics.

If we want to design a controller, which is based on vague environments, we have to specify adequate equality relations a $E_1,\dots,E_n$, F on the sets $X_1,\dots,X_n$, Y, respectively. These equality relations are used to model the indistinguishability of values as already described. Scaling factors are specially good to realize, since they have a reasonable interpretation. Small scaling factors imply a low distinguishability, i.e., in this area the control action changes only slowly with varying input values. A greater scaling factor indicates a high distinguishability, i.e., even small variations in the input might lead to greater alterations in the control action. In the field of fuzzy control we mainly have to deal with intended indistinguishability, which is not produced by difficulty in measuring exact values, but by the fact that arbitrary precision is not needed. A control expert has to provide a set of input-output tuples, i.e., tuples $((x_1^{(r)},\dots,x_n^{(r)}), y^{(r)}) \in (X_1 \times \cdots \times X_n) \times Y, (r = 1,\dots,k)$. The tuple $((x_1^{(r)},\dots,x_n^{(r)}), y^{(r)})$ simply means that $y^{(r)}$ is the appropriate output for the input $(x_1^{(r)},\ldots,x_n^{(r)})$. These k input-output tuples correspond to a partial specification of the control function, since they can be understood as the function

$$\varphi_0 : \left\{\left(x_1^{(r)},\ldots,x_n^{(r)}\right) \mid r \in \{1,\ldots,k\}\right\} \to Y,$$

$$\left(x_1^{(r)},\ldots,x_n^{(r)}\right) \mapsto y^{(r)},$$

which is a partial mapping from $X_1 \times \cdots \times X_n$ to Y. The task is now to determine an appropriate output value $\varphi(x_1,\ldots,x_n) = y \in Y$ for an arbitrary input $(x_1,\ldots,x_n) \in X_1 \times \cdots \times X_n$ from the knowledge given by the equality relations $E_1,\ldots,E_n,F$, and the partial control function φ_0. In order to solve this problem we consider the graph G_{φ_0} of φ_0 given by

$$G_{\varphi_0} \stackrel{def}{=} \left\{\left(x_1^{(r)},\ldots,x_n^{(r)},y^{(r)}\right) \mid r = 1,\ldots,k\right\}$$

which is equivalent to φ_0 and take the equality relations into account. That means, instead of considering the crisp set G_{φ_0} we take a look to the corresponding extensional hull of G_{φ_0} according to Definition 8-3 and 8-4.

Remark: If we have an ordinary mapping $\varphi : X \to Y$, we are able to derive the value $y = \varphi(x)$ by drawing the graph G_φ of φ and projecting the graph at x to the set Y.

Before we are able to calculate the extensional hull represented by the corresponding fuzzy set, we have to combine the equality relations defined on $X_1 \times \cdots \times X_n$ and Y to an equality relation E on $X_1 \times \cdots \times X_n \times Y$. Remembering Definition 8-4 we obtain

$$E((x_1,\ldots,x_n,y),(x_1',\ldots,x_n')) = \min\{E_1(x_1,x_1'),\ldots,E_n(x_n,x_n'),F(y,y')\}.$$

The extensional hull of the graph G_{φ_0} with respect to this equality relation is, according to Definition 8-3, the fuzzy set

$$\mu_{G_{\varphi_0}}(x_1,\ldots,x_n,y) = \max_{r\in\{1,\ldots,k\}}\left\{E\left(\left(x_1^{(r)},\ldots,x_n^{(r)},y^{(r)}\right),(x_1,\ldots,x_n,y)\right)\right\}$$

on $X_1 \times \cdots \times X_n \times Y$. In order to obtain an appropriate "output" for a given input tuple $(x_1,\ldots,x_n)$, we compute the projection not of the graph, but of its extensional hull at $(x_1,\ldots,x_n)$, which leads to fuzzy set

$$\mu_{\varphi_0}^{(x_1,\ldots,x_n)}(y) = \max_{r\in\{1,\ldots,k\}}\left\{\min\left\{E_1\left(x_1^{(r)},x_1\right),\ldots,E_n\left(x_n^{(r)},x_n\right),F\left(y^{(r)},y\right)\right\}\right\} \qquad (8\text{-}6)$$

on Y. If we take Definition 8-2 into account, we are able to transform (8-6) into

$$\mu_{\varphi_0}^{(x_1,\ldots,x_n)}(y) = \max_{r\in\{1,\ldots,k\}} \left\{\min\left\{\mu_{x_1^{(r)}}(x_1),\ldots,\mu_{x_n^{(r)}}(x_n),\mu_{y^{(r)}}(y)\right\}\right\}$$

because $E_1(x_1^{(r)},\cdot),\ldots,E_n(x_n^{(r)},\cdot)$, $F(y^{(r)},\cdot)$correspond to the singletons (fuzzy sets) $\mu_{x_1^{(r)}},\ldots,\mu_{x_n^{(r)}},\mu_{y^{(r)}}$., respectively. This is the output derived in the presented knowledge-based control model based on vague environments. Obviously, we obtain the same output, as we would if we considered Mamdani's fuzzy controller (before defuzzification). Thus, we are able to translate a controller based on vague environments into Mamdani's fuzzy control approach.

That means that we started with the specification of equality relations on the sets $X_1,\ldots,X_n,Y$ and a partial control mapping $\varphi_0 : X_1\times\cdots\times X_n \to Y$ in the form of the set

$$\left\{\left(\left(x_1^{(r)},\ldots,x_n^{(r)}\right),y^{(r)}\right) \mid r\in\{1,\ldots,k\}\right\}$$

Taking into account, that we are working in vague environments with respect to the equality relation, we are able to interpret the partial control mapping φ_0 as k control rules of the form

R_r : *if* ξ_1 *is approximately* $x_1^{(r)}$ *and* ... *and* ξ_n *is approximately* $x_n^{(r)}$

then η *is approximately* $y^{(r)}$, $(r=1,\ldots,k)$

where approximately $x_1^{(r)}$, ... , approximately $x_n^{(r)}$, approximately $y^{(r)}$ are represented by the singletons (fuzzy sets) $\mu_{x_1^{(r)}},\ldots,\mu_{x_n^{(r)}},\mu_{y^{(r)}}$, respectively. Taking the k control rules together with these fuzzy sets, we may define a fuzzy controller in the sense of Mamdani and obtain for an input $(x_1,\ldots,x_n)$ the fuzzy set $\mu_{x_1,\ldots,x_n}^{output}$ as output. This fuzzy set coincides with the fuzzy set $\mu_{\varphi_0}^{(x_1,\ldots,x_n)}$, which is the output of the knowledge-based controller based on vague environments.

For the answer to the question, under which conditions is one able to translate a Mandami fuzzy controller to a controller based on vague environments, the interested reader is referred to the more theoretically oriented article [6].

8.1.4 Application To Idle Speed Control

As stated in Section 8.1.1 the task of idle speed control or volumetric control is to compensate for drops in the number of revolutions by increasing the cross- section of the throttle by-pass. On the experimental vehicle a sudden drop in the engine speed might occur for different reasons:

1. switching on electrical units, like the ventilation, the lights, or the heated rear window, which load the engine via the three-phase dynamo,

2. switching on the air-conditioning system, which puts on an additional load on the engine via the air-conditioning compressor, or

3. activation of the power steering, where the hydraulic circuit pump loads the engine.

One of the main problems of this controlled system is the low torque in range of low idle speed, because in extreme situations, like simultaneous switching on of the air-conditioning system and the power steering, very strong and rapidly occurring speed drops might occur.

However, the delay of the automatic control system turns out to be the hardest problem, since it passes about 10 sequences of ignition after an alteration of the throttle by-pass, until the engine delivers a changed torque.

In addition, the inaccuracy of the engine speed information and the plurality of stochastic processes mentioned in Section 8.1 have to be taken into account.

To solve these problems of the engine idle speed controlled system, we chose an integrated concept of two controllers, where a Mamdani Fuzzy Controller (MFC) forms the basic unit and is embedded in a so-called "meta-controller" (see Figure 8.2).

The meta-controller consists of three components: data preparation, state detection (including MFC activation), and a security stage. The data preparation computes the mean of the incoming noisy data and supplies engine speed information for the evaluation of the current control situation. If the current situation requires a new control value, the state detection activates the MFC. In order to determine such a requirement, the state detection works up the modified and the original data. The security stage behind the MFC takes care of limiting the control range.

At the time of switching on the air-conditioning system, the MFC is not activated, and a pilot output value has to be chosen in order to obtain the fastest control action. The deviation dREV [rpm] of the number of revolutions to the set value and the changed gREV [rpm] of the number of revolutions between two ignitions are the two input

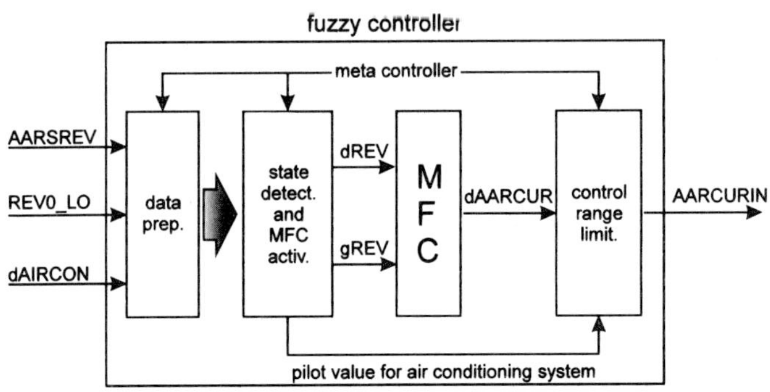

Figure 8.2 Structure of the fuzzy controller.

quantities for the MFC. The change of current dAARCUR [encoded pulse-width repetition rate] for the auxiliary air regulator serves as output quantity. Figure 8.3 shows the control surface induced by the MFC.

The inputs of the total controller are the number of revolutions REV0_LO, the set value AARSEV and the state flag dAIRCON of the air-conditioning system (on/off). The electric current AARCURIN for the auxiliary air regulator serves as output.

The controller is implemented as a C program on a 386-processor laptop. The data communication takes place through the control device of the engine management system and the RS232 laptop interface.

From the classical idea in designing a fuzzy controller, one obtains the knowledge about the underlying process by an interview of the motor engineers (experts). But our experience has shown that even experts in this field are often not able to deliver at least a restricted description of the control strategy by linguistic control rules.

For this reason the cognitive analysis was based on measurement data obtained by idle speed experiments with a real spark ignition engine of the above mentioned type.

The cognitive analysis of the available data resulted in the specification of the equality relations $E_v(v = \{dREV, gREV, dAARCUR\})$ on the domains $X^{(v)}$, induced by integrable scaling mappings $c_v: X^{(v)} \to [0, \infty)$, as we have already seen before. Beyond that, it turned out that *dREV* can assume values from the interval $X^{(dREV)} = [-70, 70]$ and that *gREV* can assume values from the interval $X^{(gREV)} = [-40, 40]$. The output variable *dAARCUR* ranges over the interval $X^{(dAARCUR)} = [-25, 25]$. The equality relations were specified by integrable scaling mappings $c_v: X^{(v)} \to [0, \infty)$ just as in Section 8.1.1. For that, the sets $X^{(v)}$ were separated into suitable areas, for which a constant scaling factor was determined. i.e., in order to avoid unintended control actions in the vicinity of zero of the input variable *gREV*, it is suggested be chosen a minor distinguishability be chosen, because in this area one has to deal with stochastic error processes. The scaling mapping c_{gREV} is, for instance, defined as follows:

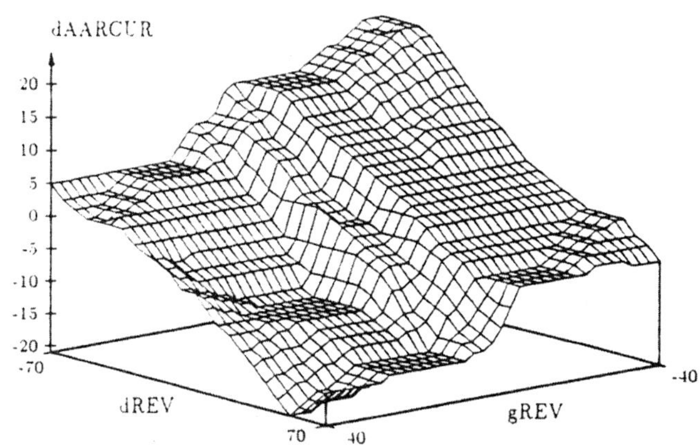

Figure 8.3 Control surface of the MFC.

$$c_{gREV}: X^{(gREV)} \to [0,\infty)$$

$$x \mapsto \begin{cases} \frac{1}{33} & if\ -40 \le x < -7 \\ 0 & if\ -7 \le x < -4 \\ 1 & if\ -4 \le x < -2 \\ 0 & if\ -2 \le x < 2 \\ 1 & if\ \ 2 \le x < 4 \\ 0 & if\ \ 4 \le x < 7 \\ \frac{1}{33} & if\ \ 7 \le x < 40 \end{cases}$$

Then, we have to specify the partial control mapping φ_0: $X^{(dREV)} \times X^{(gREV)} \to X^{(dAARCUR)}$, which assigns to certain input tuples appropriate output values coming also from the experimental data (see Figure 8.4).

By means of the equality relations E_v and the partial mapping φ_0 we are able to contruct a fuzzy controller. Figure 8.3 shows the control surface represented by a grid of supporting points. As defuzzification strategy we have used the Center Of Area method (COA). Last but not least, the final step of development is to translate the obtained controller into a Mamdani fuzzy controller as described in Section 8.1.3.

That means we have to associate each singleton with a linguistic term such as negative big (nb), negative medium (nm), negative small (ns), approximately zero (az), positive small (ps), positive medium (pm), or positive big (pb). In addition, we have to translate the partial mapping φ_0 to linguistic rules of the form

if dREV is A and gREV is B then dAARCUR is C

Figure 8.5 shows the corresponding fuzzy sets of the input variable *gREV*. The main advantage of this transformation is the opportunity of using available standard tools for an efficient implementation of Mamdani fuzzy controllers.

		gREV						
	φ_0	−40	−6	−3	0	3	6	40
dREV	−70	20	15	15	10	10	5	5
	−50	20	15	10	10	10	5	0
	−30	15	10	5	5	5	0	0
	0	5	5	0	0	0	−10	−5
	30	0	0	0	−5	−5	−10	−15
	50	0	−5	−5	−10	−15	−15	−20
	70	−5	−5	−10	−15	−15	−15	−15

Figure 8.4 The partial control mapping φ_0.

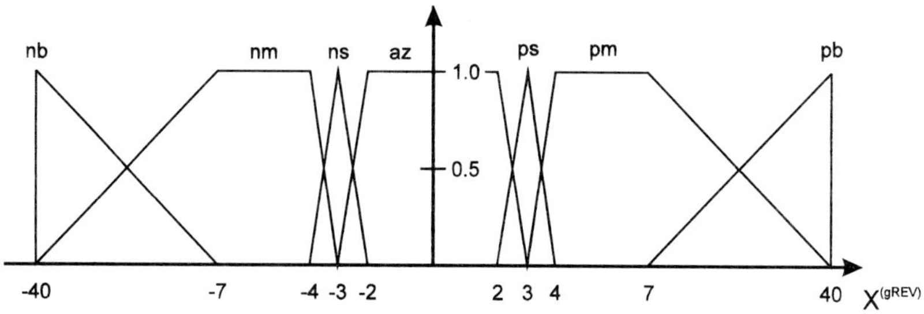

Figure 8.5 Fuzzy sets of the input variable gREV.

8.1.5 Serial Controller versus Fuzzy Controller

The following four figures show a comparison between the behavior of the serial controller and the fuzzy controller in the cases of an increased and decreased load caused by power steering or air-conditioning system, respectively. One measuring point corresponds to one stroke, i.e., the time for 180° of crank angle. AARCUROUT and AARCURIN illustrate in each case the time-dependent output of the serial controller and the fuzzy controller, respectively. Figures 8.6 and 8.7 show the situation after activating the power steering.

After a few seconds we deactivate the additional load. The over- and undershot at the point of switching on or off, respectively, cannot be regulated to disappear entirely because we have to take the delay time of the controlled system into account. Figure 8.8 and 8.9 illustrate the behavior of both controllers due to switching on the air conditioning system. All things considered, the fuzzy controller shows a very smooth control behavior compared to the serial controller. In addition, the fuzzy controller precisely and quickly reaches the desired set value and shows a very robust behavior even with slowly increasing load. The number of revolutions does not show any vibrations even after extreme changes of load.

In addition, the fuzzy control approach leads to a clean separation of states, a simple voting treatment, and a short development period.

8.1.6 Conclusion

Fuzzy control in vague environments is based on the idea of interpolating a partially specified control mapping by exploiting the information hidden in the indistinguishability in vague environments. It should be emphasized that the control rules of a fuzzy controller are not interpreted as logical implications, but as vague descriptions of crisp control actions in crisp situations. The approach of a controller based on knowledge-based interpolation provides a semantical background for fuzzy control and gives a clear explanation in which manner fuzzy sets and control rules can be interpreted.

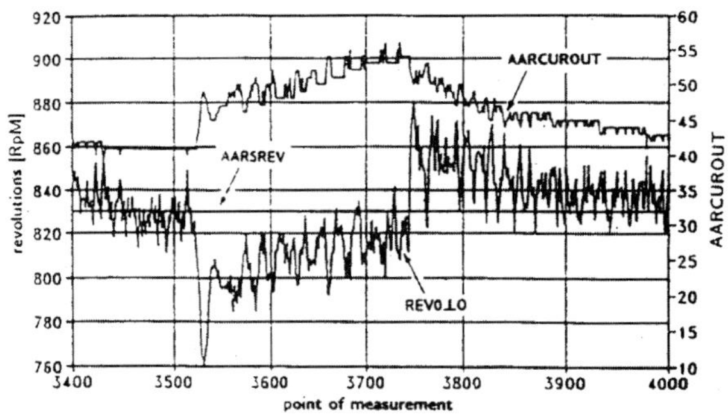

Figure 8.6 Power steering (serial controller).

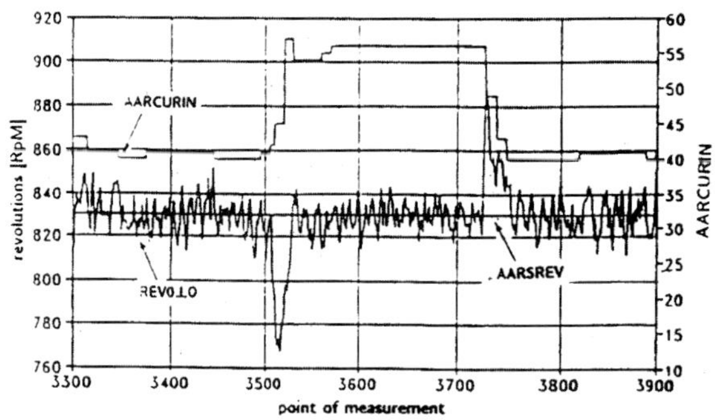

Figure 8.7 Power steering (fuzzy controller).

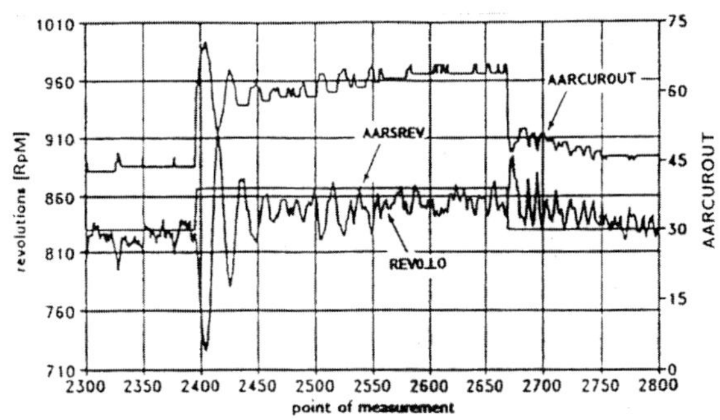

Figure 8.8 Air-conditioning system (serial controller).

The application of this approach to engine idle speed control has illustrated that the obtained fuzzy controller has a better performance of the control behavior than the serial controller.

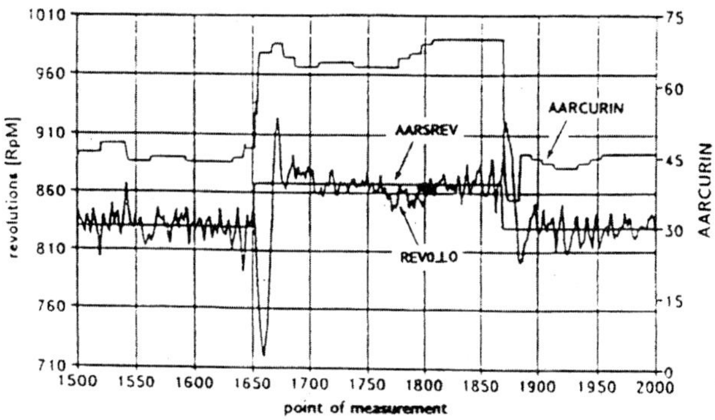

Figure 8.9 Air-conditioning system (fuzzy controller).

Additional investigations were done at Audi, Neckarsulm, where these concepts were applied to a 4200-cc, 280-hp, Audi V8 engine successfully.

8.2 FLOWING SHIFT-POINT DETERMINATION

In conventional automatic transmissions, the gear selection is done by means of gearshift diagrams, where the gearshift depends on the position of the accelerator and the vehicle's speed. There is a hysteresis between up shift and down shift in order to avoid oscillating gearshift in the case of a slightly variable vehicle speed e.g., during city traffic (see Figure 8.1). An introduction in automatic transmissions is given, for example in [8].

If the driver steps on the gas with half throttle for instance (in Figure 8.10, point A) the gearshift starts in first gear. The gearshift into second gear takes place at the intersecting point with the up shift curve. If the driving speed is suffice in the sense of the driver's

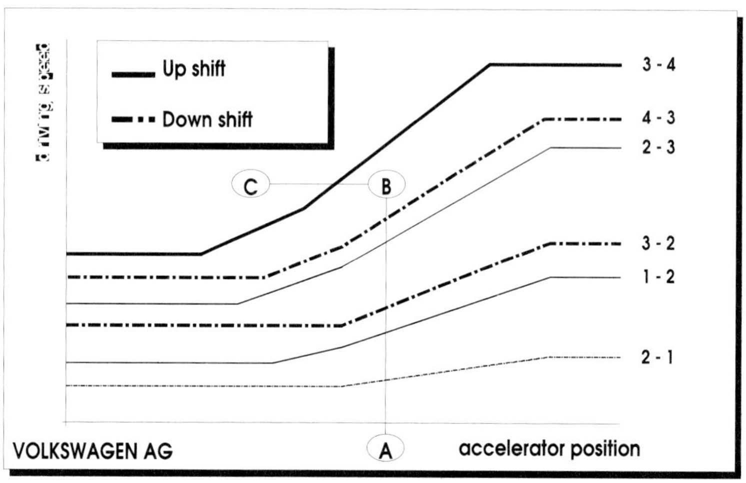

Figure 8.10 Gearshift curves of a conventional automatic transmission.

request (point B), he releases the accelerator to hold the driving speed. In this case the gearshift curve 3 → 4 will be cut and the driver obtains an up shift (point C). If the driver brakes until the car has stopped, first the driver releases the accelerator and the down shifts corresponds to the down-shift curves at zero throttle.

This adjustment leads to an engaging performance in regular situations. Nevertheless this selection of shift points is very rigid.

In order to get more flexibility, in the past there was used a division of the gearshift diagram into two different characteristic diagrams. The driver was able to switch manually between an economic and a sporting gearshift diagram by a change-over switch (see Figure 8.11 for shifting from gear 3. into the gear 4).

In the switch position "ECO," the driver obtains an early (i.e., at low speed) gearshift in order to produce a corresponding low fuel consumption.

On the other hand, in the switch position "SPORT" the driver obtains a later gearshift, i.e., at a higher level of revolutions, corresponding to more power.

The next step in the development of the automatic transmission control units at Volkswagen was the automatic change-over switch. In this system the accelerator velocity was responsible for the decision for which of the both gearshift diagrams were used.

However this procedure was still a crisp decision between the "ECO" and the "SPORT" diagrams, respectively.

The change-over switching between only two gearshift diagrams is very restrictive, i.e. some drivers will feel the "ECO" diagram as being too lazy, and other drivers will feel the "SPORT" diagram is annoying.

That is why Volkswagen has implemented a flowing shift-point determination which is based on methods of fuzzy logic. In this system, one obtains a flowing transition between the two gearshift diagram "ECO" and "SPORT" and the shift points will be adapted individually to the driver and his special driving style.

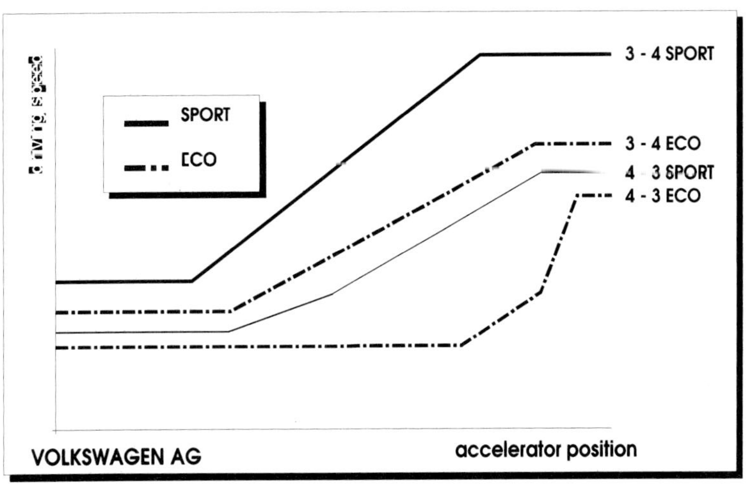

Figure 8.11 Gearshift transition 3. → 4. gear (ECO/SPORT).

8.2.1 Phases of Development

The first step was the determination of the physical quantities, which correspond to the different driving styles of individuals. A motor vehicle was equipped with a lot of sensors measuring the following quantities: vehicle speed, position of the accelerator, gear selection, engine speed, kickdown, position of the shift lever, length of acceleration of the vehicle, radial acceleration of the vehicle, vertical acceleration of the vehicle, intake manifold pressure, angle of the steering wheel, brake pressure, the flashing actuation etc. All tested driving situations were recorded by a videocamera. Before the test drives were started, all test persons were subdivided into the three groups "CALM," "REGULAR," and "SPORTING." The test drives took about one hour in the downtown traffic of Braunschweig. The evaluation of the test drives shows that only the frequency distribution of the filtered accelerator velocity was a significant quantity to describe the different types of driving styles. For the calculation of the histogram, the parts of a constant position of the accelerator (i.e., an accelerator velocity of zero) were not taken into account (e.g. waiting periods at red traffic lights). Figure 8.12 shows the histograms of the three groups of test drivers and another histogram, which reflects the driving style "NERVOUS." These histograms were transformed into fuzzy sets in order to serve as linguistic input variables for a fuzzy controller, which controls the gearshifts. One problem arises from the indistinguishability of the "SPORTING" and the "NERVOUS" groups.

A nervous driver is not able to keep his foot off the accelerator, and produces high accelerator velocities. If this driving style were interpreted as a sporting style, the driver's nervousness would be increased, because the "SPORTING" gearshift program is power oriented. In order to recognize these two different types of driving styles clearly, one has

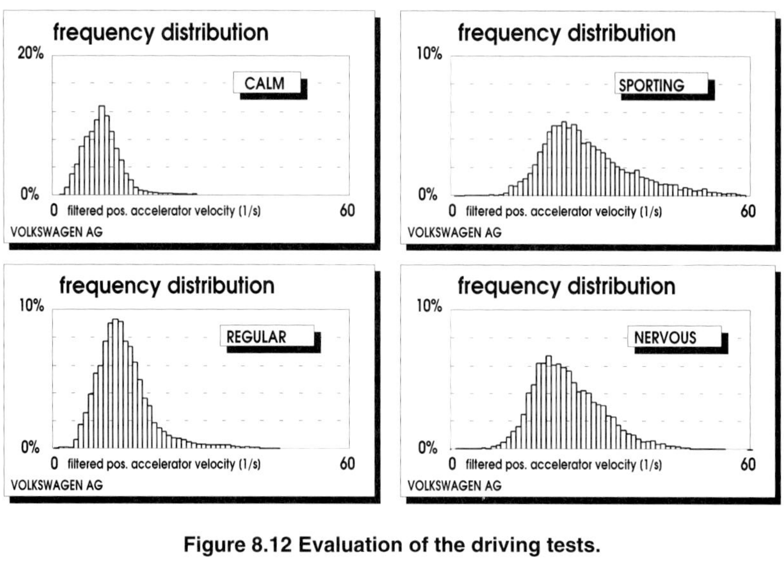

Figure 8.12 Evaluation of the driving tests.

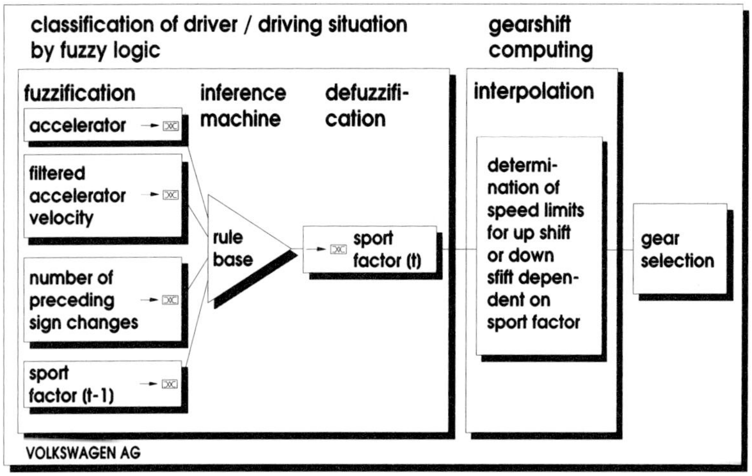

Figure 8.13 Flowing shift-point determination with fuzzy logic.

to take into account the number of changes in the accelerator direction. That means a high number of changes indicates a nervous driving style, and a small number of changes points to a sporting driving style.

8.2.2 Structure of the Fuzzy Controller

The basic structure of the driving determination is outlined in Figure 8.13. At the end there is the crisp decision, whether a gearshift takes place or not. For this decision a sport factor corresponding to the sportsmanship of the driving style is evaluated from four input variables.

The most important input variable is the filtered accelerator velocity. Therefore the filtering time depends on the changing direction of the sport factor, i.e. the increase of the sport factor takes place very fast in order to obtain a fast reaction to a high accelerator velocity, and the decrease of the sport factor takes place much slower in order to reduce the number of gear changes.

The number of direction changes of the accelerator (number of preceding sign changes of the accelerator) as input quantity serves for a distinguishability of a sporting and a nervous driver. Figure 8.14 shows the relation between the filtered positive velocity of the accelerator and the number of preceding sign changes of the accelerator velocity. At a low number of preceding sign changes, the sport factor increases linearly with respect to the increase of the filtered accelerator velocity. In the case of an additional increasing number of preceding sign changes (nervous driver), the sport factor decreases in the "ECO" direction.

The position of the accelerator in combination with the sport factor of the last evaluation step leads to the possibility of holding the sport factor constant in certain situations, e.g., if the gas is not stepped on.

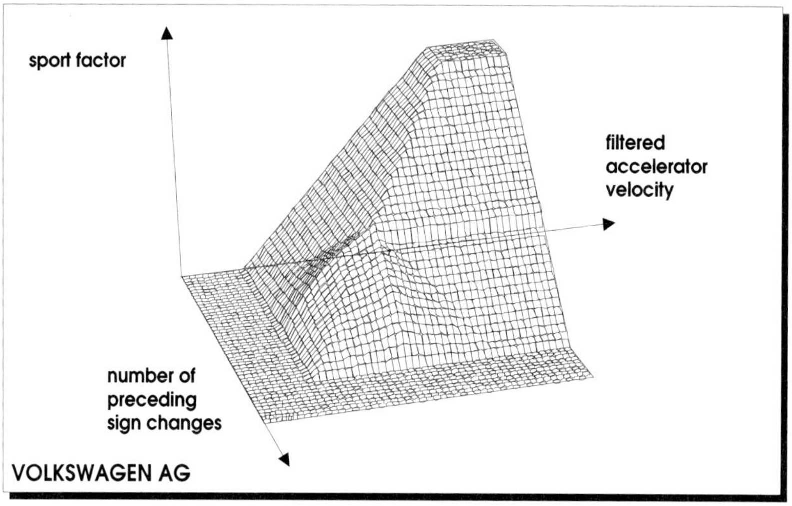

Figure 8.14 Transfer behavior.

The rule base consists of seven rules, at which the if-parts are combinations of up to four input quantities. The application of singletons in the then-parts is used, to guarantee a fast defuzzification for the on-line calculation of the fuzzy controller.

The topical shift point is determined by an interpolation between the "ECO" and the "SPORT" gearshift diagram via the sport factor.

8.2.3 The Effect of the Sport Factor

Figure 8.15 shows the influence of the sport factor in the case of gear transition from third into fourth gear. If the driver steps on the gas, for example at half throttle, and produces by it a sport factor of 82%, the up-shift point position is 82% of the difference between the "ECO" and the "SPORT" gearshift diagram, higher than it would be in the "ECO"- gearshift diagram.

Thus there is a continuously adaptive gearshift schedule between the limiting gearshift diagrams "ECO" and "SPORT," which corresponds to the sport factor. Therefore one obtains a very good adaptation to the driver's requests.

8.2.4 Conclusion

The flowing shift point determination in dependence on a calculated sport factor leads to an individual adaptation of the shift-points to the topical requests of the driver. In all this approach prefers an economic oriented shift behavior, whenever it is possible and a power oriented shift behavior, when ever it is necessary. A further advantage is that no additional sensors are needed (cost neutral). Therefore the AG4 automatic transmissions of Volkswagen motor vehicles have been adapted with the flowing shift-point determination since January 1995.

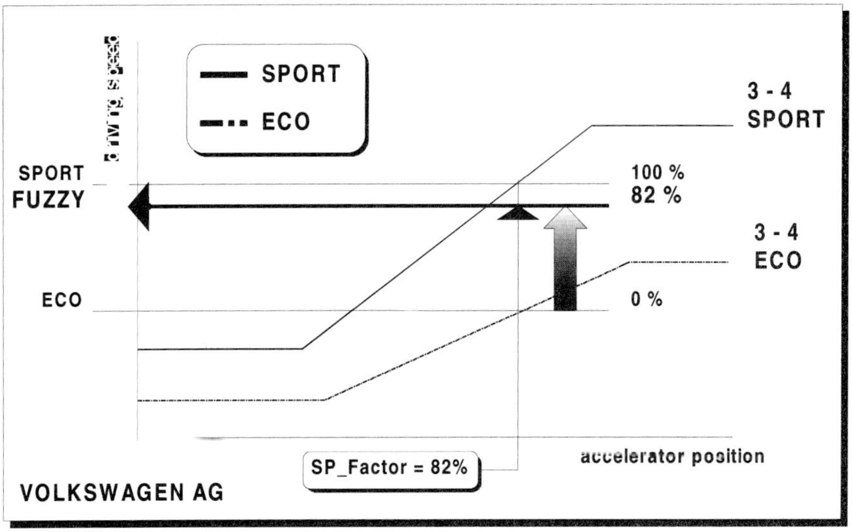

Figure 8.15 Influence of the sport factor on the gear selection.

8.3 REFERENCES

1. Abate, M. and Di Nunzio, V. *Idle Speed Control Using Optimal Regulation.* SAE 9905008, 1990.

2. Taylor, C. F., *The Internal-Combustion Engine in Theory and Practice*, Vols. I and II. MIT Press, 1985.

3. Kruse, R., Gebhardt, J. and Klawonn, F. *Foundations of Fuzzy Systems.* John Wiley & Sons, 1994.

4. Klawonn, F. and Kruse, R., "Equality Relations as a Basis for Fuzzy Control." *Fuzzy Sets and Systems.* Vol. 54, 1993, pp. 147-156.

5. Klawonn, F. and Kruse, R. "Fuzzy Control as Interpolation on the Basis of Equatilty Relations." *Proc. 2nd IEEE Int. Conf. Fuzzy Syst.*, pp. 1125-1130, San Francisco, 1993.

6. Klawonn, F., Gebhardt J. and Kruse, R., "Fuzzy Control on the Basis of Equality Relations with an Example from Idle Speed Control." *IEEE Transactions on Fuzzy Systems*, Vol. 3, No. 3, Aug.1995, pp. 336-350.

7. Lee C. C., "Fuzzy Logic in Control Systems: Fuzzy Logic Controller," part I. *IEEE Trans. Syst., Man, Cybern.* Vol. 20, 1990, pp. 404-418.

8. Förster, H. J. *Automatische Fahrzeuggetriebe*. Springer-Verlag, 1991.

9

FUZZY MAPS: MANAGING UNCERTAINTY IN SENSOR-BASED MOTION PLANNING

G. Oriolo, G. Ulivi, and M. Vendittelli
Università di Roma "La Sapienza"
ITALY

The purpose of this chapter is to present a successful application of fuzzy logic in the area of sensor-based robot motion planning. In the absence of a priori information, the robot control architecture must include a module capable of processing the uncertain sensor data in order to extract a convenient representation of the environment. Here, it is shown that fuzzy set theory provides a natural framework in which this problem can be formulated and solved. Besides, the obtained representation (*fuzzy maps*) can be effectively used for planning collision-free paths by means of graph search methods. Experimental results in real-world environments are reported.

9.1 INTRODUCTION

Mobile robotics is a very interesting field of application for fuzzy logic and, more generally, for all the soft computing techniques. This branch of Robotics aims at building autonomous mobile systems and involves many aspects of both control and perception [1,2,3]. Therefore, it provides an excellent playground for testing a variety of algorithms, ranging from low-level motor control to high-level planning and sensor data processing.

This chapter describes part of the work carried out in our laboratory, whose long-term objective is the possibly low-cost realization of an autonomous robot for navigation in an office-like environment. In particular, we will focus on two of the most important and interrelated components of the control architecture of a mobile robot, namely map building and motion planning.

In some cases, these tasks can be addressed one at a time. For example, when several views of the environment have been acquired by the sensory system, it is possible to conceive an algorithm which merges all the information into a compact representation (a map). Similarly, if an accurate map of the environment is available a priori, the path planning problem can be studied independently.

While the above problems *as such* are of practical interest in many applications, one cannot neglect the fact that, when full autonomy is pursued, the information flow is no more of the "batch" type, requiring the system components to work concurrently. In fact, no planning is possible without a map, and, on the other hand, a wide-range map cannot be reconstructed in cluttered environments without moving the robot. Therefore, the interaction between sensing and planning must be explicitly faced. This requirement rules out many elegant planning algorithms, calling for a technique that can work incrementally with a partial knowledge of the environment. Here, we present a modular solution approach: both the map building and the planning methods were originally devised to solve the off-line versions of the problems [4], but one of the design constraints was that they could be easily extended to the on-line case. This plan was successfully carried out [5], and the resulting method was implemented on a mobile platform. For the sake of clarity, the presentation will also follow this logical scheme.

The characteristics of the adopted sensors play a central role in the choice of the map building algorithm. In particular, here it is assumed that the robot is equipped with ultrasonic rangefinders, that offer a satisfactory tradeoff between cost and accuracy, together with a remarkable simplicity of use [6]. However, due to their poor angular resolution, it is very difficult to extract features from the surrounding scene. Therefore, we have chosen to represent the environment as a two-dimensional *occupancy grid*, a matrix of cells which can be in principle either "empty" or "occupied."

As the measuring process introduces a large amount of uncertainty, it is necessary to use an appropriate tool for processing the sensor data. Among the various possible choices, we elicited fuzzy logic. In this framework, the empty and the occupied space are defined as fuzzy sets over a discretized subset of $\Re^2$ (the universal set). For each cell, the degrees of membership to these sets characterize the certainty about its state resulting from the sensor measures. In particular, partial membership to both sets is possible—the principle of tertium non datur does not hold. This allows identifying areas in which the measuring process has provided insufficient or conflicting evidence about the states of the cells, in order to build a conservative representation of the environment in the form of a *gray-level* map, where each cell is characterized by a possibility of occupancy.

The planning module is in charge of generating a safe path for the robot from a start position to a goal position. Since the representation of the environment is uncertain, it is not possible to apply planning algorithms relying on a sharp distinction between the free space and the obstacles. Our solution is to characterize the risk of collision along a path by proper cost functions that correspond to different planning attitudes. The A^* class of graph search algorithms, which perform a local search driven by a suitable heuristic, has been chosen to build minimum-cost paths, because of its flexibility and efficiency.

The extension to on-line map building and navigation is relatively straightforward. In fact, the map builder is able to integrate easily the new sensory information into the current representation. Besides, while the A^*-based planner takes full advantage of the available local information, it naturally forces the robot to explore unknown areas when the goal is "out of sight," due to its global search behavior.

The chapter is organized as follows. In the next section, the necessary concepts of fuzzy logic are briefly recalled. An off-line method for building fuzzy maps from ultrasonic measures is presented in Section 9.3 and compared with existing approaches in Section 9.4. A solution method for the off-line planning problem on fuzzy maps is then outlined, while the extension to the on-line case is considered in Section 9.6.

9.2 RELEVANT TOOLS FROM FUZZY SET THEORY

From the rich theory on fuzzy sets [7,8,9], only few basic concepts are needed in order to devise a map building method.

Fuzzy sets may be easily introduced as extensions of standard crisp sets. For a *crisp set* A defined over the universal set U, *the membership function* $\mu_{\mathcal{A}}$: $U \mapsto \{0, 1\}$ identifies those elements of U that belong to A. For a fuzzy set[†] $\mathcal{A}$ defined over U, the membership function $\mu_{\mathcal{A}}$: $U \mapsto [0, 1]$ may assume any real value within the interval [0, 1], expressing the degree of membership of each element of U to $\mathcal{A}$.

Likewise, elementary fuzzy set operators (complementation, intersection, and union) can be defined as generalizations of the classical crisp set operators. Several options for the same operation are available, thus contributing to the richness and flexibility of fuzzy logic. For the map building problem, the following operators have been selected as the most appropriate according to our experience.

Set complementation is obtained by letting

$$\mu_{\bar{\mathcal{A}}}(x) = 1 - \mu_{\mathcal{A}}(x) \tag{9-1}$$

The *bounded product*

$$\mu_{\mathcal{A}\cap\mathcal{B}}(x) = \max\left(0, \mu_{\mathcal{A}}(x) + \mu_{\mathcal{B}}(x) - 1\right) \tag{9-2}$$

has been chosen as intersection operator. As for set union, we have used the class of operators introduced by Dombi [10]:

$$\mu_{\mathcal{A}\cup\mathcal{B}}(x) = \frac{1}{1 + \left[\left(\dfrac{1}{\mu_{\mathcal{A}}(x)} - 1\right)^{-\lambda} + \left(\dfrac{1}{\mu_{\mathcal{B}}(x)} - 1\right)^{-\lambda}\right]^{-\frac{1}{\lambda}}} \tag{9-3}$$

[†] Hereafter, we denote fuzzy sets by calligraphic capital letters.

with $\lambda \in (0, \infty)$. Note that:

- The Dombi operator produces "smaller" union sets† as λ is increased. This feature will be exploited for tuning the strength of the aggregation when merging the information provided by the sensors.
- The chosen triple of operators does not satisfy De Morgan's law. However, we shall make use of the intersection and the union operators in different contexts.

9.3 OFF-LINE MAP BUILDING FROM ULTRASONIC MEASURES

Assume that a series of range readings $\{r_i\}$, collected at known sensor locations, is available as input. The objective of the map building phase is to produce a fuzzy set $\mathcal{M}$, called the *fuzzy map*, that conveys information about the risk of collision for the robot in each point of the environment. The fundamental steps of the algorithm are shown in Figure 9.1. Below, each step is described in detail.

9.3.1 Range Measures Processing

Ultrasonic rangefinders generate a packet of ultrasonic waves and detect the resulting echo. The time elapsed between transmission and reception is assumed to be proportional to the distance of the obstacle. Hereafter, we will refer to the Polaroid Ultrasonic Rangefinder, a very common device [11]. The mobile robot Nomad 200™, used in our experiments in our Laboratory, has a ring of 16 such sensors.

While the interaction between ultrasonic sensors and real-world environments is very complex, there are basically three sources of uncertainty when an ultrasonic rangefinder is used for determining the position of an object. First, the measured distance r is affected by an error. The standard Polaroid Rangefinder can detect distances from 0.12 to 6.5 m with 1% accuracy over the entire range. Second, it is impossible to determine the angular position of the object that originated the echo inside the radiation cone, which has 25° of width for the Polaroid sensor. Finally, *false reflections* may occur: the sonar beam is scattered only by surfaces whose irregularities are comparable in size to the wavelength of the emitted pulse. Therefore, it may bounce off when the incidence angle is larger than a critical value which depends on the surface material; in this case the sensor reading is not significant because the beam may come back to the receiver after multiple reflections, or even get lost.

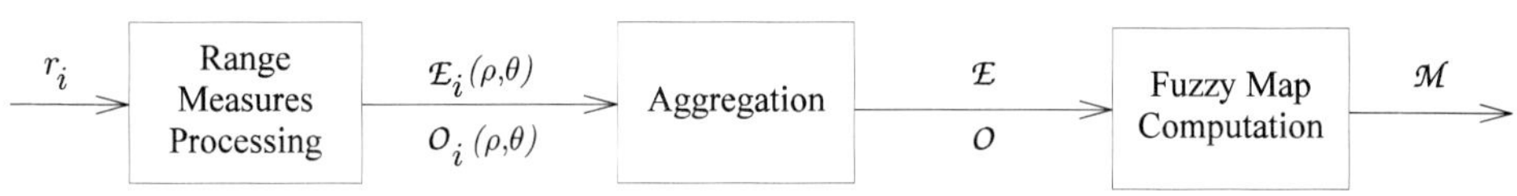

Figure 9.1 Basic steps of the map building algorithm.

† The *cardinality* of a fuzzy set $\mathcal{A}$ over U is defined[8] as $|\mathcal{A}| = \sum_{x \in U} \mu_{\mathcal{A}}(x)$

The above qualitative description can be directly used to set up a certainty model for the state of points belonging to the radiation cone. Unless a false reflection occurred, a single reading r provides the information that one or more obstacles are located somewhere along the 25° arc of circumference of radius r. Hence, there is evidence that points located in the proximity of this arc are "occupied." On the other hand, points well inside the circular sector of radius r are likely to be "empty." To model this knowledge, we introduce the two functions

$$f_{\mathcal{E}}(\rho, r) = \begin{cases} k_{\mathcal{E}} & 0 \le \rho < r - \Delta r \\ k_{\mathcal{E}}\left(\dfrac{r-\rho}{\Delta r}\right)^2 & r - \Delta r \le \rho < r \\ 0 & \rho \ge r \end{cases} \tag{9-4}$$

$$f_{O}(\rho, r) = \begin{cases} 0 & 0 \le \rho < r - \Delta r \\ k_{O}\left[1-\left(\dfrac{r-\rho}{\Delta r}\right)^2\right] & r - \Delta r \le \rho < r + \Delta r \\ 0 & \rho \ge r + \Delta r, \end{cases} \tag{9-5}$$

that describe, respectively, how the degree of certainty of the assertions "empty" and "occupied" vary with the distance ρ from the sensor, for a given range reading r. Here, $k_{\mathcal{E}}$ and k_O are two constants corresponding to the maximum values attained by the functions, and $2 \cdot \Delta r$ is the width of the area considered "proximal" to the arc of radius r. The choice of $k_{\mathcal{E}}$, k_O, and Δr, as well as other parameters to be introduced, will be detailed in Section 9.3.4.

Since the intensity of the waves decreases to zero at the borders of the radiation cone, the degree of certainty of each assertion should be higher for points close to the beam axis. This is realized by means of an *angular modulation* function

$$m_1(\vartheta) = \begin{cases} D(\vartheta) & |\vartheta| \le 12.5^\circ \\ 0 & |\vartheta| > 12.5^\circ \end{cases} \tag{9-6}$$

where $D(\vartheta)$ is the radiation directivity function which can be derived by the experimental data [11].

Finally, a *radial modulation* function is defined as

$$m_2(\rho) = 1 - \frac{1 + \tanh\left(2(\rho - \rho_v)\right)}{2} \tag{9-7}$$

in order to weaken the confidence of each assertion as the distance from the sensor increases. The parameter ρ_v plays the role of a "visibility radius", i.e., a distance at which a smooth transition occurs from certainty to uncertainty. The motivation for introducing this function is twofold. First, as false reflections give often rise to "long" readings, the use of $m_2(\rho)$ reduces the effect of erroneous measures on the map building process. Besides, narrow passages and doors appear to be obstructed if seen from a large distance, due to the sensor wide radiation angle. By varying the visibility radius according to the characteristics of the environment, it is possible to obtain a more correct detection behavior (see Section 9.3.4).

For each range measure r_i, two *local* fuzzy sets $\mathcal{E}_i$ and O_i are generated by combining the previously introduced certainty functions

$$\mathcal{E}_i(\rho,\vartheta) = f_{\mathcal{E}}(\rho, r_i) m_1(\vartheta) m_2(\rho) \tag{9-8}$$

$$O_i(\rho,\vartheta) = f_O(\rho, r_i) m_1(\vartheta) m_2(\rho) \tag{9-9}$$

The membership functions defining these two fuzzy sets represent, respectively, the certainty ascribed to the assertions "empty" and "occupied" in the radiation cone on the basis of the sensor reading r_i. Note that the above functions are expressed in polar coordinates with respect to the sensor position, and assume nonzero values only inside the radiation cone. Figure 9.2 shows the typical shape of the two sets.

9.3.2 Aggregation

The local information contained in $\mathcal{E}_i$ and O_i, which correspond to the range reading r_i, is aggregated in two *global* fuzzy sets $\mathcal{E}$ and O, respectively the set of *empty* and *occupied* points, by using the Dombi union operator (9-3)

$$\mathcal{E} = \cup_i \mathcal{E}_i \qquad O = \cup_i O_i \tag{9-10}$$

In principle, one could use different values of λ for $\mathcal{E}$ and O.

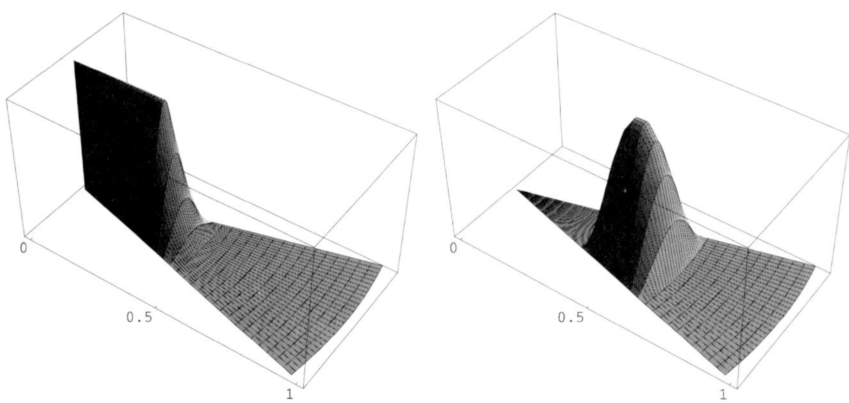

Figure 9.2 Membership functions for $\mathcal{E}_i$ (above) and O_i (below) inside the sensor radiation cone. The sensor reading is assumed to be r_i = 0.5 m.

In Equation (9-10), it is necessary to perform a conversion from polar coordinates, relative to each sensor location to absolute Cartesian coordinates. For computational reasons, the universal set $U \subset \Re^2$ is discretized into a matrix of $\beta = \sigma_1 \times \sigma_2$ square cells of side δ. As a result, two numerical values $\mu_{\mathcal{E}}(C)$ and $\mu_{\mathcal{O}}(C)$ are associated with each cell C, respectively quantifying its degree of membership to $\mathcal{E}$ and $\mathcal{O}$.

9.3.3 Computation of the Fuzzy Map

As the two sets $\mathcal{E}$ and $\mathcal{O}$ are *not* complementary, they can be combined to identify areas where the available information is either conflicting or insufficient.

The intersection of $\mathcal{E}$ and $\mathcal{O}$ is the set of *ambiguous* cells, with the corresponding membership value representing the degree of contradiction:

$$\mathcal{A} = \mathcal{E} \cap \mathcal{O}$$

while the *indeterminate* cells are those neither *empty* nor *occupied*

$$\mathcal{I} = \overline{\mathcal{E}} \cap \overline{\mathcal{O}}$$

A conservative map $\mathcal{S}$ of the *safe* cells is obtained by "discarding" the *occupied*, the *ambiguous* and the *indeterminate* cells from the *very empty* ones:

$$\mathcal{S} = \mathcal{E}^2 \cap \overline{\mathcal{O}} \cap \overline{\mathcal{A}} \cap \overline{\mathcal{I}} \qquad (9\text{-}11)$$

By squaring the value of the membership function of $\mathcal{E}$, the difference between low and high values is emphasized: according to the fuzzy logic terminology, we have applied the linguistic modifier "very" to the "empty" concept.

The fuzzy map $\mathcal{M}$ built by complementing $\mathcal{S}$ identifies *unsafe* cells that must be avoided as much as possible during the robot motion:

$$\mathcal{M} = \overline{\mathcal{S}} \qquad (9\text{-}12)$$

For compactness, the membership degree of a cell C to the fuzzy set $\mathcal{M}$ will be simply denoted by $\mu(C)$.

As already mentioned, the complementation operator (9-1) and the bounded product intersection operator (9-2) are used to perform all the above computations.

9.3.4 Choice of the Parameters

The values of the parameters of the certainty functions and of the Dombi union should be chosen on the basis of (*i*) the spatial distribution of the measure points and of the corresponding radiation cones (*ii*) the characteristics of the environment. However, a small amount of qualitative information is sufficient for successful tuning.

As already mentioned, our mobile platform is equipped with 16 sonars. To increase measured accuracy, at each position the ring is rotated twice by an angle of 7.5°: as a result, 48 range readings are obtained at each measure point, and each cell falls at least in 3 radiation cones. Moreover, overlapping may occur between radiation cones corresponding to different measure points, so that often several readings are available for the same cell. Consider a particular cell C, and assume that it belongs to the "empty" area of ν cones. Hence, each set $\mathcal{E}_i$ ($i = 1, ..., \nu$) will contribute a nonzero value to $\mu_{\mathcal{E}}(C)$, that is computed according to Equation (9-10). Immediate saturation of $\mu_{\mathcal{E}}(C)$ should be avoided, so as to require the concordance of multiple measures to build up certainty. To this end, it is first necessary to set an upper bound on the membership function of $\mathcal{E}_i$ by choosing a "small" value for $k_{\mathcal{E}}$ in Equation (9-4), and then to select accordingly an appropriate λ for the Dombi union. Typically, λ is chosen in such a way that 3 or 4 concordant readings produce a high certainty value. Similar considerations can be repeated for the choice of k_O in Equation (9-5).

As for the value of Δr in Equations (9-4–9-5), it should be selected in such a way to "augment" slightly the occupied area, providing a convenient safety margin for the planning phase.

Finally, the choice of the visibility radius ρ_v in Equation (9-7) can be made on the basis of a qualitative knowledge of the environment. For example, in an office-like area (as in our experiments) it is reasonable to ask that the map building algorithm correctly detects passages of the size of a door, say 0.7 m. To this end, the measure point must be sufficiently close to the opening, so that the 25° radiation cone does not intersect the door features. A simple geometric construction shows that the maximum admissible distance is 1.5 m. In this case, it would be therefore necessary to choose $\rho_v < 1.5$ m.

9.3.5 Off-Line Map Building: Experimental Results

As a testbed for the proposed map building and path planning methods, we used the Nomad 200™ mobile robot produced by Nomadic Technologies. Nomad has a cylindrical shape with an approximate radius of 0.23 m, and a kinematic model equivalent to a unicycle. The upper turret, which carries the 16 Polaroid ultrasonic sensors, may be independently rotated. The robot control software runs under UNIX on an IBM RISC 6000 that communicates with Nomad through a radio link. All the algorithms were implemented in the C language.

The experiment area was a rectangle of 18 × 12 m, containing a corridor and a room of our Department. To define the fuzzy map $\mathcal{M}$, this area was discretized into a matrix of 180 × 120 square cells of side $\delta = 0.1$ m. Halfway along the corridor there is an intersection, where a small obstacle is present. The open space is delimited by smooth surfaces (walls and closed glass cabinets) with poor diffraction properties, an adverse condition for ultrasonic sensing. Nomad was driven to 43 measure points, whose position was determined a priori. At each measure point, a complete sensing procedure was performed, providing 16 × 3 = 48 range readings.

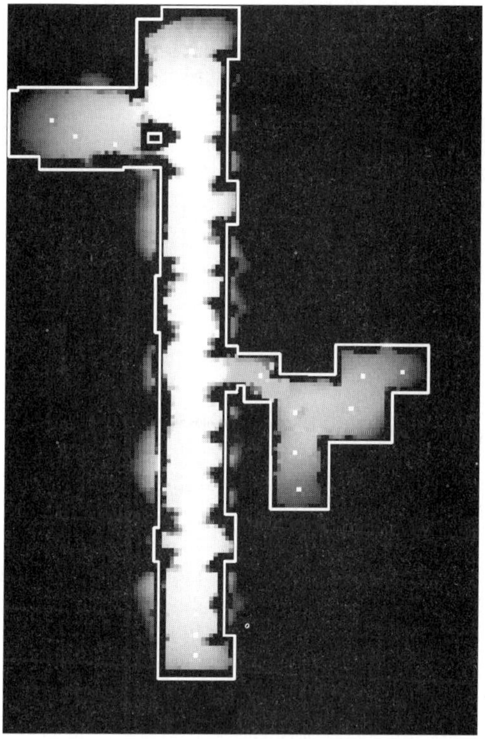

Figure 9.3 Plot of the fuzzy map $\mathcal{M}$ computed off-line from experimental data; darker areas correspond to higher values of μ. The actual profile of the corridor, the room, and the obstacle are superimposed. White spots indicate the measure points.

For the map building algorithm, we have chosen the following set of parameter values: $k_{\mathcal{E}} = 0.1$, $k_O = 0.25$, $\Delta r = 0.15$ m, $\rho_v = 1.2$ m, and $\lambda = 0.4$. Figure 9.3 shows the plot of the resulting fuzzy map $\mathcal{M}$. The total time necessary to compute $\mathcal{M}$ from the range readings was approximately 13 s. Note the satisfactory accordance of the map with the actual boundary of the open space. The small gray areas extending beyond the corridor walls are due to false reflections occurring for large angles of incidence. Nevertheless, the algorithm estimated accurately the profile of the walls, by incorporating the range readings obtained for smaller incidence angles.

9.4 COMPARISON WITH OTHER APPROACHES

After describing the method we are currently using in our Laboratory, a comparison with other possible approaches is in order. All kinds of *uncertainty calculi* are indeed candidates for the considered application. In this section, the proposed Fuzzy Logic (FL) approach is compared with two well-known methods to manage uncertainty, namely Probability Theory (PT) and Belief Measures (BM).

The debate on the most appropriate tool for reasoning in the presence of uncertainty is long-standing and very active [12,13,14]. Our conclusions are to be understood as limited in scope to the map building problem and to the described algorithms.

The PT approach is the most established among the three considered. If all the necessary hypotheses are satisfied, the obtained results have the reliability and the elegance of the underlying theorems. However, in an effort to devise a computationally feasible algorithm, one may be forced to make unrealistic simplifying assumptions in the problem formulation. On the other hand, both the FL and the BM approaches need less constraining hypotheses and prove to be more flexible tools, although additional preliminary work is required to select operators and parameters.

9.4.1 Probabilistic Approach

One of the best known algorithms for building occupancy grids from ultrasonic measures is due to Elfes [15]. The environment map is represented as an *occupancy grid*, containing probabilistic estimates of the occupancy state of the cells. In particular, the random variable $s(C)$ describing the state of cell C can assume only two values, namely E (empty) and O (occupied). The occupancy grid is modeled as a discrete-state Markov field, a realization of which consists in the estimation, for each cell C, of the conditional probability $P[s(C) = \mathsf{O} \mid \{r_1,\ldots, r_n\}]$, i.e., the probability of cell C being occupied, given that a series of range readings $\{r_1,\ldots, r_n\}$ has occurred. This conditional probability is updated by properly applying the sequential version of Bayes theorem

$$P\big[s(C)=\mathsf{O}\big|\{r_1,\ldots r_n\}\big]=\frac{p\big[r_n\big|s(C)=\mathsf{O}\big]P\big[s(C)=\mathsf{O}\big|\{r_1,\ldots r_{n-1}\}\big]}{\sum_{s(C)}p\big[r_n\big|s(C)\big]P\big[s(C)\big|\{r_1,\ldots r_{n-1}\}\big]}$$

In this formula:

- $p[r_n \mid s(C) = \mathsf{O}]$ and $p[r_n \mid s(C) = \mathsf{E}]$ are obtained through the application of Kolmogorov theorem—a cumbersome computation. In doing this, it is necessary to use a stochastic model of the sensor in the form of a probability density function $p[r \mid z]$, characterizing the probability of measuring r given that the closest obstacle to the sensor is located at a distance z.
- $P[s(C) = \mathsf{O} \mid \{r_1,\ldots, r_{n-1}\}]$ and $P[s(C) = \mathsf{E} \mid \{r_1,\ldots, r_{n-1}\}]$ are the previous estimates of the cell state. For $n = 1$, it is necessary to use a priori estimates of the probabilities $P[s(C) = \mathsf{O}]$ and $P[s(C) = \mathsf{E}]$, satisfying the constraint

$$P\big[s(C)=\mathsf{E}\big]+P\big[s(C)=\mathsf{O}\big]=1$$

To keep the problem tractable, it is necessary to make the assumption that the random variables $s(\cdot)$ are independent (zero-order Markov field), i.e., that no relationship exists between the states of two cells, even if they are adjacent. Besides, the a priori probabilities are typically estimated with the *maximum entropy* assumption, or

$$P[s(C) = \mathsf{E}] = P[s(C) = \mathsf{O}] = 1/2.$$

Most criticisms raised about the Bayesian approach question the validity of these two assumptions. As for the independence assumption, it has been observed that it may induce large errors even in the presence of a slight degree of dependence between the random variables [13]. This is the case of map building, since the occupied cells are not evenly distributed, but concentrated in clusters (obstacles). Furthermore, with the maximum entropy assumption the convergence of the Bayesian updating procedure toward an acceptable characterization of the occupancy grid requires a large number of measures.

The above hypotheses bring forth paradoxical consequences even in a simple one-dimensional case. Assume that the environment is a half-line, with the sensor located at the origin. The cells are line segments of length δ. In view of the maximum entropy assumption, the a priori probability of two adjacent cells being occupied is $0.5^2 = 0.25$, of three is 0.125, and so on. Hence, the a priori probability that an obstacle exists depends on its size expressed in terms of δ.

A second consequence is that the a priori probability of a given sensor reading (which is used in the computation of $p[r \mid s(C)]$) depends on the measured distance. In fact, for the ultrasonic sensor to detect the presence of an obstacle at cell C_j, cells $C_1,\ldots, C_{j-1}$ must be empty and cell C_j must be occupied. The states of cells $C_{j+1},\ldots$ is immaterial. The probability of this reading is 0.5^j, implying that the most probable measured distance is 0 (the sensor is inside an obstacle!).

This odd behavior of a priori probabilities is certainly disturbing. While it is known [16]. that with the sequential Bayes formula the effect of a priori information vanishes asymptotically as the number of measures increases, in practice convergence is so slow—particularly for finer discretizations—that such theoretical issue has no relevance.

A final remark is in order about the choice of the sensor model $p[r \mid z]$, which is often assumed to be Gaussian. This is a reasonable choice in the absence of false reflections, but these do happen and are difficult to detect as *outliers* for the measuring process. As the simulation in Section 9.4.3 will show, the Bayesian estimation formula is very sensitive to outliers, and the resulting map building algorithm is prone to serious errors. In principle, it should be possible to make the method more robust by using more complicated probability density functions to model the sensor [1,15]. However, this requires a careful calibration of the model through several preliminary experiments to be performed in the environment.

9.4.2 Belief Measures Approach

In the FL approach described in Section 9.3, we express the uncertainty on the state of each cell through its membership degree to the fuzzy sets $\mathcal{E}$ and $\mathcal{O}$. A more direct way to model this uncertainty may be derived from the theory of belief measures [8,9]. For example, a related technique has been used for stereo vision [17].

Consider the power set $\psi = \{U, O, E, \varnothing\}$, where U is the universal set (the environment discretized in cells), E and O are the sets of empty and occupied cells, and $\varnothing$ is the empty set. With this approach, all sets are considered to be crisp, a formulation which fits better the nature of the problem. For each set $A \in \psi$, one can define a *belief measure* Bel(A) to characterize the degree of belief that each element of the universal set belongs to A. For each cell, our belief about its state is then represented by {Bel(U), Bel(E), Bel(O), Bel($\varnothing$)}. At the beginning of the measuring process, the initial belief for all cells is {1, 0, 0, 0}, i.e., complete ignorance.

To qualify as a belief measure, a set function must satisfy natural axioms concerning boundary conditions, monotonicity, continuity, and additivity. One way to define a belief measure Bel(A) is through the *basic mass assignment* $m(A)$ (also called *basic probability assignment* according to Dempster-Shafer terminology). The value of $m(A)$ for a particular cell C represents the degree of evidence that C belongs to A, independently from any evidence that it belongs to a (strict) subset of A. For example, for a cell C that is known to be certainly empty, we have $m(U) = 0$ even if $m(E) = 1$ and $E \subseteq U$. A basic mass assignment $m(\cdot)$ must satisfy the following properties:

$$m(\varnothing) = 0 \qquad \sum_{A \in \psi} m(A) = 1 \tag{9-13}$$

Given a basic assignment, the corresponding belief measures are obtained as

$$\begin{aligned}
\text{Bel}(U) &= m(U) + m(O) + m(E) + m(\varnothing) \\
\text{Bel}(E) &= m(E) + m(\varnothing) \\
\text{Bel}(O) &= m(O) + m(\varnothing) \\
\text{Bel}(\varnothing) &= m(\varnothing)
\end{aligned}$$

As the measuring process provides evidence about the environment, mass is moved from U to E and O according to Dempster-Shafer rule. In particular, the value of the basic mass assignment after the $(i+1)$th measure is computed from its value after the ith measure as

$$\begin{aligned}
m_{i+1}(U) &= \frac{1}{1-K_i}\left[m_i(U)m_s(U)\right] \\
m_{i+1}(O) &= \frac{1}{1-K_i}\left[m_i(O)m_s(O) + m_i(O)m_s(U) + m_i(U)m_s(O)\right] \\
m_{i+1}(E) &= \frac{1}{1-K_i}\left[m_i(E)m_s(E) + m_i(E)m_s(U) + m_i(U)m_s(E)\right]
\end{aligned}$$

being

$$K_i(O) = m_i(O) \cdot m_s(E) + m_i(E) \cdot m_s(O)$$

Here, $m_s(\cdot)$ are the *sensor basic mass assignments*, expressing the degrees of evidence provided by a single sensor measurement. The functions $m_s(E)$ and $m_s(O)$ can be chosen on the basis of considerations similar to those made in Section 9.3.1 to derive the membership functions of $\mathcal{E}$ and $\mathcal{O}$, while $m_s(U)$ is obtained by Equation (9-13).

To obtain a conservative representation of risk from this formulation, one can use the *plausibility measure* Pl(O), which characterizes cells for which no evidence shows that they are safe. In the present formulation, being E and O crisp sets, one has

$$\mathrm{Pl}(O) = \mathrm{Pl}(\overline{E}) = 1 - \mathrm{Bel}(E) = m(E).$$

We are currently investigating a modification of the above approach based on the studies of Smets [18]. Basically, it consists in relaxing the requirement that $m(\varnothing) = 0$, so as to regard $\varnothing$ as a "should-be-empty set," containing all the contradictions coming from the measuring process. Preliminary results show slightly better performance than the standard BM approach.

9.4.3 Simulation Results

The three above approaches have been applied to a simulated one-dimensional environment, chosen as a half-line of total length 2 m. The cell size is 0.05 m, and the sensor is located at the origin. An obstacle is present at 0.6 m from the origin. We have assumed that the rangefinder provides a succession of ten range readings $\{r_1, \ldots, r_{10}\}$, with $r_1 = \ldots = r_9 = 0.6$ m, and $r_{10} = 1.2$ m. Similar measure patterns often occur, due to the phenomenon of false reflections. While detecting and rejecting the outlier value $r_{10} =$ 1.2m appears to be easy in this case, the situation is different in real-world sensing, when measures are taken from various positions.

The fuzzy maps $\mathcal{M}$ obtained with the FL method after 1, 3, 5, and 10 range readings are shown in Figure 9.4(a) (the latter is the final map). *Safe* cells were defined through the simpler formula $\mathcal{S} = \mathcal{E} \cap \overline{\mathcal{O}}$ in place of the complete expression (9-11), while $\mathcal{M} = \overline{\mathcal{S}}$ as before. Cells located close the obstacle are considered to be unsafe from the very first readings, while the area close to the sensor becomes increasingly safe as successive measures confirm the obstacle position. Cells beyond the detected obstacle are considered to be unsafe. This cautious attitude is consistent with our final goal, that is to use the obtained map for path planning. The method appears to be fairly robust with respect to outliers, as the tenth reading does not modify substantially the certainty about the existence of an obstacle at 0.6 m.

We have then applied the PT method. Figure 9.4(b) shows the profiles of the occupancy probabilities after 1, 3, 5, 7, 9, and 10 readings. Note that the method correctly uses the first nine readings to build up certainty about the occupancy state of cells located in the vicinity of 0.6 m. However, when the tenth reading is processed, all the previous knowledge is erased and the final occupancy grid takes into account only the last range reading r_{10}. Besides, the PT method is somewhat too confident: one reading is sufficient to bring to zero the probability that the area between 0 and 0.4 m is empty.

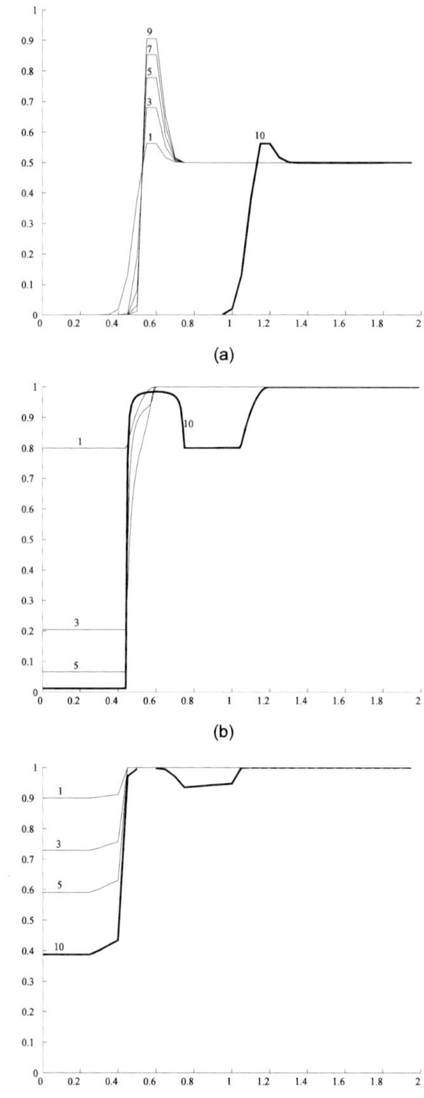

Figure 9.4 Map building for a simple one-dimensional case: FL approach (a); PT approach (b); BM approach (c).

Finally, Figure 9.4(c) shows the results obtained using the BM approach that are similar to those of the FL approach. It appears that this method is slower in recognizing the empty spaces, but this behavior is likely to be adjustable by changing the values of the algorithm parameters.

9.5 OFF-LINE PATH PLANNING ON FUZZY MAPS

Assume that a fuzzy map $\mathcal{M}$ of the environment has been reconstructed from the available sensor readings by using the map building method so far described.[†] We shall now address the problem of robot path planning on $\mathcal{M}$: given a start cell S and a goal cell G, specifying respectively the initial and the desired final position of the robot, find a *collision-free* path from S to G (a path is a sequence of adjacent cells). For simplicity, the case of a *point* robot will be considered first; however, this assumption is not restrictive and will be removed later.

One difficulty with the above formulation of the path planning problem is that it makes use of a crisp concept ("collision-free") which cannot be defined on a fuzzy map. In fact, the latter does not convey a separation between the free and the occupied space: in principle, all cells of $\mathcal{M}$ are admissible for planning. Thus. to devise a planner one needs to specify:

1. a *safety* (or a *risk*) *criterion* that can be computed for a given path;
2. an *algorithm* for finding the safest path connecting S to G.

Depending on the choice of the criterion and of the algorithm, several planning techniques can be obtained.

A straightforward approach is to avoid cells with large values of μ (the membership degree to the map of *unsafe* cells $\mathcal{M}$) to minimize the risk of traversing cells corresponding to detected obstacles. Following this idea, two possible path risk functions for a path are introduced next, together with an algorithm for finding minimum-risk paths.

9.5.1 Risk Functions

A first risk function for a path P can be defined as

$$g_1(P) = \sum_{C_i \in P} \mu^2(C_i),$$

i.e., a measure of the integral risk along the path. By squaring the value of each cell membership degree, we intend to make the inclusion of unsafe cells less likely.

A second possibility is to let

$$g_2(P) = \max_{C_i \in P} \mu(C_i)$$

that is the maximum risk encountered along the path.

[†] As a matter of fact, the planning technique presented in this section can be applied to any gray-level map, independently from the chosen map building method.

While the minimization of g_1 or g_2 is naturally appealing, it can also be given a more formal significance. In fact, both these risk functions may be interpreted as *energy measures* [7]. of the fuzzy set $\mathcal{M}$ over the path P, characterizing the distance of P from an *ideally* collision-free path, i.e., a path for which $g_1 = 0$ ($i = 1, 2$). In particular [19], g_1 and g_2 are respectively the *power* and the *height* of the fuzzy set $\mathcal{M}$ over P.

The two risk functions measure the aforementioned distance using different metrics. Since the value of g_1 depends on the path length through the number of addends, a path minimizing g_1 will yield a tradeoff between length and risk. The minimization of g_2 will give more conservative results, even if the final path may be longer. In some instances, however, the use of g_1 may yield safer paths. For example, assume that S is located in a zone of $\mathcal{M}$ with uniform high values of μ. An optimal path with respect to g_2 might be unsatisfactory, since the maximum value of μ will probably be attained in the proximity of S. From there on, any subpath leading to G is admissible, if it does not increase the value of g_2. Minimization of g_1 will perform better, since this risk function depends on the *whole* path from S to G.

9.5.2 Planning via the A^* Algorithm

By associating to $\mathcal{M}$ a graph whose nodes are the cells of the map and adjacent cells are connected by arcs, the problem of planning a minimum-risk path is easily transformed into a graph search problem. In order to solve the latter, the A^* algorithm can be adopted. Under mild assumptions, this algorithm is *complete*, i.e., it is guaranteed to return a minimum-cost path whenever one exists, and to return failure otherwise. It also allows the incorporation of heuristic information when available, resulting in an efficient search. We refer to the original reference [20] for details.

Given a graph whose arcs are labeled by costs, define the cost of a path from a node S to a node G as the sum of the costs of its arcs. A^* makes use of an evaluation function $e(N)$ associated with each node N that estimates the cost of the optimal path from S to G constrained to go through N as the sum of two parts:

$$e(N) = g(N) + h(N)$$

where $g(N)$ is the cost of the current path from S to N, and $h(N)$ is a *heuristic* estimate of the cost of the optimal path from N to G. A^* is complete under the condition

$$0 \le h(N) \le h^*(N), \forall N \tag{9-14}$$

where $h^*(N)$ is the *actual* cost of the minimum-cost path from N to G. A heuristic function that satisfies the condition (9-14) is said to be *admissible*. When no estimate of $h^*(N)$ is available, it is possible to set $h(N) = 0$, resulting in a *noninformed* A^*. If in addition the heuristic function is *locally consistent*, i.e., for any pair of adjacent nodes (N, N') connected by an arc of cost $k(N, N')$ it is

$$0 \le h(N) \le h(N') + k(N, N')$$

Then whenever a node N is selected for expansion by the algorithm, the current path from S to N is already optimal. Again, $h(N) = 0$ is trivially consistent.

The use of A^* for determining minimum-g_1 paths is straightforward. The cost of an arc joining two adjacent cells C_i and C_j is defined as

$$k_1(C_i, C_j) = \mu^2(C_j)$$

so that the cost of a path P coincides with $g_1(P)$, except for the additive constant $\mu^2(S)$. As for the heuristic function, we can let

$$h_1(C_j) = d(C_j) \cdot \mu^2_{\min}$$

Here, $d(C_j)$ is the minimum possible number of cells that compose a subpath from C_j to G (its expression depends on the adjacency definition on the map), and $\mu_{\min}$ is the minimum value of μ over $\mathcal{M}$. The above heuristic function is clearly admissible. Note that, in order to obtain an *informed* A^*, it must be $\mu_{\min} > 0$. The resulting version of the A^* algorithm will be denoted by A_1^*.

The use of A^* for determining minimum-g_2 paths is somewhat nonstandard, the difficulty coming from the nonadditive nature of the function g_2. It is necessary to define the arc cost $k_2(C_i, C_j)$ in such a way that, adding the costs of all arcs of a path, the result is g_2. To this end, let

$$k_2(C_i, C_j) = \begin{cases} \mu(C_j) - \hat{\mu}(C_i) & \text{if } \mu(C_j) > \hat{\mu}(C_i) \\ 0 & \text{else,} \end{cases}$$

where $\hat{\mu}(C_i)$ denotes the maximum value of μ encountered on the *optimal* path from S to C_i. If a locally consistent heuristic function is used, then when C_i is expanded, the current path from S to C_i is optimal and $\hat{\mu}(C_i) = g(C_i)$. Thus, even if the arc costs are not known a priori, they are uniquely defined and computable as the algorithm proceeds.

For the heuristic function to be admissible, it is necessary to set $h_2(C_j) = 0$, since it may happen that $g(G) = g(C_j)$ (the maximum value of μ on the path is attained on the subpath from S to C_j). However, a noninformed A^* may be very inefficient, due to the large number of *ties* that can occur. To cope with this problem, we have used a tie-resolution strategy privileging those cells whose distance $d(\cdot)$ from G is smaller. With this modification, A^* works like a *depth-first* search method whenever this is possible without increasing the value of g_2 The algorithm just described will be indicated by A_2^*.

At this point, we can remove the point robot assumption and take into account the actual robot dimensions with a simple modification. Assume that the robot can be approximated by a circle of radius γ, whose center is located at (the center of) cell C. Since each bitmap cell has side δ ($\delta < \gamma$), the robot body will be contained in a square of $\eta \times \eta$ cells centered at C, where $\eta = 2(\text{round}(\gamma/\delta) + 0.5)$. An *augmented* fuzzy map $\mathcal{M}_a$

can be built by assigning to each cell C a membership degree $\mu_a(C)$ defined as the maximum value of μ attained in the square of $\eta \times \eta$ cells centered at C. Planning for a point in $\mathcal{M}_a$ is equivalent to plan for the actual robot in $\mathcal{M}$. Such procedure may be implemented as a pre-processing phase on the map $\mathcal{M}$, but this is not necessary. In fact, it is sufficient to modify A^* so that the value of $\mu_a(C)$ is computed only when cell C is visited.

For a graph with a arcs and n nodes, the time complexity of A^* is $O(a \log n)$. Since our representation is a bitmap of β cells, the number of arcs is $a = 2\beta^2 - 3\beta + 1$ when 2-adjacency is used. The resulting time complexity is $O(\beta^2 \log \beta)$

9.5.3 Off-Line Path Planning: Experimental Results

The proposed path planning method has been applied to the map presented in Section 9.3.5. To take into account the actual Nomad dimensions, the augmentation procedure described at the end of the previous section was performed with $\gamma = 0.23$ m, giving $\eta = 5$.

The results of the proposed planning method for a trial case are reported in Figure 9.5, while details on the solution paths are given in Table 9.1. The start cell S is located inside the room, while the goal cell G is at the end of the left branch of the corridor. The straight-line distance between S and G is approximately 12 m. Note that the path obtained with A_2^* is more dangerous in the integral sense (the value of g_1 is roughly tripled with respect to the path obtained with A_1^*), but the maximum risk along the path is the same for both paths. This value is attained in correspondence of the narrow passage between the room and the corridor (recall that planning is performed on the augmented map $\mathcal{M}_a$.)

In this case the algorithm A_2^*, although non-informed, is much faster than A_1^*. This is due to the tie-resolution strategy, that forces a depth-first behavior when possible. As a drawback, while 2-adjacency was chosen for A_1^*, we had to use 1-adjacency for A_2^*, to prevent the path from grazing the walls.

The time necessary to compute a solution path may vary remarkably. While time efficiency was not stressed in our implementation, a considerable increase in speed could be obtained by using a sub-optimal [21]. version of A^*. A_2^* is faster in returning a solution path, provided that no backtracking phase is necessary. On the other hand, A_1^* generally produces safer and smoother paths.

9.6 ON-LINE MAP BUILDING AND NAVIGATION

After having developed a method for planning paths on fuzzy maps, we can turn to the problem of autonomous robot motion in unknown environments, which can be approached in different ways. If the robot must repeatedly move in a static workspace, it may be convenient to perform a preliminary exploration step in which the environment is accurately reconstructed [2]. The high cost of such action will be compensated by the ease of planning on a globally known map. Instead, when the scene characteristics change frequently a purely reactive behavior [22,23]. may be appropriate. A whole range of intermediate approaches may be conceived, with different balance between planning and

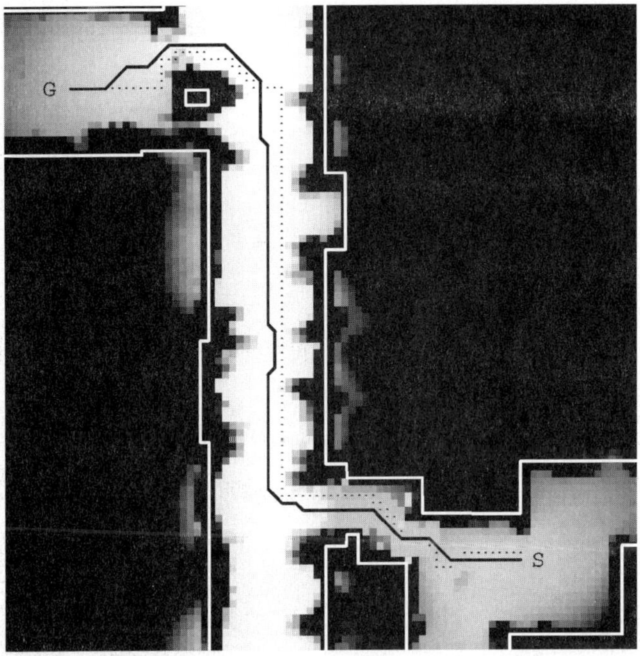

Figure 9.5 Off-line planning on the map of Figure 9.3: paths generated by A_1^* (continuous) and A_2^* (dotted) for a trial case.

Table 9.1 Details on the Solution Paths.

	g_1	g_2	Cells in the Path	Expanded Cells	Computing Time (sec)
A_1^*	5.34	0.50	121	2394	6.21
A_2^*	17.13	0.50	145	378	0.08

sensing [24]. The solution method proposed here makes use of two basic processes, *perception* and *navigation*, interleaved during the task execution.

In the perception process, the robot collects data from its sensors, builds a local representation of the surrounding scene and updates the current fuzzy map by integrating this information. In the navigation process, the planner is in charge of generating a path from the current position to the goal that is *locally* safe: i.e., it should avoid collisions in the area so far explored. The robot follows the planned path up to the boundary of the explored area, where a new perception process takes place. The proximity sensors are always active to detect unexpected obstructing obstacles.

9.6.1 The Perception Process

The perception process is composed by two consecutive phases: *sensing* and *map building* (see Figure 9.6). During the process the robot position is constant.

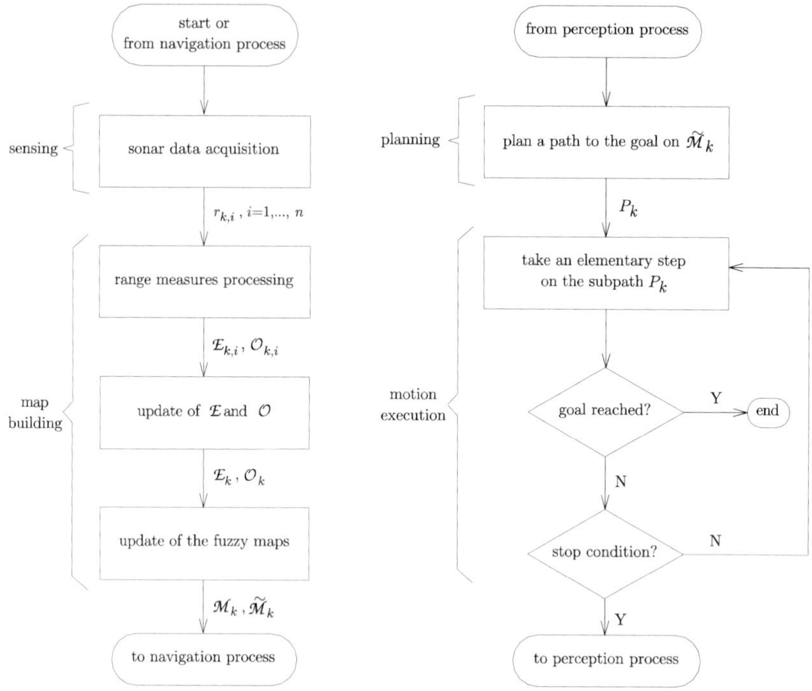

Figure 9.6 On-line map building and navigation: the *k*th perception process (left); the *k*th navigation process (right).

Sensing. As already mentioned, in our implementation the sonar ring of Nomad undergoes two consecutive rotations of 7.5° at each perception point: as a result, n = 48 *new* range readings are available for the map building phase. Note that since the sensor ring rotates independently from the body, the robot wheels do not move during this process. As a consequence, no odometry error is accumulated during sensing.

Map building. The inputs to the map building phase are the n range measures coming from the ultrasonic sensors. The outputs are two updated *fuzzy maps* $\mathcal{M}$ and $\tilde{\mathcal{M}}$. Both convey information about the risk of collision at each point of the environment, but in $\mathcal{M}$ the unexplored areas are regarded as dangerous, while they are safe in $\tilde{\mathcal{M}}$. Planning is performed on the latter, while the former is used during motion execution. The fundamental steps of map building are basically the same as described in Section 9.3. However, some modifications are in order.

During the kth perception process, for each range measure $r_{k,i}$ (i = 1,..., n), two local fuzzy sets $\mathcal{E}_{k,i}$ and $O_{k,i}$ are generated, similarly to Equations (9-8–9-9), as

$$\mathcal{E}_{k,i}(\rho,\vartheta) = f_{\mathcal{E}}(\rho, r_{k,i})m_1(\vartheta)m_1(\rho)$$
$$O_{k,i}(\rho,\vartheta) = f_O(\rho, r_{k,i})m_1(\vartheta)m_1(\rho)$$

The information contained in $\mathcal{E}_{k,i}$ and $\mathcal{O}_{k,i}$ is then used to update the two global fuzzy sets of *empty* and *occupied* cells:

$$\mathcal{E}_k = \mathcal{E}_{k-1} \cup \left(\bigcup_{i=1}^{n} \mathcal{E}_{k,i} \right) \qquad \mathcal{O}_k = \mathcal{O}_{k-1} \cup \left(\bigcup_{i=1}^{n} \mathcal{O}_{k,i} \right)$$

In turn, the above sets are used to update the fuzzy map $\mathcal{M}$, which is computed according to Equation (9-12), and the auxiliary fuzzy map $\tilde{\mathcal{M}}$ defined through

$$\tilde{\mathcal{M}}_k = \left(\overline{\tilde{S}_k} \right)$$

where

$$\tilde{\mathcal{S}} = \mathcal{E}_k^2 \cap \overline{\mathcal{O}}_k \cap \overline{\mathcal{A}}_k \cup \mathcal{I}_k$$

Note that the indeterminate cells are "included" in $\tilde{\mathcal{S}}$, whereas they are "discarded" to obtain $\mathcal{S}$ (see Equation (9-11)). As before, the membership degrees of cell C to the sets $\mathcal{M}$ and $\tilde{\mathcal{M}}$ are denoted by $\mu(C)$ and $\tilde{\mu}(C)$, respectively.

It is not necessary to compute the fuzzy maps from scratch during each perception process. In fact, due to the choice of $m_2(\rho)$, only limited portions of $\mathcal{E}$ and $\mathcal{O}$ are updated. Therefore, also $\mathcal{M}$ and $\tilde{\mathcal{M}}$ need to be updated only partially. In practice, the time needed for map building is negligible if compared to the sensing phase.

9.6.2 The Navigation Process

When the perception process is completed, the fuzzy maps $\mathcal{M}$ and $\tilde{\mathcal{M}}$ are available for navigation. The navigation process is composed by two sequential phases: *planning* and *motion execution* (see Figure 9.6).

During the k-th planning phase, a subpath P_k is produced from the *current* robot position to the goal G by applying the path planning technique of Section 9.5 on $\tilde{\mathcal{M}}_k$. For example, if the integral risk along the subpath is of concern, the algorithm A_1^* is applied to minimize the risk function

$$g_1(P_k) = \sum_{C_i \in P_k} \tilde{\mu}_k^2(C_i)$$

It should be emphasized that the planner makes use of the fuzzy map $\tilde{\mathcal{M}}_k$, in which the *indeterminate* cells are not regarded as dangerous. The rationale for this choice is simple: in order to reach the goal, the robot will have to traverse areas that at the beginning of the motion task were indeterminate. Thus, the planner must be allowed to propose subpaths going through such areas; during the motion execution phase the robot will follow these subpaths as long as they are safe on $\mathcal{M}_k$, and will activate a new perception process *before* entering an indeterminate area.

Once the planner has produced a subpath P_k on $\tilde{\mathcal{M}}_k$ from the current robot position to the goal, the motion execution phase takes place. The robot follows P_k until one of two *stop conditions* is verified:

1. The robot proximity sensors—which are always active—detect the presence of an obstacle closer than a minimum clearance. The choice of a small clearance is more hazardous, but allows the robot to go across narrow passages. In our implementation, a ring of infrared sensor was used to this end. A clearance of 12 cm was selected, that is the minimum detectable range for the sonar.

2. The robot is entering a specified α-*cut* of $\mathcal{M}_k$, i.e., the subset of cells whose membership function μ_k is larger than α. Since *indeterminate* cells are identified by high values of μ_k, the robot will stop if the planned subpath enters an indeterminate area. Our experiments were performed with $\alpha = 0.9$.

If any of these is met, the robot stops and performs a new perception process.

9.6.3 On-line Case: Experimental Results

To test the on-line navigation method we have used a cluttered area of 7.5×10 m, shown in Figure 9.7(a). The start and goal positions are denoted by S and G, respectively. As before, the environment has been approximated by a bitmap of square cells with a side of 0.1 m. The planner minimizes g_1 on a map obtained growing $\tilde{\mathcal{M}}$ to allow for the robot actual size. Few trials were necessary to tune the various parameters. For map building, we set $k_O = 0.25$, $k_E = 0.2$, $\Delta r = 0.15$ m, and $\rho_v = 1.5$ m.

The partial (10 perceptions) and the final (23 perceptions) maps $\tilde{\mathcal{M}}$ are reported in Figs. 9.7(b-c). Perception points are numbered in the order of visit. Darker areas indicate cells that are unsafe for planning; indeterminate cells outside the reliability area are safer than the empty cells inside the same area. As a consequence, the robot displays a pronounced *exploratory* attitude.

The robot path is reasonably efficient. Nomad tries to follow a direct path to G until (location 7) it detects the presence of a wall. Thereafter, it tries to find a convenient passage driven by the heuristic information in the A^* planner. Since there is no opening on the left, the robot loops around the obstacle and crosses the door on the right (location 14), that has already been detected. From there on, navigation proceeds smoothly towards the goal.

The final map is fairly accurate: this satisfactory behavior was confirmed by several experiments. Failure was reported in a few very complicated situations, essentially due to the ultrasonic sensors inherent limitations.

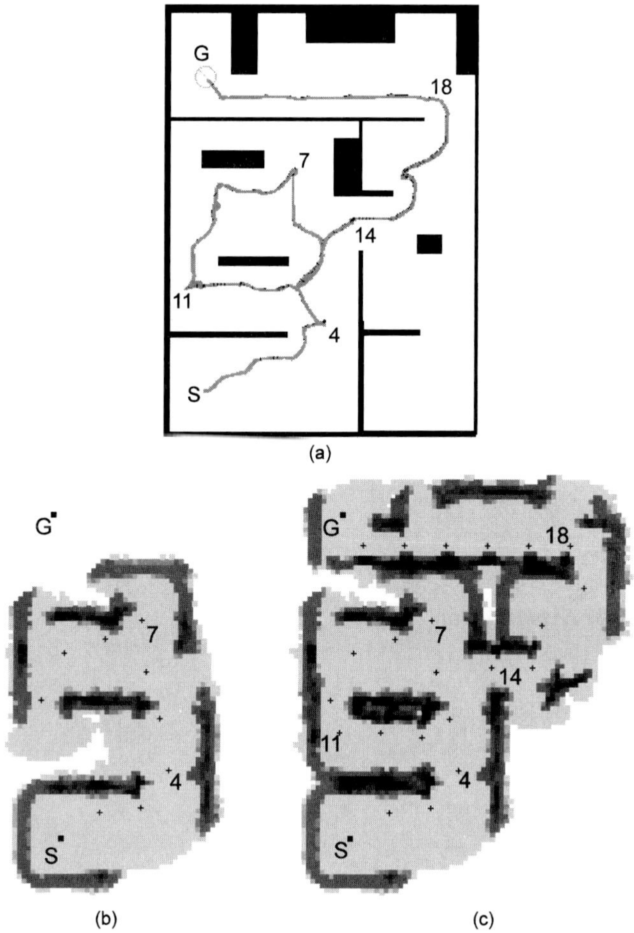

Figure 9.7 On-line map building and navigation in a trial case: the path executed by the robot (a); map $\tilde{\mathcal{M}}$ after 10 perceptions (b); after 23 perceptions (c).

9.7 CONCLUSION

Fuzzy logic, as a tool for managing information affected by uncertainty, finds a natural application in the field of perception-based mobile robotics. By virtue of its less constraining axiomatic foundation with respect to more conventional approaches (e.g., probability theory), it provides a framework in which it is relatively easy to take into account complex interactions between sensors and real-world environments, and to devise effective map building methods. Also the synthesis of planning algorithms may benefit from the simplicity and the flexibility of the representation obtained by this approach.

9.8 REFERENCES

1. Durrant-Whyte, H. F. *Integration, Coordination and Control of Multi-Sensor Robot Systems*. Kluwer Academic Publishers, 1988.

2. Cox, I. J. and Wilfong, G. T., eds. *Autonomous Robot Vehicles*. Springer-Verlag, 1990.

3. Latombe, J.-C. *Robot Motion Planning*. Kluwer Academic Publishers, 1991.

4. Oriolo, G., Ulivi, G. and Vendittelli, M. "Motion Planning with Uncertainty: Navigation on Fuzzy Maps." *Proc. 4th IFAC Symp. on Robot Control (SY.RO.CO. '94)*, I, pp. 71-78, Capri, 1994.

5. Oriolo, G., Ulivi, G. and Vendittelli, M. "On-Line Map Building and Navigation for Autonomous Mobile Robots." *Proc. 1995 IEEE Int. Conf. on Robotics and Automation*, pp. 2900-2906, Nagoya, Japan, 1995.

6. Leonard, J. J. and Durrant-Whyte, H. F. *Directed Sonar Sensing for Mobile Robot Navigation*. Kluwer Academic Publishers, 1992.

7. Bandemer, H. and Näther, W. *Fuzzy Data Analysis*. Kluwer Academic Publishers, 1992.

8. Klir, G. J., and T. A. Folger. *Fuzzy Sets, Uncertainty and Information*. Prentice Hall, 1988.

9. Zimmermann, H.-J. *Fuzzy Set Theory—and Its Applications*. Kluwer Academic Publishers, 1991.

10. Dombi, J. "A General Class of Fuzzy Operators, the De Morgan Class of Fuzzy Operators and Fuzziness Measures Induced by Fuzzy Operators." *Fuzzy Sets and Systems*. Vol. 8, 1982, pp. 149-163.

11. Polaroid Corporation. *Ultrasonic Ranging System*. 1987.

12. Quinlan, J. R. "Consistency and Plausible Reasoning." *Proc. 8th Int. Joint Conf. on Artificial Intelligence*, pp. 137-144, Karlsruhe. D, 1983.

13. Henkind, S. J. and Harrison, M. C. "An Analysis of Four Uncertainty Calculi." *IEEE Trans. on Systems, Man, and Cybernetics*. Vol. 18, No. 5, 1988, pp. 700-714.

14. Bonissone, P. P. "Reasoning, Plausible." *The Encyclopedia of Artificial Intelligence* (Shapiro, S., ed.). 2nd Ed., John Wiley and Sons, 1991, pp. 1307-1322.

15. Elfes, A. "Occupancy Grids: A Stochastic Spatial Representation for Active Robot Perception." In *Autonomous Mobile Robots: Perception, Mapping, and Navigation* (Iyengar, S. S., and A. Elfes, eds.). IEEE Computer Society Press, 1991, pp. 60-71.

16. Hill, S. D. and Spall, J. C. "Sensitivity of a Bayesian Analysis to the Prior Distribution." *IEEE Trans. on Systems, Man, and Cybernetics*. Vol. 24, No. 2, 1994, pp. 216-222.

17. Tirumalai, A. P., Schunk, B. G. and Jain, R. C.. "Evidential Reasoning for Building Environment Maps." *IEEE Trans. on Systems, Man, and Cybernetics*. Vol. 25, No. 1, 1995, pp. 10-20.

18. Dubois, D., Wellmann, N. P., D'Ambrosio, B. and Smets, Ph. "The Nature of Unnormalized Beliefs Encountered in the Transferable Belief Model." In *Proc. 8th Conf. on Uncertainty in Artificial Intelligence*, San Matteo, CA, 1992, pp. 292-297.

19. De Luca, A. and S. Termini. "Entropy and Energy Measures of a Fuzzy Set." In *Advances in Fuzzy Sets and Applications* (Gupta, M. M., R. K. Ragade, and R. R. Yager, eds.). North-Holland, 1979,pp. 321-338.

20. Hart, P. E., Nilsson, N. J. and Raphael, B. "A Formal Basis for the Heuristic Determination of Minimum Cost Paths." *IEEE Trans. on Systems, Science, and Cybernetics*. Vol. SSC-4, No. 2, 1968,pp. 100-107.

21. Warren, C. W. "Fast Path Planning Using Modified A^* Method." *Proc. 1993 IEEE Int. Conf. on Robotics and Automation*, Atlanta, GA, 1993, pp. 662-667.

22. Khatib, O. "Real-Time Obstacle Avoidance for Manipulators and Mobile Robots." *Int. J. of Robotics Research*. Vol. 5, No. 1, 1986, pp. 90-98.

23. Skewis, T., and Lumelsky, V. "Experiments With a Mobile Robot Operating in a Cluttered Unknown Environment." *J. of Robotic Systems*. Vol. 11, pp. 281-300, 1994.

24. Borenstein, J., and Koren, Y. "The Vector Field Histogram—Fast Obstacle Avoidance for Mobile Robots." *IEEE Trans. on Robotics and Automation*. Vol. 7, No. 3, pp. 278-288, 1991.

10

FUZZY MEASUREMENT OF MANUFACTURING FLEXIBILITY

N. C. Tsourveloudis and Y. A. Phillis
Technical University of Crete
GREECE

Flexibility is a concept describing a host of inherent properties of manufacturing systems. Due to its fuzziness, it cannot be modeled satisfactorily in any algebraic form. In this chapter, a new approach is provided to compute flexibility using fuzzy logic in the context of expert systems. The approach is applied in the determination of flexibility for machines, routing, material handling systems, products, and manufacturing systems.

10.1 INTRODUCTION

As the life cycle of consumer products decreases and market needs continuously change, organizations recognize the importance of manufacturing flexibility in facing the uncertainty of their environment. Manufacturing flexibility has been conventionally associated with a company's ability to absorb quickly and economically external changes. Although managers have a clear view of the benefits of flexibility, they cannot systematically use it in the decision-making process because it is not properly defined and measured in a quantitative way. Indeed, a universally acceptable method for such measurement is lacking [1,2], mainly because of the multidimensionality and the inherent vagueness of the notion.

Flexibility could be approached from a number of different angles. One such important angle examines the economic benefits of flexibility (see for example [3]–[6]).

Olhager [7] notes that a general relationship between profitability and manufacturing flexibility cannot be developed, again due to the vague character of the latter. In contrast to the above works which suggest indirect measurements of flexibility, many researchers examine the operational characteristics of the notion. Browne et al. [8] analyzed flexibility by defining eight distinct types which proved to be useful in understanding and measuring flexibility. The types of flexibility we review below are based on the comprehensive taxonomy of Sethi and Sethi [1], which constitutes an enlargement of the classification reported in [8].

1. *Machine flexibility* can be described by the ease of making changes among various operations required to produce a set of products. It is traditionally measured by the number of different operations which a machine performs and the time needed to switch from one operation to another.

2. *Routing flexibility* is the ability of a production system to manufacture a part using several alternative routes in the system, and it is determined by the number of such potential routes and the back-up machinery in case of breakdowns.

3. *Material handling flexibility* is the ability of a transportation system to move efficiently many different part types from one point to another. It can be measured by the number, diversity, and transportation time of workpieces.

4. *Product flexibility* is called the ease with which the part mix can be changed in order to manufacture or assemble new products. Quantitatively it is measured by the time or the cost needed to switch from one part mix to another.

5. *Operation flexibility* of a part refers to the ease of changing the sequence of the operations required to produce this part and it can be measured by the number of different processing schedules.

6. *Process flexibility* measures the ability of a manufacturing system to produce several part-types without reconfiguration. An index of this flexibility is the number of part types that can be simultaneously processed by the system.

7. *Volume flexibility* is the ability of a system to operate profitably at different throughput levels. It is quantified by the range of volumes with which the system runs profitably.

8. *Expansion flexibility* refers to a system's capability of being modularly built and expanded. It can be measured by the time or cost required for the system's expansion to a given capacity.

9. *Production flexibility* is the set of part types that a system can produce without major investments in machinery and it is measured by the size of this set.

10. *Market flexibility* is the ability to absorb changes by adapting to the market orders. Time and cost required for the adaptation can be considered as measures of market flexibility.

Machine flexibility can be regarded as the ratio of the setup time and the corresponding manufacturing time for a set of products processed by the machine [9]. Brill and Mandelbaum [10] assess the same type of flexibility via the efficiency with which a machine performs a number of tasks. Their measurement can be easily extended to a group of machines or to an industry. In practice, however, it is more important to compare flexibilities rather than compute them on an absolute scale. Indeed such a scale would be arbitrary and its usefulness limited to a manager, whereas comparisons of flexibilities of similar machines on a given scale or even larger production systems are of paramount importance to engineers and decision makers. With this in mind, Abdel-Malek and Wolf [11] developed an algorithm which produces a comparative measurement of flexibility by examining the functional parameters of the systems under study. Wharton and White [12], Gupta and Somers [13], used a statistical analysis of the expert's opinions to determine the factors affecting flexibility as well as their importance in the assessment of flexibility. In [12] and [13] the necessary data come out of questionnaires answered by managers.

By representing the workstations with nodes and the various transitions with arrows, many researchers developed measures in the context of graph theory (see for example [14]). A Petri net model is suggested in [15] which contains events such as product arrivals, start-end of operations etc., and situations such as a machine is available, or a part is waiting to be processed. Flexibility, here, is measured by the time the system needs to settle down after a disturbance. Kochikar and Narendran [16] use directed graphs in order to define measures for machine, routing, product, and material handling flexibilities. The overall flexibility is given as the inner product of weights and individual flexibilities.

Measures of routing flexibility, which are based on information theory and entropy, were introduced in the work of Yao [17]. In [18], Kumar reports fifteen axioms that a flexibility measure should satisfy and suggests measures in analogy to entropy measures, as flexibility reflects the freedom with which various choices can be made. Recently, Benjaafar et al. [19], proposed a measurement also in the context of information theory.

Other schemes can be found in Zelenovic [20], Buzacott [21], Lim [22], Dooner [23], and De Groote [24].

As already mentioned, a universally accepted framework for measuring flexibility is hard to define, while the concept is filled with uncertainty which is of also unspecified dimension. Below we provide a number of reasons to explain the difficulties in establishing such framework [25,26].

1. There exists no effective method for the synthesis of :
    a) the functional parameters affecting each type of flexibility.

b) the flexibility types which are observed on different hierarchical levels and are crucial in the determination of manufacturing flexibility.

As an example consider two manufacturing systems with setup times s_1 and s_2 respectively. Let $s_1 > s_2$ which means that the second system is more flexible in the sense of setups. Furthermore, the products which the second system can produce are a subset of the products of the first system. As a result, the first system appears now to be more flexible than the second. Clearly, correct conclusions should be based on the knowledge of the overall effect of these two parameters.

2. There is no correspondence between flexibility and the physical characteristics of the production system. For example, the ability of a system to recover after malfunction (routing flexibility) is associated either with redundant machinery (identical parallel machines), or with the existence of versatile workstations which are capable of performing many different operations, substituting for nonoperational machines.

3. Flexibility may be studied in different hierarchical levels demanding data for its measurement which are not accurately defined, such as the number of operations that a workstation can perform or the rerouting ability of a system.

4. Certain operational characteristics have contradictory effects on flexibility. The existence of common parts in the assembly of final products increases, for instance, product flexibility, while, at the same time, it may decrease operations flexibility.

In this chapter we present a knowledge-based framework for the determination of manufacturing flexibility. Our measurement scheme depends on existing expertise regarding the factors that influence flexibility. The fundamental assumptions of the method are:

1. Manufacturing flexibility is an inherently fuzzy notion which can be measured by the synthesis of its constituents. This synthesis is achieved better with logical rather than algebraic operations.

2. A measurement should always take into account the particular characteristics of the installation under study.

10.2 FLEXIBILITY MEASUREMENT AND FUZZY LOGIC

Although one may observe flexibility characteristics in all manufacturing systems, the benefits of the notion are obvious in the context of Flexible Manufacturing Systems (FMSs). An FMS is a production system consisting of a set of numerically controlled machining centers which are connected through an automated material handling system. FMSs are capable of processing a wide variety of parts with negligible setup delays. Furthermore, parts and tools can be quickly positioned at the machines thanks to an exchange-storage system which accompanies every FMS. A detailed discussion on FMS hardware can be found in [27].

In real world applications, precise data concerning flexibility factors of FMSs are not available or very hard to be extracted. In addition, decision makers prefer natural language expressions such as "high," "low," "average," etc., rather than numerical values in assessing flexibility. Fuzzy logic offers a systematic base in dealing with situations which are ambiguous or not well defined. Indeed, the uncertainty in expressions such as "low flexibility" or "high utilization," which are frequently encountered in the flexibility literature, is fuzziness.

Manufacturing flexibility (MF) is a multidimensional, but also vague notion, which for reasons of analysis has been broken down by the researchers into several distinct types. More than fifty overlapping types are defined in the literature, but some of them are widely accepted as the most important for the explanation and determination of manufacturing systems flexibility. In the sequel we deal with a knowledge-based measuring scheme concerning machine routing, material handling, and product flexibilities.

10.2.1 Machine Flexibility

Machine flexibility (F_M) is the simplest kind of flexibility that can be defined in a manufacturing system and constitutes a necessary building block for the assessment of the total flexibility. Although it is mainly determined by the hardware installation existing, it is quite difficult to be computed analytically. For example, low changeover time or high machine versatility improves F_M, but an analytical formula which involves the simultaneous effect of the operational parameters is lacking. Here, we tackle this problem by introducing a rule-based measurement with which a sufficient amount of knowledge about how F_M parameters are logically combined to obtain an assessment is described in the form of fuzzy IF < Antecedents > THEN < Consequent > expert rules. Knowledge acquisition can be achieved by using various methodologies, as for example questionnaires ([12] and [13]) or existing surveys etc.

The observable components we define as the linguistic variables of F_M are:

1. Setup or changeover time S required for various preparations, such as tool and part positioning and release, software changes, etc. In many cases machines are controlled by a central computer and the corresponding software changeover time is very small or negligible. Setup time represents the ability of a machine to absorb efficiently changes of the production process and it influences flexibility heavily when the batch sizes are small.

2. Variety of operations that a machine is capable of performing which will be called versatility V of the machine. Processes with different tools and conditions (longitudinal or form turning, twist drilling or boring, external or internal grinding, etc.) are also considered to be operations. Versatility may be associated with the physical characteristics of a machine such as the number of motion axes, maximum accuracy, range of cutting speeds, number of fixtures, etc. Finally, the quantity and the diversity of workpieces are strongly connected with versatility.

3. The range of adjustments or the adjustability R of a machine. This parameter is defined as the size of working space and is related to the maximum and minimum dimensions of the parts that the machine can handle.

In order to evaluate these parameters, which are among the most common aspects of F_M, we need information concerning:

1. The size of product batches, since this is an index of the influence of setup time in flexibility.

2. The product variety which characterizes the versatility of a machine.

3. The necessary variety of product sizes in the determination of adjustability.

4. The values we assign to F_M and the linguistic variables of setup S, versatility V and adjustability R are Low (L), About Low (AL), Average (A), High (H), About High (AH).

The rules which represent the expert knowledge on how the antecedents affect machine flexibility are of the following form:

If Setup time is S
  AND Versatility is V
  AND Adjustability is R
THEN
Machine flexibility is F_M

and in compact form

$$\text{If } S \text{ AND } V \text{ AND } R \text{ THEN } F_M \quad \text{or} \quad S \text{ AND } V \text{ AND } R \rightarrow F_M \tag{10-1}$$

The rule base contains $5^3 = 125$ rules. The linguistic values of S, V, R, and F_M which represent the inexact knowledge of the experts, are fuzzy sets with membership functions shown in Figure 10.1. More importantly, logical connectives are also fuzzy. Generally, the representation of managerial knowledge by linguistic rules performs better when there are no units of measurement concerning the attributes of the system and no quantitative criteria for the values of such attributes [28]. If the AND operators were crisp as in two-valued logic, the result of the statement "S AND V AND R" would be unambiguous. Here, however, we are faced with the problem of expressing such ambiguities. In our measurement scheme, machine flexibility is given as the conclusion of an approximate reasoning method. Informally, approximate reasoning is the process by which, given an observation about the parameters of a system, a conclusion is deduced from the existing

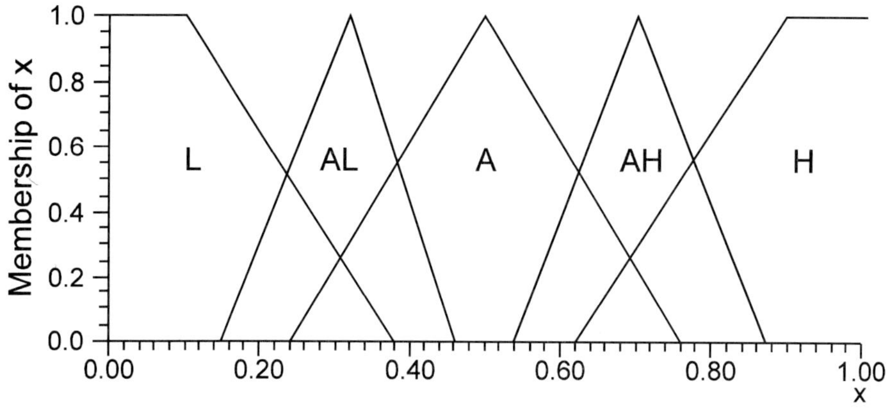

Figure 10.1 Membership functions of the linguistic variables.

rule base. In particular let $C = S$ AND V AND R and S', V', R' be real world observations about setup, versatility, and adjustability respectively. Machine flexibility is computed by

$$F'_M = f(C', C \rightarrow F_M) \tag{10-2}$$

where f is an approximate reasoning method, and $C' = S'$ AND V' AND R'. Using the generalized modus ponens, we obtain:

$$F'_M = C' \circ C \rightarrow F_M \tag{10-3}$$

where $\circ$ is the *Compositional Rule Of Inference* (CRI) proposed by Zadeh [29].

In the membership function domain and for the CRI with max-min operators, machine flexibility is

$$\mu_{F'_M}(y_i) = \vee_i [\mu_{C'}(x_i) \wedge \mu_{C \rightarrow F_M}(x_i, y_i)], \; i = 1, \ldots, I \;\; j = 1, \ldots, J \tag{10-4}$$

where $\mu_{F'_M}(y_i)$ is the membership grade of the jth element of F'_M, i.e., the inferred membership value of y_j, $\mu_{C'}(x_i)$ is the membership grade of the ith element of C', i.e., the observed membership value of x_i, and $\mu_{C \rightarrow F_M}(x_i, y_i)$ is the membership grade of the ijth element of the implication relation matrix.

10.2.2 Routing Flexibility

Routing flexibility (F_R) refers to the ability of a manufacturing system to produce a part using alternate paths. This ability allows for a quick reaction to unexpected events such as machine breakdowns and it minimizes the effect of interruptions of the production process. Two forms of F_R have been identified [8]:

1. Potential routing flexibility, when part routes are specified from the processing schedule, but parts can be dynamically rerouted during a breakdown, and

2. Actual routing flexibility, when identical parts are processed through different routes, independently of breakdown situations.

The benefits from the existence of routing flexibility are well understood among researchers but there is still some confusion as to its definition [1]. Routing flexibility appears in the literature in the different guises of scheduling, operations, process and, more often, manufacturing flexibility. It is achieved when the system consists of interchangeable and multipurpose machines together with a material handling system, rescheduling control software, and redundancy in machines, tools, and processes. These requirements demand a high investment, often making the production prohibitively costly.

Several measures of F_R have been proposed in the literature. Most of them are functions of the different available routes through which products should be processed (see for example Chatterjee et al. [30], Das and Nagendra [31]).

In our view, routing flexibility is an inherent property of the manufacturing system which expresses the ability of the system to respond to unanticipated internal changes and variations. We are mainly motivated by the fact that F_R arises from the existence of interchangeable machines capable of performing similar operations. The ability to handle breakdowns, which is the main characteristic of F_R, exists if each operation can be performed on more than one machines [8]. Furthermore, we recognize that a key prerequisite in measuring F_R is the ability of a machine to substitute for another.

The linguistic variables of concern in assessing routing flexibility are:

1. Operation Commonality CO, which expresses the number of common operations that a group of machines can perform in order to produce a set of parts.

2. Substitutability SB, which is defined as the facility of a system to reroute and reschedule effectively under failure conditions. It is an index of substitution quality among machines, which may also be used to characterize some built-in capabilities of the system (real-time scheduling or available transportation links). Substitutability is associated with the material handling system and the layout of the machines.

A crisp measure of F_R which is based on the same variables and pairwise comparison between machines is the following [26]:

$$F_R = \frac{2}{N(N-1)} \sum_{i=1}^{N-1} \sum_{j=i+1}^{N} \frac{CO_{ij} SB_{ij}}{O_i + O_j - CO_{ij}} \tag{10-5}$$

where N is the number of machines; O_i the number of tasks that machine i performs; O_j is the number of operations that machine j can perform; CO_{ij} is the number of common operations for machines i and j; and SB_{ij} the substitutability or equivalently the index of the quality of substitution.

Equation (10-5) becomes very complicated when it simultaneously compares a group of machines, as it is the common practice. Furthermore, variables as *substitutability* and *operations commonality* are better evaluated by using linguistic rather than numerical values.

The rule base of our prototype contains statements IF THEN of the form:

IF Operations Commonality is *CO*
AND Substitutability is *SB*
THEN
Routing Flexibility is F_R

Equivalently,

$$\text{IF } CO \text{ AND } SB \text{ THEN } F_R \tag{10-6}$$

Obviously, in the above rule *CO*, *SB*, and F_R represent the linguistic values taken by Operations Commonality, Substitutability and Routing Flexibility, respectively. These values, as in the case of machine flexibility, are Low (*L*), About Low (*AL*), Average (*A*), About High (*AH*), High (*H*).

It is known that there are various representations of conjugation and implication connectives. In addition, one may use infinitely many t-norm-conorm operators in these connectives [32] in order to express the meaning of (10-6).

Here however, we utilize the interval-valued representation of AND and IF-THEN connectives suggested by Turksen [33]. This leads to the construction of lower and upper bounds for each logical connective, based on Disjunctive and Conjunctive Normal Forms of the linguistic values (Table 10.1), denoted DNF and CNF respectively. The main advantage of this representation is that, by defining a bounded region for the membership function of a statement, such as "*CO* AND *SB*," we capture the dispersion of imprecise knowledge acquired by the experts. Generally, if *LC*(*) is a logical connective for (*), then

$$\text{DNF}\left[LC(*)\right] \subseteq LC(*) \subseteq \text{CNF}\left[LC(*)\right] \tag{10-7}$$

or in the membership functions domain

$$\mu_{\text{DNF}}\ LC(*) \leq \mu LC(*) \leq \mu_{\text{CNF}}\ LC(*)$$

where DNF stands for "Disjunctive Normal Form" and CNF for "Conjunctive Normal Form" as already mentioned.

In Table 10.1 $\vee$, $\wedge$ denote maximum and minimum operators respectively, while A^c denotes the standard complement of a fuzzy set *A*, i.e.,

Table 10.1 Interval-valued representation of connectives			
Title	Verbal Expression	DNF	CNF
Conjunction	*A* AND *B*	$A \wedge B$	$(A \vee B) \wedge (A \vee B^c) \wedge (A^c \vee B)$
Implication	IF *A* THEN *B*	$(A \wedge B) \vee (A \wedge B^c) \vee A^c (\wedge B)$	$A^c \vee B$

$$\mu_{A^c}(x) = 1 - \mu_A(x), \quad \forall x \in X$$

where X is the universal set.

The procedure for assessing routing flexibility, using the interval-valued representation of connectives, is as follows. We compute the bounds of the connective AND:

1. The lower bound $\mathrm{L}^{[2]}_{\mathrm{AND}}$ for the two antecedent rule (CO AND $SB \to F_R$) is (see Table 10.1):

$$\mathrm{L}^{[2]}_{\mathrm{AND}} = \mathrm{DNF}\ (CO \text{ AND } SB) = CO \wedge SB \tag{10-8}$$

   and the corresponding upper bound

$$\begin{aligned} \mathrm{U}^{[2]}_{\mathrm{AND}} &= \mathrm{CNF}\ (CO \text{ AND } SB) \\ &= (CO \wedge SB) \wedge (CO \vee SB) \wedge \left(CO \vee SB^c\right) \wedge \left(CO^c \vee SB\right) \end{aligned} \tag{10-9}$$

In order to simplify the notation of the implication bounds, we define the center of the region determined by (10-8) and (10-9)

$$C_{\mathrm{AND}} = \frac{1}{2}\left(\mathrm{L}^{[2]}_{\mathrm{AND}} + \mathrm{U}^{[2]}_{\mathrm{AND}}\right) \tag{10-10}$$

Now the bounds of the implication $C_{\mathrm{AND}} \to F_R$, are

$$\begin{aligned} \mathrm{L}_{(\to)} &= \mathrm{DNF}\left(C_{\mathrm{AND}} \to F_R\right) \\ &= \left(C_{\mathrm{AND}} \wedge F_R\right) \vee \left(C_{\mathrm{AND}} \wedge F_R\right) \vee \left(C^c_{\mathrm{AND}} \wedge F_R\right) \vee \left(C^c_{\mathrm{AND}} \wedge F^c_R\right) \end{aligned} \tag{10-11}$$

and

$$\mathrm{U}_{(\to)} = \mathrm{CNF}\left(C_{\mathrm{AND}} \to F_R\right) = C^c_{\mathrm{AND}} \vee F_R \tag{10-12}$$

Following (10-10), the center of implication is

$$C_{(\rightarrow)} = \frac{1}{2}\left(\mathrm{L}_{(\rightarrow)} + \mathrm{U}_{(\rightarrow)}\right) \tag{10-13}$$

As a matter of notational convenience, we denote membership functions of a set by the set itself. For example $\mu_{CO}(x)$ becomes simply *CO*.

In general, the observations of linguistic variables CO' and SB' differ from the predefined scaled values defined in the 25 rules of the rule base. That is, a finer scale may be needed, which on the other hand would raise the number of rules. In this case, routing flexibility is the conclusion of an approximate reasoning method. In particular, and for the compositional rule of inference, F_R' is given by

$$F_R' = C_{\mathrm{AND}}' \circ C_{(\rightarrow)} \tag{10-14}$$

where $\circ$ is the max-min composition and C_{AND}' is the new center of the AND connective which results from the statement "CO' AND SB'".

10.2.3 Material Handling and Product Flexibilities

A Material Handling System (MHS) is the key component of a flexible manufacturing system. MHS flexibility has been defined as the ability to move quickly several types of parts for proper positioning and processing through the manufacturing system [1]. In contrast to previously examined cases, the development of MHS flexibility measures has been limited. Stecke and Browne [34] explored the impact that several types of material handling equipment have on flexibility. They have reported linguistic assessments of flexibilities defined in [8] concerning the following types of equipment: belt and powered roller conveyors, monorails, power-and-free conveyors, towline carts, and automated guided vehicles. Other measurement schemes can be found in [30] as well as in [35].

The linguistic variables we define for the knowledge-based measurement of MHS flexibility (F_{MHS}) are:

1. Rerouting factor *B*, which indicates the ability of a material handing system to change traveling paths automatically or with small setup delay and cost. Rerouting ability is a necessary property for the establishment of routing flexibility.

2. The variety of loads *P* which a material handling system carries, for example, workpieces, tools, jigs, fixtures, etc. It is restricted by the volume, dimension, and weight requirements of the load.

3. Maximum transfer speed *C*. It is associated with the weight of products, as well as the frequency of transportation. The membership functions of the relative linguistic values are shown in Figure 10.2.

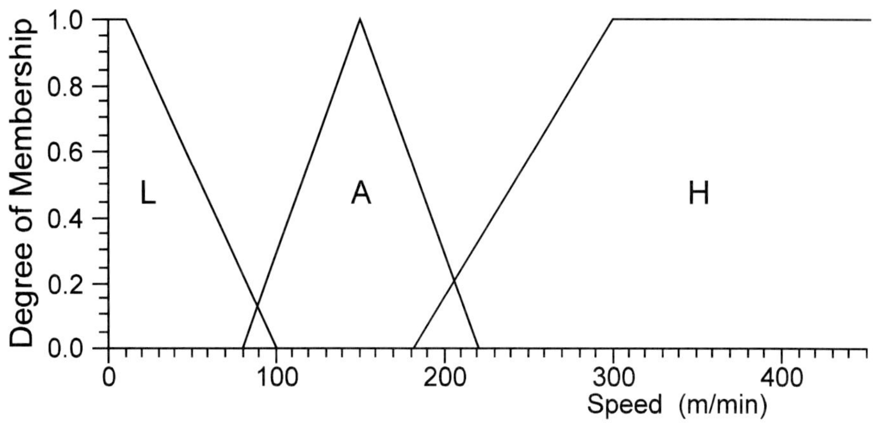

Figure 10.2 The linguistic values of transfer speed (*C*).

4. Number of connected elements M, for example machines or buffers.

The expert rules concerning F_{MHS} are of the following form:

IF Rerouting Factor is B
AND Load Variety is P
AND Transfer Speed is C
AND Number of Connected Elements is M
THEN
Material Handling Flexibility is F_{MHS}

or

$$B \text{ AND } P \text{ AND } C \text{ AND } M \rightarrow F_{MHS} \tag{10-15}$$

In the same manner, the rules of Product Flexibility (F_P) are:

IF Part Variety is V_p
AND Changeover Time is T
AND Part Commonality is C_p
THEN
Product Flexibility is F_p

or compactly

$$V_p \text{ AND } T \text{ and } C_p \rightarrow F_p \tag{10-16}$$

where:

1. Part variety V_p is associated with the number of new products that the manufacturing system can produce in a time period.

2. Changeover time T is the total time required for preparations in order to produce a new product mix. It expresses the ability of a system to absorb market variations.

3. Part commonality C_p refers to the number of common parts used in the assembly of a final product. It measures the ability of introducing new products fast and economically.

Both F_{MHS} and F_p are the conclusions of an approximate reasoning schema. In previous sections we discussed CRI with max-min operators, which is the first known approximate reasoning method. This methods is the most frequently used in the literature and it is suggested from a structural perspective. In real-life applications, suitable conjunction and implication must be selected in order to achieve an inference with desirable properties. In the software implementation of our prototype, however, which is presented in [26], one may select various methods of approximate reasoning as well as operators. Having noted this, we will continue our presentation of the knowledge-based measurement of manufacturing flexibility.

10.3 MANUFACTURING FLEXIBILITY MEASUREMENT

Up to now, we have discussed the measurement of four flexibility types, which are observed in different hierarchical levels of an organization. In this section, manufacturing flexibility (*MF*) is the logical synthesis of the types already examined (Figure 10.3). This is expressed via If-Then rules, as follows

IF Machine Flexibility is F_M
  AND Routing Flexibility is F_R
  AND MHS Flexibility is F_{MHS}
  AND Product Flexibility is F_p
THEN
  Manufacturing Flexibility is *MF*

or

$$F_M \text{ AND } F_R \text{ AND } F_{MHS} \text{ AND } F_p \rightarrow MF \qquad (10\text{-}17)$$

It is known that there are many ways to represent fuzziness in a conditional statement. Here, we demonstrate the use of the interval representation of connectives, because this mechanism provides the largest region of uncertainty.

The lower bound of the AND connective for (10-17) is:

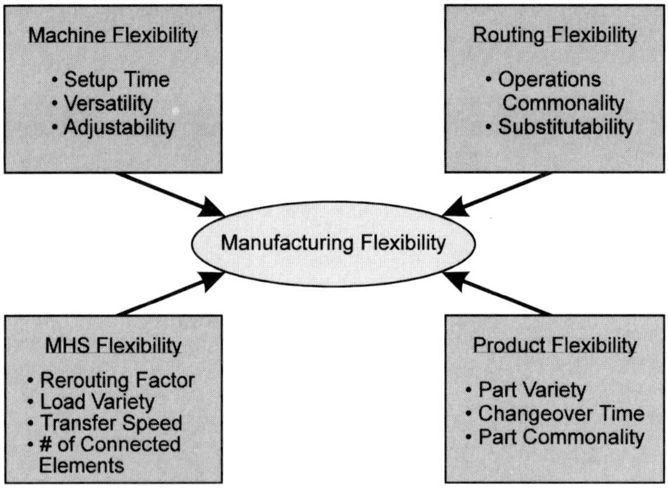

Figure 10.3 Manufacturing flexibility as the synthesis of distinct types.

$$\begin{aligned} \mathrm{L}_{\mathrm{AND}}^{[4]} &= \mathrm{DNF}\left\{\mathrm{DNF}\left[\mathrm{DNF}(F_M \text{ AND } F_R)\text{AND } F_{MHS}\right]\text{AND } F_p\right\} \\ &= F_M \wedge F_R \wedge F_{MHS} \wedge F_P \end{aligned} \tag{10-18}$$

as DNF (*) $\subseteq$ CNF (*) by definition. We will compute the upper bound of the same connective. Obviously $\mathrm{L}_{\mathrm{AND}}^{[2]} = \mathrm{DNF}(F_M \text{ AND } F_R) \subseteq \mathrm{CNF}(F_M \text{ AND } F_R) = \mathrm{U}_{\mathrm{AND}}^{[2]}$, and

$$\mathrm{DNF}\left(\mathrm{L}_{\mathrm{AND}}^{[2]} \text{ AND } F_{MHS}\right) \subseteq \mathrm{CNF}\left(\mathrm{L}_{\mathrm{AND}}^{[2]} \text{ AND } F_{MHS}\right)$$

$$\mathrm{DNF}\left(\mathrm{U}_{\mathrm{AND}}^{[2]} \text{ AND } F_{MHS}\right) \subseteq \mathrm{CNF}\left(\mathrm{U}_{\mathrm{AND}}^{[2]} \text{ AND } F_{MHS}\right)$$

Furthermore, we know that

$$\mathrm{CNF}\left(\mathrm{L}_{\mathrm{AND}}^{[2]} \text{ AND } F_{MHS}\right) = \left(\mathrm{L}_{\mathrm{AND}}^{[2]} \vee F_{MHS}\right) \wedge \left(\mathrm{L}_{\mathrm{AND}}^{[2]} \vee F_{MHS}^{c}\right) \wedge \left(\mathrm{L}_{\mathrm{AND}}^{[2]c} \vee F_{MHS}\right) \tag{10-19}$$

$$\mathrm{CNF}\left(\mathrm{U}_{\mathrm{AND}}^{[2]} \text{ AND } F_{MHS}\right) = \left(\mathrm{U}_{\mathrm{AND}}^{[2]} \vee F_{MHS}\right) \wedge \left(\mathrm{U}_{\mathrm{AND}}^{[2]} \vee F_{MHS}^{c}\right) \wedge \left(\mathrm{U}_{\mathrm{AND}}^{[2]c} \vee F_{MHS}\right) \tag{10-20}$$

In the general case, we cannot decide whether (10-19) is smaller or larger than (10-20), and since we are looking for the upper bound we compute the maximum of these two:

$$\mathrm{U}_{\mathrm{AND}}^{[3]} = \vee\left[\mathrm{CNF}\left(\mathrm{L}_{\mathrm{AND}}^{[2]} \text{ AND } F_{MHS}\right), \mathrm{CNF}\left(\mathrm{U}_{\mathrm{AND}}^{[2]} \text{ AND } F_{MHS}\right)\right] \tag{10-21}$$

Analogously, the upper bound of AND in rule (10-17) is

$$\begin{aligned}\mathrm{U}_{\mathrm{AND}}^{[4]} &= \vee\left[\mathrm{CNF}\left(\mathrm{L}_{\mathrm{AND}}^{[3]} \text{ AND } F_P\right), \mathrm{CNF}\left(\mathrm{U}_{\mathrm{AND}}^{[3]} \text{ AND } F_P\right)\right] \\ &= \vee\left\{\left[\left(F_M \wedge F_R \wedge F_{MHS}\right) \vee F_P\right] \wedge \left[\left(F_M \wedge F_R \wedge F_{MHS}\right) \vee F_P^c\right]\right. \\ &\quad \wedge \left[\left(F_M \wedge F_R \wedge F_{MHS}\right)^c \vee F_P\right], \\ &\quad \left.\left(\mathrm{U}_{\mathrm{AND}}^{[3]} \vee F_P\right) \wedge \left(\mathrm{U}_{\mathrm{AND}}^{[3]} \vee F_P^c\right) \wedge \left(\mathrm{U}_{\mathrm{AND}}^{[3]c} \vee F_P\right)\right\}\end{aligned} \tag{10-22}$$

where

$$\begin{aligned}\mathrm{U}_{\mathrm{AND}}^{[3]} &= \vee\left\{\left[\left(F_M \wedge F_R\right) \vee F_{MHS}\right] \wedge \left[\left(F_M \wedge F_R\right) \vee F_{MHS}^c\right] \wedge \left[\left(F_M \wedge F_R\right)^c \vee F_{MHS}\right]\right. \\ &\quad \left\{\left[\left(\left(F_M \vee F_R\right) \wedge \left(F_M \vee F_R^c\right) \wedge \left(F_M^c \vee F_R\right)\right) \vee F_{MHS}\right]\right. \\ &\quad \wedge \left[\left(\left(F_M \vee F_R\right) \wedge \left(F_M \vee F_R^c\right) \wedge \left(F_M^c \vee F_R\right)\right) \vee F_{MHS}^c\right] \\ &\quad \left.\left.\wedge \left[\left(\left(F_M \vee F_R\right) \wedge \left(F_M \vee F_R^c\right) \wedge \left(F_M^c \vee F_R\right)\right)^c \vee F_{MHS}\right]\right\}\right\}\end{aligned} \tag{10-23}$$

In order to simplify the approximate reasoning process we compute the center of the AND connective

$$C_{\mathrm{AND}} = \frac{1}{2}\left(\mathrm{L}_{\mathrm{AND}}^{[4]} + \mathrm{U}_{\mathrm{AND}}^{[4]}\right) \tag{10-24}$$

Note that C_{AND} is a point-valued fuzzy set as opposed to an interval-valued set. The lower bound of the implication now is

$$\mathrm{L}_{(\to)} = \mathrm{DNF}\left(C_{\mathrm{AND}} \to MF\right) = \left(C_{\mathrm{AND}} \wedge MF\right) \vee \left(C_{\mathrm{AND}}^c \wedge MF\right) \vee \left(C_{\mathrm{AND}}^c \wedge MF^c\right) \tag{10-25}$$

and the upper bound

$$\mathrm{U}_{(\to)} = \mathrm{CNF}\left(C_{\mathrm{AND}} \to MF\right) = C_{\mathrm{AND}}^c \vee MF \tag{10-26}$$

Suppose next that, instead of the five linguistic values *L*, *AL*, *A*, *AH*, and *H*, we make an observation with finer gradation of the variables of interest. Let the new observations be F_M', F_R', F_{MHS}', F_P' and consequently the center of AND would yield C_{AND}'. Then the bounds of manufacturing flexibility are

$$MF_L' = C_{\mathrm{AND}}' \circ \mathrm{L}_{(\to)} \tag{10-27}$$

and

$$MF'_U = C'_{\text{AND}} \circ \text{U}_{(\rightarrow)} \tag{10-28}$$

where MF'_L, MF'_U are the lower and the upper bounds of MF respectively, and $\circ$ represents the CRI with max-min operators. An exhaustive discussion on approximate reasoning with interval-valued fuzzy sets can be found in [36]. We illustrate the above procedure with an example.

Suppose that for a given input, the rule which fires is

IF $< F_M$ is Average $>$
AND $< F_R$ is Low $>$
AND $< F_{MHS}$ is High $>$
AND $< F_P$ is High $>$
THEN
$< MF$ is About High $>$

or

$$A \text{ AND } L \text{ AND } H \text{ AND } H \rightarrow AH \tag{10-29}$$

The membership functions of the linguistic values of concern are shown in Table 10.2, where $\mu(x)/x$ represents the membership grade $\mu(x)$ of point x.

We first compute the bounds of the AND connective. From (10-18) we obtain

$$\begin{aligned} \text{L}^{[4]}_{\text{AND}} &= \text{DNF}(A \text{ AND } L \text{ AND } H \text{ AND } H) \\ &= (0/.1,\ 0/.15,\ .1/.25,\ .5/.35,\ .75/.45,\ 1/.5,\ .5/.65,\ .1/.75,\ 0/.85,\ 0/.9) \\ &\wedge (1/.1,\ .8/.15,\ .5/.25,\ .1/.35,\ 0/.45,\ 0/.5,\ 0/.65,\ 0/.75,\ 0/.85,\ 0/.9) \\ &\wedge (0/.1,\ 0/.15,\ 0/.25,\ 0/.35,\ 0/.45,\ 0/.5,\ .1/.65,\ .5/.75,\ .8/.85,\ 1/.9) \\ &\wedge (0/.1,\ 0/.15,\ 0/.25,\ 0/.35,\ 0/.45,\ 0/.5,\ .1/.65,\ .5/.75,\ .8/.85,\ 1/.9) \\ &= (0 / x),\ \text{where } x \in [0,\ 1] \end{aligned}$$

Table 10.2 The discrete membership functions of the linguistic values

Low:	(1/.1, .8/.15, .5/.25, .1/.35, 0/.45, 0/.5, 0/.65, 0/.65, 0/.75, 0/.85, 0/.9)
About *Low*:	(0/.1, 0/.15, .4/.25, 1/.35, .4/.45, 0/.5, 0/.65, 0/.75, 0/.85, 0/.9)
Average:	(0/.1, 0/.15, .1/.25, .5/.35, .75/.45, 1/.5, .5/.65, .1/.75, 0/.85, 0/.9)
About *High*:	(0/.1, 0/.15, 0/.25, 0/.35, 0/.45, 0/.5, .4/.65, 1/.75, .4/.85, 0/.9)
High:	(0/.1, 0/.15, 0/.25, 0/.35, 0/.45, 0/.5, .1/.65, .5/.75, .8/.85, 1/.9)

The upper bound from (10-22) is

$$\mathrm{U}_{\mathrm{AND}}^{[4]} = \vee\left[\mathrm{CNF}\left(\mathrm{L}_{\mathrm{AND}}^{[3]} \text{ AND } H\right), \mathrm{CNF}\left(\mathrm{U}_{\mathrm{AND}}^{[3]} \text{ AND } H\right)\right] \tag{10-30}$$

where

$$\begin{aligned}\mathrm{CNF}\left(\mathrm{L}_{\mathrm{AND}}^{[3]} \text{ AND } H\right) &= \left[(A \wedge L \wedge H) \vee H\right] \wedge \left[(A \wedge L \wedge H) \vee H^{c}\right] \wedge \left[(A \wedge L \wedge H)^{c} \vee H\right] \\ &= (0/.1,\ 0/.15,\ 0/.25,\ 0/.35,\ 0/.45,\ 0/.5,\ .1/.65,\ .5/.75,\ 8/.85,\ 1/.9) \\ &\wedge (1/.1,\ 1/.15,\ 1/.25,\ 1/.35,\ 1/.45,\ 1/.5,\ .9/.65,\ .5/.75,\ .2/.85,\ 0/.9) \\ &\wedge (1/.1,\ 1/.15,\ 1/.25,\ 1/.35,\ 1/.45,\ 1/.5,\ 1/.65,\ 1/.75,\ 1/.85,\ 1/.9) \\ &= (0/.1,\ 0/.15,\ 0/.25,\ 0/.35,\ 0/.45,\ 0/.5,\ .1/.65,\ .5/.75,\ .2/.85,\ 0/.9)\end{aligned}$$

$$\begin{aligned}\mathrm{CNF}\left(\mathrm{U}_{\mathrm{AND}}^{[3]} \text{ AND } H\right) &= \left(\mathrm{U}_{\mathrm{AND}}^{[3]} \vee H\right) \wedge \left(\mathrm{U}_{\mathrm{AND}}^{[3]} \vee H^{c}\right) \wedge \left(\mathrm{U}_{\mathrm{AND}}^{[3]c} \vee H\right) \\ &= (0/.1,\ .2/.15,\ .5/.25,\ .5/.35,\ .25/.45,\ 0/.5,\ .5/.65,\ .5/.75,\ .2/.85,\ 0/.9)\end{aligned}$$

Equation (10-30) yields

$$U_{\mathrm{AND}}^{[4]} = (.2/.15,\ .5/.25,\ .5/.35,\ .25/.45,\ .5/.65,\ .5/.75,\ .2/.85)$$

The center of the AND connective is directly from (10-24)

$$C_{\mathrm{AND}} = (.1/.15,\ .25/.25,\ .25/.35,\ .125/.45,\ .25/.65,\ .25/.75,\ .22/.85)$$

A sufficient condition for the implication to be meaningful is given in [37] according to which C_{AND} should be normalized. Dividing C_{AND} by 0.25, which is the maximum membership grade, we obtain

$$C_{\mathrm{AND}} = (.4/.15,\ .1/.25,\ .1/.35,\ .5/.45,\ .1/.65,\ .1/.75,\ .4/.85)$$

We now compute the implication relation matrices in order to extend the range of expert advice concerning *MF*. The bounds of implication involve the normalized C_{AND} and the knowledge provided by the selected rule. From (10-25) and (10-26) we have

$$\mathrm{L}_{(\to)} = \mathrm{DNF}\left(C_{\mathrm{AND}} \to AH\right) = \left(C_{\mathrm{AND}} \wedge AH\right) \vee \left(C_{\mathrm{AND}}^{c} \wedge AH\right) \vee \left(C_{\mathrm{AND}}^{c} \wedge AH^{c}\right) \tag{10-31}$$

$$\mathrm{U}_{(\to)} = \mathrm{CNF}\left(C_{\mathrm{AND}} \to AH\right) = C_{\mathrm{AND}}^{c} \vee AH \tag{10-32}$$

where C_{AND} = (.4/.15, .1/.25, .1/.35, .5/.45, .1/.65, .1/.75, .4/.85) and AH =(.4/.65, 1/.75, .4/.85). All values of the bounds will be arranged in two 7×3 relation matrices, where the number of rows equals the number of values for C_{AND} and the number of columns equals the number of values of AH. Thus

$$L_{(\rightarrow)} = \text{DNF}(C_{\text{AND}} \rightarrow AH) = (C_{\text{AND}} \wedge AH) \vee (C^c_{\text{AND}} \wedge AH) \vee (C^c_{\text{AND}} \wedge AH^c) \quad (10\text{-}31)$$

$$U_{(\rightarrow)} = \text{CNF}(C_{\text{AND}} \rightarrow AH) = C^c_{\text{AND}} \vee AH \quad (10\text{-}32)$$

where C_{AND} = (.4/.15, .1/.25, .1/.35, .5/.45, .1/.65, .1/.75, .4/.85) and AH =(.4/.65, 1/.75, .4/.85). All values of the bounds will be arranged in two 7×3 relation matrices, where the number of rows equals the number of values for C_{AND} and the number of columns equals the number of values of AH. Thus

$$L_{(\rightarrow)} = \begin{bmatrix} .6 & .4 & .6 \\ .6 & 1 & .6 \\ .6 & 1 & .6 \\ .5 & .5 & .5 \\ .6 & 1 & .6 \\ .6 & 1 & .6 \\ .6 & .4 & .6 \end{bmatrix} \text{ and } U_{(\rightarrow)} = \begin{bmatrix} .6 & .4 & .6 \\ 1 & 1 & 1 \\ 1 & 1 & 1 \\ .6 & .5 & .6 \\ 1 & 1 & 1 \\ 1 & 1 & 1 \\ .6 & .4 & .6 \end{bmatrix}$$

where , for example, L_{11} =.6 = (.4 ∧ .4) ∨ (.6 ∧ .4) ∨ (.6 ∧ .6), and U_{11}= .6 = (.6 ∨ .4). Now, consider the compositional rule of inference. By using the above matrices and the center of the AND connective, we obtain the membership function of manufacturing flexibility. Suppose that we have a slightly different input with C'_{AND} = (.5/.15, 1/.25, .9/.35, .6/.45, 1/.65, 1/.75, .6/.85). The bounds of flexibility are

$$MF'_L = C'_{\text{AND}} \circ L_{(\rightarrow)} = [.5 \;\; 1 \;\; .9 \;\; .6 \;\; 1 \;\; 1 \;\; .6] \circ \begin{bmatrix} .6 & .4 & .6 \\ .6 & 1 & .6 \\ .6 & 1 & .6 \\ .5 & .5 & .5 \\ .6 & 1 & .6 \\ .6 & 1 & .6 \\ .6 & .4 & .6 \end{bmatrix} = (.6/.65 \;\; 1/.75 \;\; .6/.85), \quad (10\text{-}33)$$

The methodology we have described is very useful in making flexibility comparisons among alternate systems, and, therefore, it is an important tool for a manager in choosing a specific manufacturing configuration.

10.4 CONCLUSION

A novel methodology for the computation of manufacturing flexibility was presented using fuzzy logic in conjunction with expert systems. Such a methodology appears to be the natural context for this problem, since the concept of flexibility is not amenable to precise mathematical modeling. Indeed, flexibility means different things to different people and its assessment is not unique. Our approach in essence makes use of such different opinions and ideas through an expert system which operates via fuzzy logic.

The results of the computation facilitate comparisons among alternative manufacturing configurations, thus enabling managers and engineers to make proper choices. Up until now, such choices and decisions were based on algebraic formulas of questionable validity. We use interval-valued implication which in a sense encompasses a wide range of uncertainty. Any other type of implication can be used with no major difficulty, depending on specific requirements.

The model for manufacturing flexibility, which is the epitome of all individual flexibilities (see Figure 10.3), utilizes four components. It is easy to expand the model to any number of contributing components, if they are deemed important. Also, each individual flexibility is assessed from a number of attributes which, without difficulty, can also be expanded. Such expansions for practical systems is the subject of future research.

10.5 REFERENCES

1. Sethi, A. and Sethi, S. "Flexibility in Manufacturing: A Survey." *International Journal of Flexible Manufacturing Systems*. Vol. 2, 1990, pp. 289-328.

2. Gupta, Y. and Goyal, S. "Flexibility of Manufacturing Systems: Concepts and Measurements." *European Journal of Operational Research*. Vol. 43, 1989, pp. 119-135.

3. Kulatilaka, N. "Valuing the Flexibility of Flexible Manufacturing Systems." *IEEE Transactions on Engineering Management*. Vol. 78, no. 3, 1988, pp. 250-257.

4. Hutchinson, G.K. and Sinha, D. "Quantification of the Value of Flexibility." *Journal of Manufacturing Systems*. Vol. 8, No. 1, 1989, pp. 47-56.

5. Kulatilaka, N. and Marks, S. E. "The Strategic Value of Flexibility: Reducing the Ability to Compromise." *American Economic Review*. Vol. 78, No. 3, 1988, pp. 574-580.

6. Son, Y. K. and Park, C. S. "Economic Measure of Productivity, Quality and Flexibility in Advanced Manufacturing Systems." *Journal of Manufacturing Systems.* Vol. 6, No. 3, 1984, pp. 193-207.

7. Olhager, J. "Manufacturing Flexibility and Profitability." *International Journal of Production Economics.* Vol. 30-31, 1993, pp. 67-78.

8. Browne, J., Dubois, D., Rathmill, K., Sethi, S. P. and Stecke, K. E. "Classification of Flexible Manufacturing Systems." *The FMS Magazine.* Vol. 2, No. 2, 1984, pp. 114-117.

9. Falkner, C. H. "Flexibility in Manufacturing Plants." *Proceedings of 2nd ORSA/TIMS Conference on Flexible Manufacturing Systems*, Ann Arbor, MI (Stecke, K. E. and Suri, R., eds.). Elsevier, 1986, pp. 95-106.

10. Brill, P. H. and Mandelbaum, M. "On Measures of Flexibility in Manufacturing Systems." *International Journal of Production Research.* Vol. 27, No. 5, 1989, pp. 747-756.

11. Abdel-Malek, L. and Wolf, C. "Evaluating Flexibility of Alternative FMS Designs—A Comparative Measure." *International Journal of Production Economics.* Vol. 23, No. 1, 1991, pp. 3-10.

12. Wharton, T. J. and White, E. M. "Flexibility and Automation: Patterns of Evolution." *Operations Management Review.* Vol. 6, No. 3, 1988, pp. 1-8.

13. Gupta, Y. P. and Somers, T. M. "The Measurement of Manufacturing Flexibility." *European Journal of Operational Research.* Vol. 60, 1992, pp. 166-182.

14. Afentakis, P. "A Model for Layout Design in FMS." *Flexible Manufacturing Systems: Methods and Studies.* (Kusiak, A., ed.), Elsevier, 1986, pp. 127-139.

15. Barad, M. and Sipper, D. "Flexibility in Manufacturing Systems: Definitions and Petri Net Modeling." *International Journal of Production Research.* Vol. 26, No. 2, 1988, pp. 237-248.

16. Kochikar, V. P. and Narendran, T. T. "A Framework for Assessing the Flexibility of Manufacturing Systems." *International Journal of Production Research.* Vol. 30, No. 12, 1992, pp. 2873-2895.

17. Yao, D.D. "Material and Information Flows in Flexible Manufacturing Systems." *Material Flow.* Vol. 3, 1985, pp. 143-149.

18. Kumar, V. "Entropic Measures of Manufacturing Flexibility." *International Journal of Production Research.* Vol. 25, No. 7, 1987, pp. 957-966.

19. Benjaafar, S., Morin, T. L. and Talavage, J. J. "The Strategic Value of Flexibility in Sequential Decision Making." *European Journal of Operational Research.* Vol. 82, 1995, pp. 438-457.

20. Zelenovic, D. M. "Flexibility: A Condition for Effective Production Systems." *International Journal of Production Research.* Vol. 20, No. 3, 1982, pp. 319-337.

21. Buzacott, J.A. "The Fundamental Principles of Flexibility in Manufacturing Systems." *Proceedings of 1st International Conference on Flexible Manufacturing Systems*, Brighton, UK. Elsevier, 1982, pp. 23-30.

22. Lim, S.M. "Flexible Manufacturing Systems and Manufacturing Flexibility in the United Kingdom." *International Journal of Operations and Production Management.* Vol. 7, No. 6, 1987, pp. 44-54.

23. Dooner, M. "Conceptual Modeling of Manufacturing Flexibility." *International Journal of Computer Integrated Manufacturing.* Vol. 4, No. 3, 1991, pp. 135-144.

24. De Groote, X. "The Flexibility of Production Processes: A General Framework." *Management Science.* Vol. 40, No. 7, 1994, pp. 933-945.

25. Gupta, D., and Buzacott, J. A. "A Framework for Understanding Flexibility of Manufacturing Systems." *Journal of Manufacturing Systems.* Vol. 8, No. 2, 1989, pp. 89-97.

26. Tsourveloudis, N.C. Manufacturing flexibility measurement: A fuzzy logic approach. Ph.D Dissertation (in Greek). Department of Production Engineering and Management, Technical University of Crete, 1995.

27. Greenwood, N.R. *Implementing Flexible Manufacturing Systems,* Macmillan, 1988.

28. Zadeh, L.A. "The Role of Fuzzy Logic in the Management of Uncertainty in Expert Systems." *Fuzzy Sets and Systems.* Vol. 11, 1983, pp. 199-227.

29. Zadeh, L.A. "Outline of a New Approach to the Analysis of Complex Systems and Decision Processes." *IEEE Transactions on Systems, Man, and Cybernetics.* Vol. 3, No. 1, 1973, pp. 28-44.

30. Chatterjee, A., Cohen, M. A., Maxwell, W. L. and Miller, L. W. "Manufacturing Flexibility: Models and Measurements." *Proceedings of 1st ORSA/TIMS Special Interest Conference on FMS*, Ann Arbor, MI. (Stecke, K. E. and Suri, R., eds.), Elsevier, 1984.

31. Das, S. K. and Nagendra, P. "Investigations Into the Impact of Flexibility on Manufacturing Performance." *International Journal of Production Research.* Vol. 31, No. 10, 1984, pp. 2337-2354.

32. Schweitzer, B. and Sklar, A. *Probabilistic Metric Spaces.* North Holland, 1983.

33. Turksen, I. B. "Interval Valued Fuzzy Sets Based on Normal Forms." *Fuzzy Sets and Systems.* Vol. 20, 1986, pp. 191-210.

34. Stecke, K. E. and Browne, J. "Variations in Flexible Manufacturing Systems According to the Relevant Types of Automated Materials Handling." *Material Flow.* Vol. 2, 1985, pp 179-185.

35. Noble, J. S. and Tanchoco, J. M. A. "A Framework for Material Handling System Justification." *International Journal of Production Research.* Vol. 31, No. 1, 1993, pp. 81-106.

36. Turksen, I. B. "Four Methods of Approximate Reasoning with Interval-Valued Fuzzy Sets." *International Journal of Approximate Reasoning.* Vol. 3, 1989, pp. 121-142.

37. Turksen, I.B. and Yao, D. D. W. "Representation of Connectives in Fuzzy Reasoning: The View Through Normal Forms." *IEEE Transactions on Systems, Man, and Cybernetics.* Vol. 14, No. 1, 1984, pp. 146-151.

11

NEO-FUZZY NEURON AND ITS LEARNING ALGORITHMS WITH APPLICATIONS TO THE MODELING OF NONLINEAR DYNAMICAL SYSTEMS

T. Yamakawa* and E. Uchino*
Kyushu Institute of Technology
JAPAN

In this chapter a new type of a fuzzy neuron, i.e. a neo-fuzzy neuron (a fusion of a fuzzy logic and a neural network) is presented. Some types of learning algorithms of this neo-fuzzy neuron are also discussed, and some actual applications are given.

11.1 INTRODUCTION

The attempt was made to modify a conventional neuron model by using a fuzzy logic in order to obtain a novel neuron model, which can be better adapted to a study of the behavior of the nonlinear dynamical systems imprecisely defined by virtue of their high degree of complexities. S. C. Lee and E. T. Lee [1] modified the "all-or-nothing" activity of the MacCulloch-Pitts model [2] to have the output equal to zero in non firing state and some positive number $0 < \mu_i \leq 1$ in firing. They also presented a general synthesis

* The authors are also with the Fuzzy Logic Systems Institute (FLSI), Iizuka, Fukuoka 820, Japan.

procedure for realizing arbitrary fuzzy automaton by using their fuzzy neurons [1]. One of the authors presented another type of fuzzy neuron [3], where the linear synaptic connection is replaced with a nonlinear one characterized by a membership function labeled as "tightly connected," "loosely connected," and so on. The excitatory connection and the inhibitory connection are represented by a fuzzy logic intersection and by a fuzzy logic complement, respectively. This fuzzy neuron was successfully employed to construct a hardware system which achieves a handwritten character recognition within 1 μsec [4]. A design algorithm for this fuzzy neuron was also established by an example-based learning [5].

In this chapter a new type of a fuzzy neuron model is presented, whose synaptic transfer characteristic is nonlinear. Its nonlinear synaptic characteristic is realized by a set of fuzzy implication rules, and thus its network characteristic is quite flexible. The neo-fuzzy neuron has both the characteristics of the fuzzy systems and the neural networks, i.e., the representational ability of the former and the learning ability of the latter.

The type of fuzzy rules employed here is the one with complementary triangular membership functions in the antecedent, and singletons in the consequent. Some types of learning algorithms to assign the parameters in the fuzzy rules are also discussed. It is then guaranteed that the global minimum in the error space can be reached by the present algorithm. The neo-fuzzy neuron is, so to speak, a kind of a realization of a general nonlinear functional link network, whose nonlinearity is characterized by a set of fuzzy linguistic if-then rules.

The computer simulations have shown that the neo-fuzzy neuron has a high capability of generalization, and the learning speed and its accuracy have been drastically improved compared with the conventional neural networks. The attempt was also made to apply the neo-fuzzy neuron to the modeling of the nonlinear chaotic systems. The system of concern was modeled adaptively by simply feeding to the proposed neo-fuzzy neuron the input and the output data of the system. The accuracy of modeling was excellent, and it could predict well the chaotic behavior of the system. The experimental results are promising for a practical use.

The neo-fuzzy neuron proposed in this chapter is very simple in its structure and its learning, but flexible enough to achieve the desired action. This is an important point, because it should be as simple as possible when it is implemented by hardware. This work is the first step before implementing it by a Large Scale Integration (LSI) chip, which is the final goal of our research.

11.2 ARCHITECTURE OF A NEO-FUZZY NEURON

The basic architecture of a neo-fuzzy neuron is shown in Figure 11.1. It consists of an input node, a nonlinear synapse, and a summing node. An input x_i goes through a nonlinear synapse f_i; then the outputs of each nonlinear synapse are summed up to produce the output y of the neo-fuzzy neuron. The aggregation of the synaptic signals is achieved by just an algebraic sum, and the soma does not exhibit a sigmoid function at all. Thus, the output of this neo-fuzzy neuron is given by:

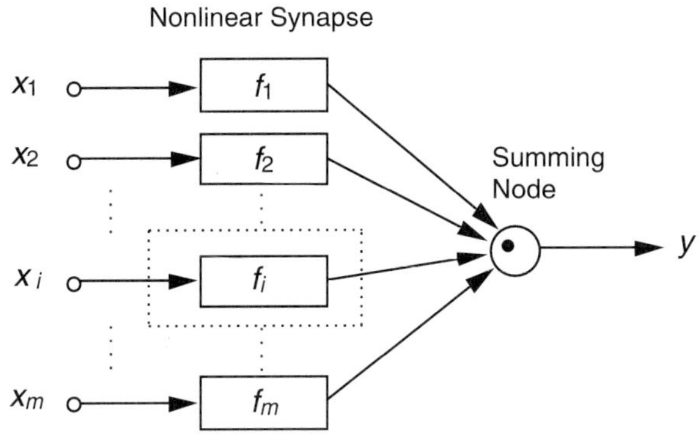

Figure 11.1. General structure of a neo-fuzzy neuron.

$$y = \sum_{i=1}^{m} f_i(x_i), \tag{11-1}$$

where $f_i(\cdot)$ is an output of each nonlinear synapse. The nonlinear function $f_i(\cdot)$ can be easily realized by a set of fuzzy rules, e.g., with triangular membership functions in the antecedent and with singletons in the consequent, which corresponds mathematically to an approximation of a nonlinear function by piecewise linear functions. Other types of fuzzy rules can of course be employed.

Figure 11.2 shows the internal mechanism of each nonlinear synapse. The fuzzy rules employed here are simple ones with triangular membership functions in the antecedent and with singletons in the consequent for the ease of hardware implementation. The input space for each signal x_i is divided into several fuzzy segments, which are characterized by the membership functions across the input space of x_i as illustrated in Figure 11.3. Each membership function μ_{ij} is followed by each weight w_{ij}, which corresponds to a singleton in the consequent of each fuzzy rule.

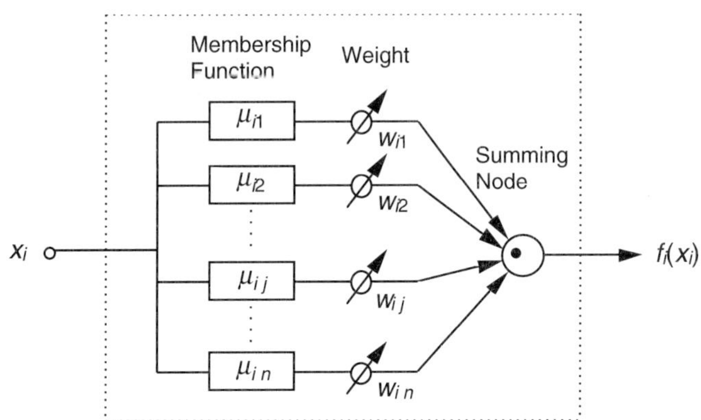

Figure 11.2. Structure of a nonlinear synapse.

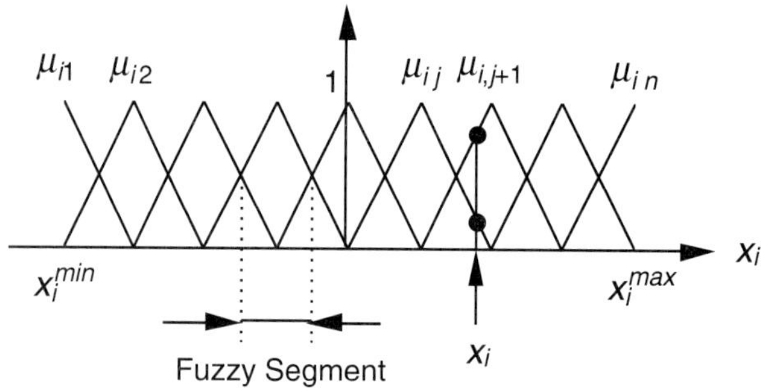

Figure 11.3. Complementary membership functions assigned to cover the input space. x_i^{min} and x_i^{max} denote the minimum and the maximum values of an input x_i, respectively.

Mapping from x_i to $f_i(x_i)$ is achieved by the following fuzzy inference:

$$\text{If } x_i \text{ is } A_{ij} \text{ then the output is } w_{ij} \tag{11-2}$$

where A_{ij} is a fuzzy set whose membership function is $\mu_{ij}(\cdot)$. w_{ij} is a singleton. The membership functions are so arranged that they cover a whole fluctuation range of an input signal as shown in Figure 11.3. The fuzzy inference adopted here is a Mamdani type fuzzy inference [6] and a center of gravity is taken for defuzzification as shown in Figure 11.4.

It should be noted again that each membership function in the antecedent is triangular and complementary with each other. In other words, an input signal x_i activates only two neighboring membership functions simultaneously, and the sum of grades of these two membership functions is always equal to 1, i.e., $\mu_{ij}(x_i) + \mu_{i,j+1}(x_i) = 1$. Thus the defuzzification taking a center of gravity does not need an arithmetic division, which is very convenient for a hardware implementation. The output is then given by the following simple form:

$$f_i(x_i) = \frac{\sum_{j=1}^{n} \mu_{ij}(x_i) w_{ij}}{\sum_{j=1}^{n} \mu_{ij}(x_i)} = \frac{\mu_{ij}(x_i) w_{ij} + \mu_{i,j+1}(x_i) w_{i,j+1}}{\mu_{ij}(x_i) + \mu_{i,j+1}(x_i)}$$
$$= \mu_{ij}(x_i) w_{ij} + \mu_{i,j+1}(x_i) w_{i,j+1} \tag{11-3}$$

The weight w_{ij} is an adaptive parameter which is assigned by an iterative learning to train the network to achieve a desired behavior. Some of these learning algorithms are described in the next section.

The neo-fuzzy neuron has a single-layer network structure, and the soma does not have a nonlinearity but the synapse does instead. This enables the high modeling capability of nonlinearity only by a single-layer structure. Furthermore, as discussed in the next

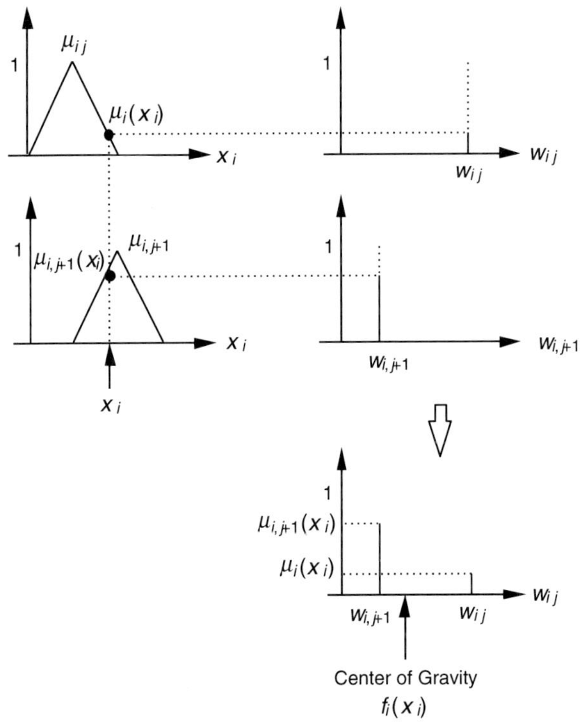

Figure 11.4. Fuzzy inference procedure achieved in the nonlinear synapse.

section, it has a global minimum in the error-weight space. This also enables the speedy convergence to the minimum point, and high learning accuracy is obtained.

The information on the structure of the nonlinear dynamical system of concern is stored in each nonlinear synapse of the neo-fuzzy neuron. Furthermore, because of the transparent structure of the network (each nonlinear synapse of the neo-fuzzy neuron is constituted by a set of fuzzy rules), we can easily estimate what synapse has mainly contributed to the present output. This is essentially important for the analysis of the complicated nonlinear dynamical system.

11.3 LEARNING ALGORITHMS AND THEIR PERFORMANCE

Let y_k be an actual network output, and $\tilde{y}_k$ a desired network output at time k (sometimes called as training data), respectively. The actual output y_k often differs from the desired one $\tilde{y}_k$. Thus we have to train the network to produce the desired output by adjusting a weight w_{ij}. In this section some learning algorithms for weight w_{ij} are presented, and their performance is examined.

11.3.1 Batch Learning

Now we define the mean-squared error to be minimized over the learning (training) cycle as follows:

$$E = \frac{1}{2}\sum_{k=1}^{N_m}\left(y_k - \tilde{y}_k\right)^2 \tag{11-4}$$

where N_m is a number of training data. The gradient descent method [7] is employed here to find a point which gives the minimum value of E. Thus the renewal of weight w_{ij} is given by:

$$\Delta w_{ij} = -\alpha\frac{\partial E}{\partial w_{ij}} = -\alpha\sum_{k=1}^{N_m}\left(y_k - \tilde{y}_k\right)\mu_{ij}\left(x_{ik}\right) \tag{11-5}$$

where x_{ik} is an input signal applied to the ith nonlinear synapse of the neo-fuzzy neuron at time k, and α is a learning rate empirically obtained. As Equation (11-5) is cumulative for instantaneous errors, the updating of weights is achieved only after applying N_m training patterns. Then this learning is called batch learning. In this algorithm, the initial values of the weights are all assigned to be zero, and hence they do not contribute to the output until they are updated.

11.3.2 Stepwise Learning

Practically speaking, as for the updating of Equation (11-5) we can employ the following approximated algorithm in case when α is small [8,9]:

$$\Delta w_{ij} = -\alpha\left(y_k - \tilde{y}_k\right)\mu_{ij}\left(x_{ik}\right) \tag{11-6}$$

In this case the updating of weights is achieved every time a training pattern is applied. This is called a stepwise learning. In this learning algorithm only those weights that contribute to the output at time k are updated. This drastically reduces the amount of learning, and also the effect of learning interference is almost completely reduced. In other words, the learning of the neo-fuzzy neuron is local, i.e., the learning in some parts of the network does not affect the learning of the other parts of the network. In this algorithm all the initial values of the weights can also be assigned to be zero.

11.3.3 Direct Search Learning with Use of the Second-Order Interpolation

The value of a learning rate must be carefully chosen. Proper value leads to a speedy convergence, or otherwise it may cause an oscillation. Further, it cannot be determined uniquely independent of the problem of concern. Therefore, in order to realize speedy and accurate learning, the learning rate must be settled adaptively and at the same time optimally for each case.

In this section, we find adaptively at each learning step the suboptimum value of a learning rate α with the aid of the Lagrange's second-order interpolation. Now define the weight vector at time t by $W(t) = [w_{11}(t),..., w_{ij}(t),... w_{mn}(t)]$. The optimum value of the learning rate α_{opt} at time t is the one which realizes:

$$\min_{\alpha} E\left(W(t) - \alpha \frac{\partial E(W(t))}{\partial W(t)}\right) \tag{11-7}$$

where E is the error criterion of Equation (11-4). It can be shown that the error criterion E is a unimodal function of α. The procedure to find the exact optimum value of α which realizes Equation (11-7) is not impossible but takes much computing time. In this section, we consider finding the sub-optimum value of α which gives approximately the minimum value of Equation (11-7).

The basic principle of this searching algorithm is very simple. First we find three points α_1, α_2, and α_3 ($\alpha_1 < \alpha_2 < \alpha_3$) in the error-$\alpha$ space, where the optimum value α_{opt} exists in the interval $[\alpha_1, \alpha_3]$. Then, we approximate the $E(\alpha)$ error function by the second-order curve by using Lagrange's interpolation method [10]. It is then easy to find analytically the point, which is the sub-optimum value of α, that gives the minimum value of the approximated second-order error function $E'(\alpha)$. The general algorithm is as follows.

(0) Set an initial interval $I_0 = [\alpha_0, \alpha_0 + h]$.

(i) In case $E(\alpha_0) > E(\alpha_0 + h)$, then renew the initial interval I_0 by $I_1 = [\alpha_0 + h, \alpha_0 + \lambda h]$ ($\lambda > 1$). If it is still $E(\alpha_0 + h) > E(\alpha_0 + \lambda h)$, renew I_1 by I_2 where $I_n = [\alpha_0 + \lambda^{n-1}h, \alpha_0 + \lambda^n h]$ ($n \geq 1$). Continue this renewal process, i.e., the renewal of I_{n-1} by I_n, until it becomes $E(\alpha_0 + \lambda^{n-1}h) < E(\alpha_0 + \lambda^n h)$. In another initial case of $E(\alpha_0) < E(\alpha_0 + h)$, the renewal process of interval is achieved in the counter direction. That is, $I_{-(n-1)}$ is renewed by I_{-n} until it becomes $E(\alpha_0 + h - \lambda^n h) > E(\alpha_0 + h - \lambda^{n-1}h)$, where $I_{-n} = [\alpha_0 + h - \lambda^n h, \alpha_0 + h - \lambda^{n-1}h]$ ($n \geq 1$).

(ii) It can be easily observed that the optimum point α_{opt} exists in $I = I_{n-1} \cup I_n$ (or $I = I_{-n} \cup I_{-(n-1)}$). Let $I = I_{n-1} \cup I_n = [\alpha_1, \alpha_2] \cup [\alpha_2, \alpha_3]$ ($\alpha_1 < \alpha_2 < \alpha_3$) (or $I_{-n} \cup I_{-(n-1)} = [\alpha_1, \alpha_2] \cup [\alpha_2, \alpha_3]$ ($\alpha_1 < \alpha_2 < \alpha_3$)) and $E_i = E(\alpha_i)$ ($i = 1, 2, 3$). The second-order curve which passes through the points $(\alpha_i, E_i)(i = 1, 2, 3)$ can be easily formulated, and the point α_{s-opt} which gives the minimum value of this approximated second-order curve is found analytically as:

$$\alpha_{s-opt} = \frac{(\alpha_2^2 - \alpha_3^2)E_1 + (\alpha_3^2 - \alpha_1^2)E_2 + (\alpha_1^2 - \alpha_2^2)E_3}{2(\alpha_2 - \alpha_3)E_1 + (\alpha_3 - \alpha_1)E_2 + (\alpha_1 - \alpha_2)E_3} \tag{11-8}$$

which is the suboptimum value of α that realizes approximately the minimum value of Equation (11-7).

The updating of weight $W(t)$ is achieved for each set of input-output training patterns.

11.3.4 Stepwise Learning with Variable Learning Rate

In the usual stepwise learning algorithm (see e.g., Section 11.3.2) the learning rate is fixed to a certain value through the whole learning cycle. The weight vector comes up near to the optimum point as the learning goes on, but sometimes oscillates around the optimum point. The bigger the learning rate is, the bigger the oscillation becomes.

In this section, we give a practical method to choose a better learning rate. In this learning algorithm, several learning rates are prepared in advance, and they are all examined in the first learning cycle. The learning rate which has given the best performance is selected among them, and it is used later on for the learning process. And, at the same time, the following error ratio:

$$R(t+1) = \left| \frac{E(W(t+1) - W(t))}{E(W(t))} \right| \tag{11-9}$$

is further checked at every learning. When it becomes less than a pre-fixed value r, its learning rate is reduced according to some rule, e.g., replace α by $\rho\alpha$, where α is the present value of a learning rate and ρ is a reduction ratio. By repeating this process, as the learning goes on, the learning ratio is reduced little by little, and then finally the minimum error can be reached.

11.3.5 Gradient Descent Learning with Momentum Term

In the gradient descent learning, what value to select for the learning rate is an important and difficult problem. The learning performance is greatly affected by the value of the learning rate α. It is known that a large value of α may realize a rapid learning speed, while on the other hand, it may easily cause an oscillation around the optimum point. In order to suppress this oscillation, an additional momentum term [7] is introduced, and then the learning algorithm for the neo-fuzzy neuron becomes as follows:

$$\Delta w_{ij}(t+1) = -\alpha \sum_{k=1}^{N_m} (y_k - \tilde{y}_k)\mu_{ij}(x_{ik}) + \beta \Delta w_{ij}(t) \tag{11-10}$$

where $\Delta w_{ij}(t)$ is a renewal of weight at time t, and $\beta (0 < \beta < 1)$ is called a momentum parameter. There is not any guideline to determine analytically its value, but it is often determined empirically.

11.3.6 Performance Comparison

In this section, the learning algorithms proposed so far in the previous sections for the neo-fuzzy neuron are examined, and their performance is compared.

A test signal for learning is generated by:

$$x_{k+1} = f(x_k) = 4x_k(1 - x_k) \tag{11-11}$$

with an initial value $x_0 = 0.6$. This nonlinear dynamical system is known as a logistic map which produces a chaotic signal. A single-input synapse neo-fuzzy neuron is employed for this experiment. The synapse has 12 fuzzy segments across the input space [0, 1] characterized by 12 membership functions followed necessarily by 12 weights to be adjusted. 50 chaotic data are generated with use of Equation (11-11) and they are used for training the network.

The signal data x_k at time k is applied to the neo-fuzzy neuron as an input, and then the actual output of the neo-fuzzy neuron is compared with the signal x_{k+1} (desired network output), and then the weights are changed (updated) in order for the network to produce the desired output. This is repeated for a set of 50 signal data, which is defined as one cycle of learning. 100 cycles of learning has been achieved with the initial weights being zero. The root-mean-square error $E_{RMS}(\equiv \sqrt{E(x_k - \tilde{x}_k)^2})$, where $\tilde{x}_k$ is a desired output, is evaluated for each learning algorithm. The experimental results are shown in Figure 11.5. The horizontal axis is the learning time (executed on Sun Sparc Station 1).

Learning performance is also compared with the conventional neural network of the 3-layer type. Figure 11.6 shows the decay of RMS errors versus the learning time. The notation NFN*i* means the neo-fuzzy neuron with *i* fuzzy segments with the direct search learning algorithm. NN*i* means the neural network with *i* hidden layers with the error back propagation learning algorithm [11]. The drastically improved learning speed and the learning performance of the neo-fuzzy neuron can be seen in Figure 11.6. Figure 11.7 shows the identified nonlinear mechanism of Equation (11-11).

11.4 IDENTIFICATION OF NONLINEAR DYNAMICAL SYSTEMS

The neo-fuzzy neuron is applied to the identification problem of nonlinear dynamical systems. Here we employ a neo-fuzzy neuron shown in Figure 11.8 with cascaded delay elements at an input part. This type of neo-fuzzy neuron can identify the structure of the nonlinear dynamical system by using its time series data. In order to show its effectiveness we test with the following 3rd-order chaotic nonlinear dynamical system [12]:

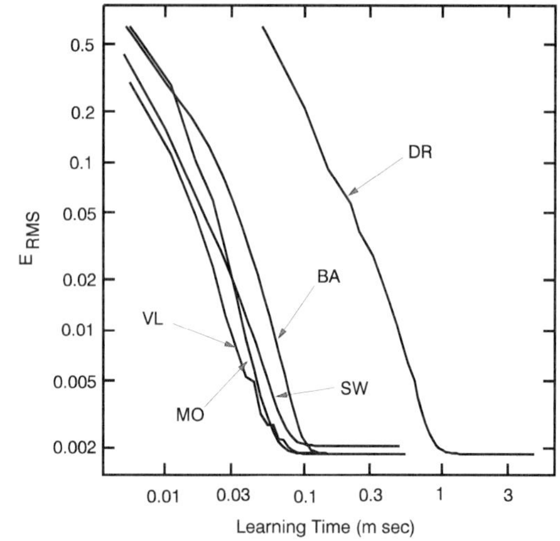

Figure 11.5. Performance comparison for several types of algorithms. BA, Batch learning with $\alpha = 0.3$; SW, stepwise learning with $\alpha = 0.3$; DR, direct search learning with $\lambda = 2$ and $h = 0.05$; VL, stepwise learning with variable learning rates of $\alpha = 0.05, 0.1, 0.2, 0.5, 1.0, 2.0$ with $r = 0.1$; MO, gradient descent learning with momentum term with $\alpha = 0.3$ and $\beta = 0.2$.

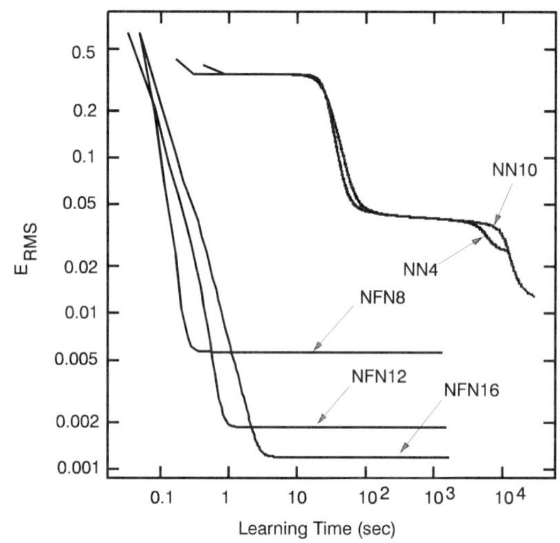

Figure 11.6. Performance comparison with the neo-fuzzy neuron and the conventional neural network. NFN*i*, neo-fuzzy neuron with *i* fuzzy segments; NN*i*, layered neural network with *i* hidden layers.

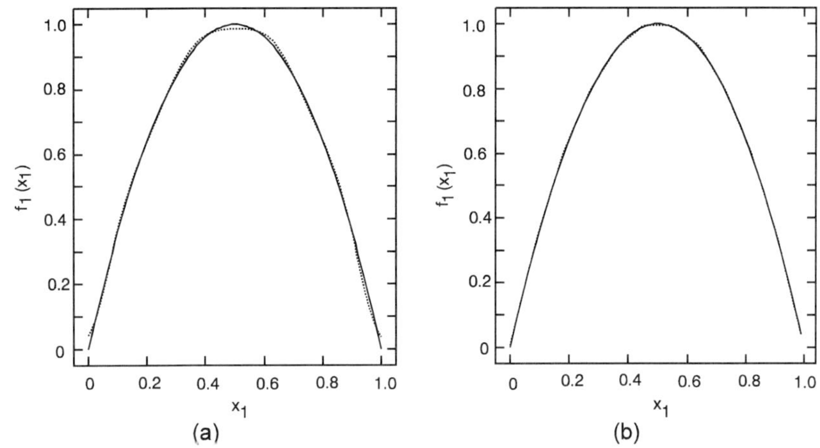

Figure 11.7. Identified nonlinear mechanism. (a) Conventional neural network (———, true nonlinear function; - - - - - - - -, identified function (NN10)); (b) neo-fuzzy neuron (———, true nonlinear function; - - - - - - - -, identified function (NFN12)).

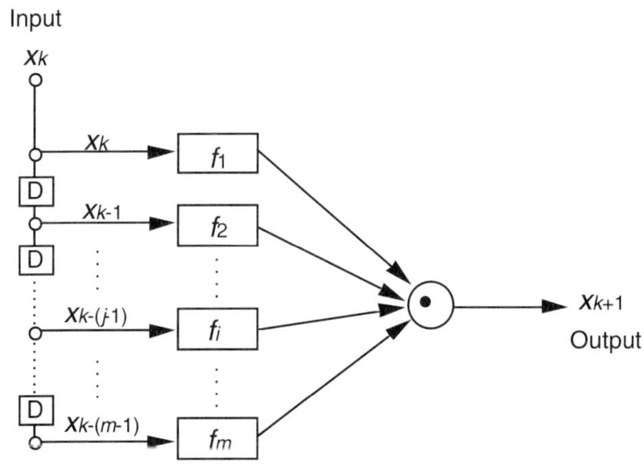

Figure 11.8. Neo-fuzzy neuron with cascaded delay elements.

$$\begin{aligned} x_{k+1} &= f_0(x_k) + f_1(x_{k-1}) + f_2(x_{k-2}) \\ &= \frac{5x_k}{1+x_k^2} - 0.5x_k - 0.5x_{k-1} + 0.5x_{k-2} \end{aligned} \tag{11-12}$$

with initial values $x_0 = 0.733$, $x_1 = 0.234$, and $x_2 = 0.973$. A sample time series pattern of a chaotic test signal generated by Equation (11-12) is shown in Figure 11.9, which is quite chaotic and thus it is very difficult to identify or predict.

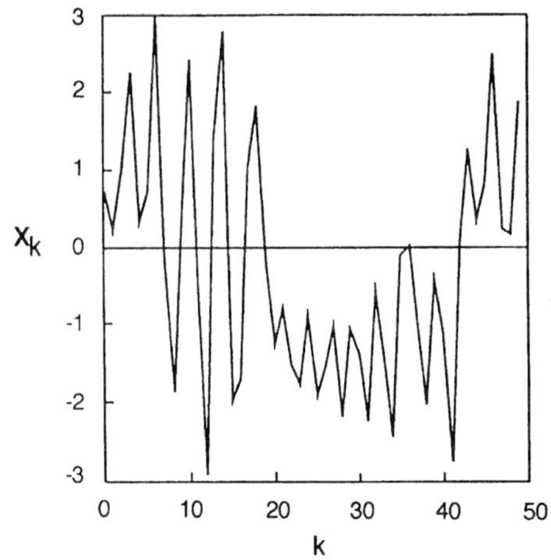

Figure 11.9. A sample chaotic test signal generated by Equation (11-12).

We have used the neo-fuzzy neuron model of Figure 11.8 with 1 to 5 nonlinear synapses (m = 1~5), in order to study how precisely they can identify the system of concern. The number of fuzzy segments in each input space x_i is 12, and thus 12 membership functions are assigned to each nonlinear synapse. The updating of weight is achieved by stepwise learning discussed in Section 11.3.2. The initial values of weights are assigned to zero, and in this experiment the learning rate is set as $\alpha = 0.2$.

The time series data of x_k (k = 0~49) are sequentially applied to the input terminal, and they are shifted downwards one by one through the cascaded delay chain. The desired output of the network is taken from the sample time series x_k. For example, in case of a 3-synapses neo-fuzzy neuron, the desired output for the input x_{k-2}, x_{k-1}, x_k, is x_{k+1}. One cycle of learning is defined as feeding a complete set of training data x_k (k = 0~49) to the neo-fuzzy neuron. Figure 11.10 shows the decay of RMS errors versus the learning cycle for a 1- to 5-synapse neo-fuzzy neuron model. It is observed that a 3-synapse neo-fuzzy neuron is not necessarily best for the 3rd order dynamical system, but a better identification is obtained by the neo-fuzzy neuron model with more synapses.

The effect of the learning rate α on the learning speed was also examined. The experimental results are shown in Figure 11.11. It is observed that when α is small, the learning speed is low; on the other hand, a big α causes an oscillatory learning. It is estimated that the optimum α exists around 0.2, and in this case the learning of 20 cycles is enough to get a reasonably small error.

The fuzzy inference with triangular membership functions in the antecedent and singletons in the consequent will exhibit a linear interpolation, and thus any nonlinear function is approximated by a piecewise linear function. This essentially means some error is produced. This will be improved in the extended neo-fuzzy neuron in Section 11.6.

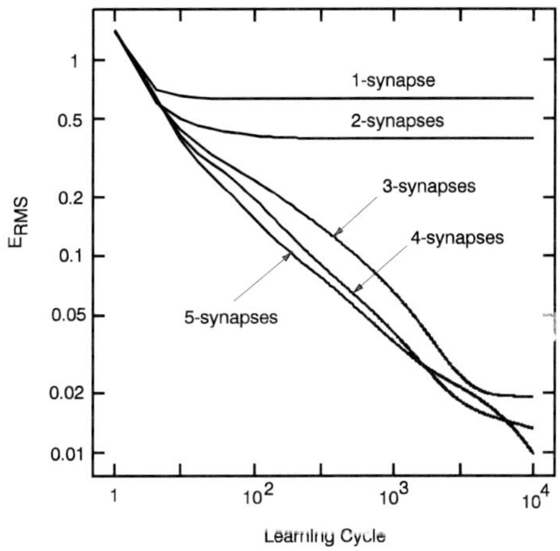

Figure 11.10. Learning performance for 1- to 5-synapse neo-fuzzy neuron model.

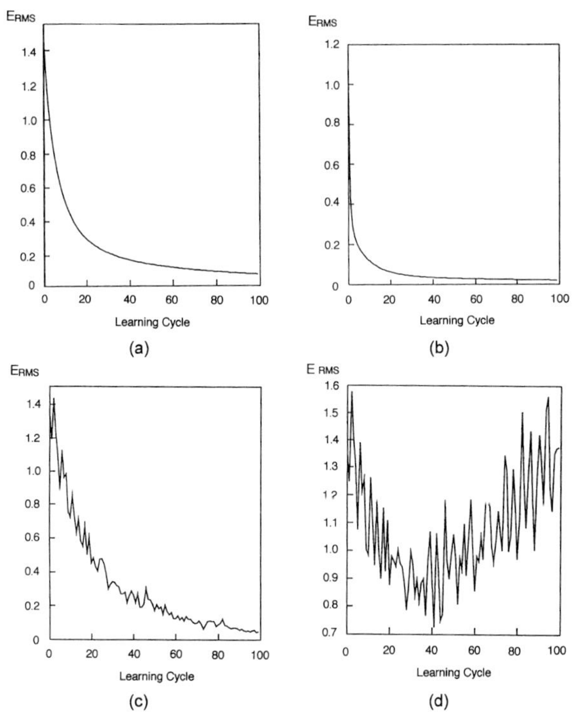

Figure 11.11 The effect of the learning rate α on the learning speed. (a) $\alpha = 0.02$; (b) $\alpha = 0.2$; (c) $\alpha = 0.56$; (d) $\alpha = 0.57$.

11.5 PREDICTION OF NONLINEAR DYNAMICAL SYSTEMS

Once the learning of the neo-fuzzy neuron with a delay chain, see Figure 11.8, has been completed, it can predict the future behavior of the system. There are several modes of prediction according to the structure of the neo-fuzzy neuron.

11.5.1 Single-Step Ahead Prediction

This is done by the network configuration of Figure 11.8. The time series sequence obtained from the unknown dynamical system of concern is applied to the input terminal. Then it goes through the delay chain to produce the network inputs x_k, x_{k-1}, ..., $x_{k-(m-1)}$. The training of this network is achieved by comparing the actual output of the network and the desired one. After training the network, the allocation of the present data to the input terminal provides the one-step-ahead prediction at the output terminal of the network. This is a single-step-ahead prediction.

11.5.2 Multi-Steps Ahead Prediction

A small modification of the network configuration of Figure 11.8 enables a multi-steps-ahead prediction. Now consider the configuration of Figure 11.12 with $p - 1$ additional delay elements. The training of this network is achieved in the same way as in the case of a single-step ahead prediction. After training, the allocation of the present data to the input of the first synapse of the neo-fuzzy neuron provides the p-steps-ahead prediction at the output terminal. This is a multi-steps-ahead prediction.

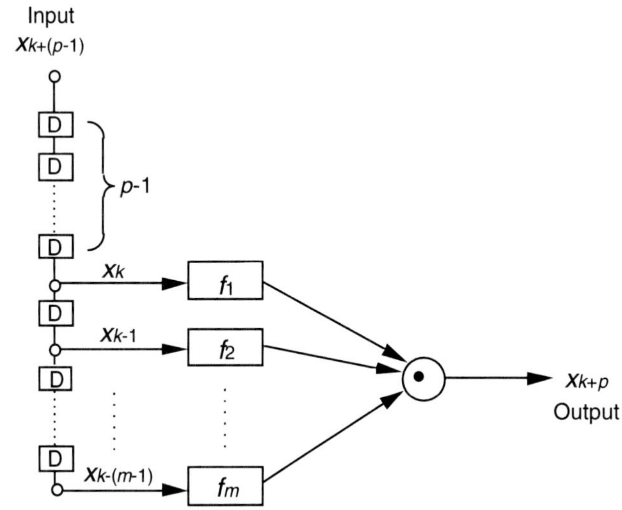

Figure 11.12. Network configuration for a multi-steps-ahead prediction.

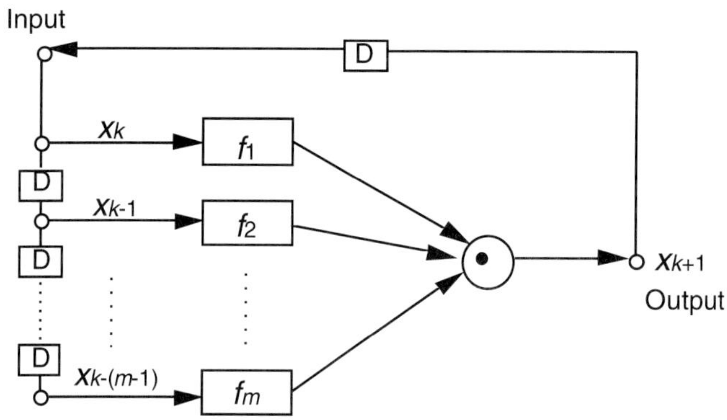

Figure 11.13 Network configuration for a sequential prediction.

11.5.3 Sequential Prediction

After training the network, if we feed back the output of the neo-fuzzy neuron direct to the input terminal, sequential prediction is achieved. Figure 11.13 shows this network configuration.

In this type of sequential prediction, the weights are fixed at the final values obtained by learning. They are not updated during the prediction. It can be easily seen that a very small modeling error can cause a large prediction error even in the short-term prediction. We dare use in Section 11.6 this type of prediction to check the modeling accuracy. This becomes a very severe check.

11.5.4 Adaptive Prediction

In order to get a good prediction in the long term, the weights of the network must be updated adaptively by the new data. A new signal data can be obtained after one-step-ahead prediction; then it can be used as the training data for the next step prediction. Every time we get a new signal data, the network is retrained and the weights are updated adaptively.

Figure 11.14 shows the experimental results of an adaptive prediction for the chaotic signal generated by Equation (11-12). The network is first trained with the past 50 data by 100 learning cycles. Then every time a new signal data comes, the network is retrained by 50 learning cycles. If the modeling accuracy is the same, this gives in general a better prediction than the sequential prediction in Section 11.5.3.

11.6 EXTENSION OF A NEO-FUZZY NEURON

In order to get a better generalization capability, a neo-fuzzy neuron is extended. Concretely, learning is made here both on the membership functions in the antecedent and on the singletons in the consequent, i.e., the shapes of the membership functions in the antecedent and the positions of the singletons in the consequent are both adjusted adaptively.

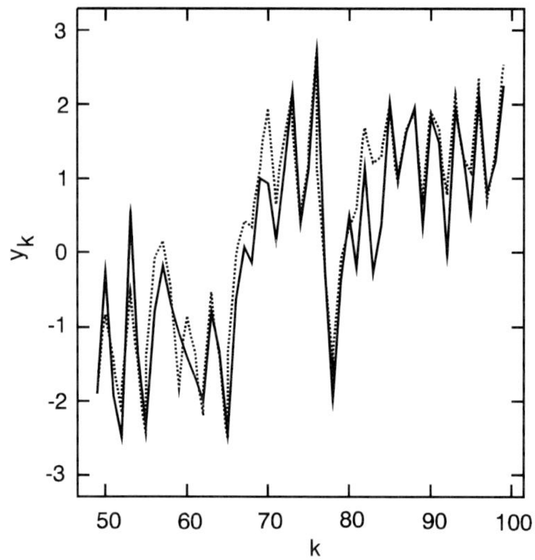

Figure 11.14 Adaptive prediction of a chaotic signal by the neo-fuzzy neuron.

11.6.1 Structure of the Extended Neo-Fuzzy Neuron

In what follows, attention is paid to a single nonlinear synapse for the ease of mathematical expression. Extension to the multi-synapse neuron is straightforward. The structure of the extended neo-fuzzy neuron is shown in Figure 11.15. In this model, a height h_i of the membership function, an x-coordinate a_i which gives the peak of the membership function in the antecedent, and a position w_i (weight) of a singleton in the consequent are adjusted adaptively by learning.

Now consider the similar error function as Equation (11-4) is minimized over the learning cycle. Then, the renewal of weight w_i and the renewal of height h_i are given as follows in the same way as Equation (11-6) is derived:

$$\Delta w_i = -\alpha(y_k - \tilde{y}_k)\frac{h_i\mu_i(x_k)}{h_i\mu_i(x_k) + h_{i+1}\mu_{i+1}(x_k)} \tag{11-13}$$

$$\Delta h_i = -\beta(y_k - \tilde{y}_k)\frac{(w_i - y_k)\mu_i(x_k)}{h_i\mu_i(x_k) + h_{i+1}\mu_{i+1}(x_k)} \tag{11-14}$$

where α and β are learning rates, which are empirically determined. Equations (11-13) and (11-14) are stepwise learning algorithm for w_i and h_i.

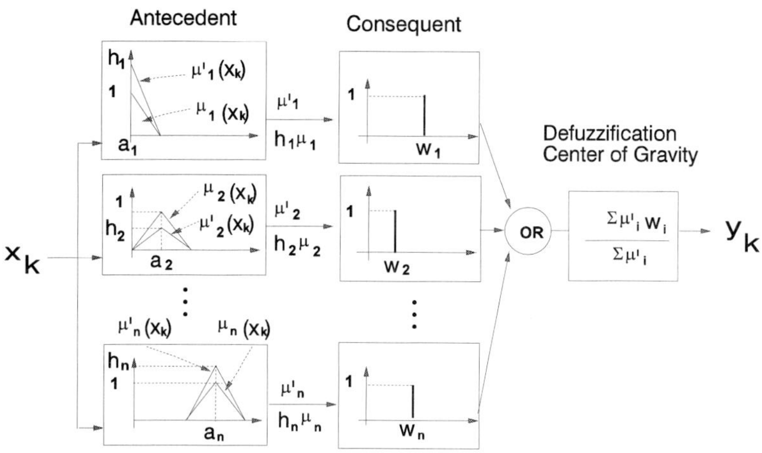

(a) Nonlinear synapse of the extended neo-fuzzy neuron.

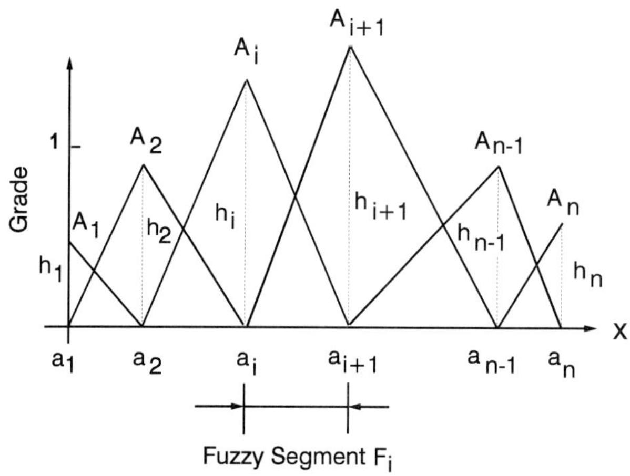

(b) Parameters assigned on the membership functions in the antecedent.

Figure 11.15. Structure of the extended neo-fuzzy neuron model.

The learning of a_i is carried out as follows. First, we define the mean error E_i at each fuzzy segment F_i ($\equiv [a_i, a_{i+1}]$) as follows:

$$E_i = \sum_{k=1}^{N_m} \frac{p(x_k)|y_k - \hat{y}_k|}{n(F_i)} \tag{11-15}$$

where $p(x_k)$ is defined as:

$$p(x_k) = \begin{cases} 0, & x_k < a_i \\ 1, & a_i \le x_k < a_{i+1} \\ 0, & a_{i+1} \le x_k \end{cases} \tag{11-16}$$

where $n(F_i)$ is the number of input x_k that hit the fuzzy segment F_i. Then we calculate the renewal of a_i by:

$$\Delta a_i = \gamma(E_i - E_{i-1}) \tag{11-17}$$

where γ is a learning rate. In this learning a_1 and a_n (lower and upper bounds of a fluctuation range of x_k) are fixed.

11.6.2 Performance of the Extended Neo-Fuzzy Neuron

In the next, the extended neo-fuzzy neuron is applied to the identification and prediction problem of a chaotic signal. The same dynamical system of Equation (11-12) with the same initial values is adopted to create a time series signal which is used for the experiment. The signal generated by Equation (11-12) is quite chaotic. Figure 11.16 shows the identified nonlinear function in each synapse in the extended neo-fuzzy neuron with 3-synapses. Figure 11.17 shows the comparisons among the prediction results by the neo-fuzzy neuron model and by the extended neo-fuzzy neuron models of Type 1 (only the height h_i of the membership function in the antecedent and the position w_i of the singleton in the consequent are adjusted adaptively) and Type 2 (in addition to the

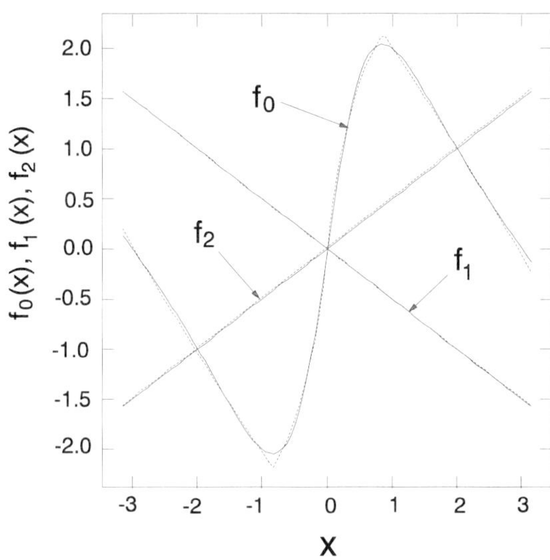

Figure 11.16 Identified nonlinear dynamical system of Equation (11-12). ———, true curve; · · · · · · · ·, identified curve.

adaptation of h_i and w_i, the position a_i of the membership function in the antecedent is further adjusted adaptively). We can see from these results that the prediction ability of the Type 2 model is excellent.

Space limitation precludes inclusion of the practical applications of the extended neo-fuzzy neuron. For example, in [13] an application to the identification problem of the acoustic sound insulation system can be found. In [14] an application to the medical image processing can be found. In this application, the positions of the landmarks in the roentgenographic cephalogram, which are used for the actual orthodontic treatments, have been identified with good accuracy. The applications to other actual problems are now under study.

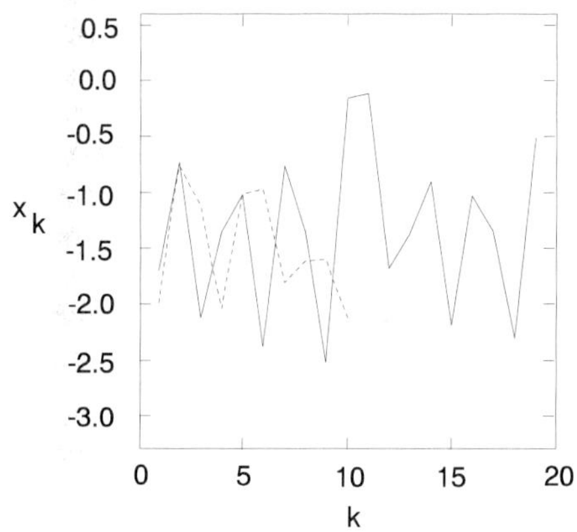

(a) Prediction results by the neo-fuzzy neuron model.

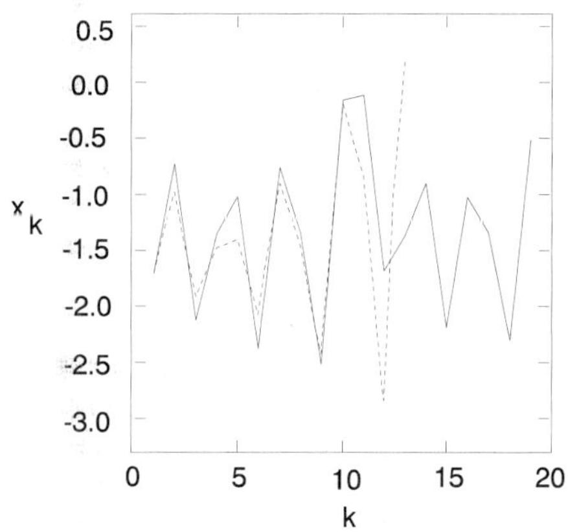

(b) Prediction results by the extended neo-fuzzy neuron model of Type 1.

Figure 11.17. Comparisons of prediction results.
———, true chaotic signal; ········, prediction by the respective model.

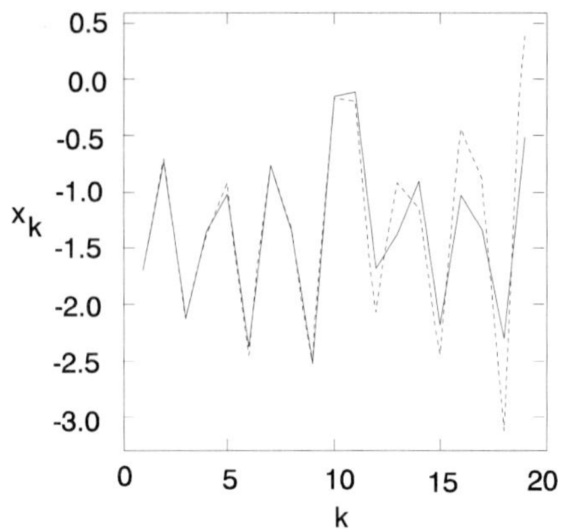

(c) Prediction results by the extended neo-fuzzy neuron model of Type 2.

Figure 11.17 (continued)

11.7 CONCLUSIONS

A structure and some learning algorithms of a neo-fuzzy neuron have been described in this chapter. The remarkable features of this neo-fuzzy neuron are (i) the nonlinearity is shifted into the synapse, i.e. nonlinear synapse, in order to get a high description capability only by single-layer networks; (ii) the nonlinearity of the synapse is realized by a set of fuzzy linguistic if-then rules, which gives a flexibility of description; (iii) the convergence to the global minimum is guaranteed in the error-weight space under arbitrary initial conditions because of the single-layer network structure; (iv) it shows a drastic improvement in learning speed and accuracy compared with the conventional neural networks.

The neo-fuzzy neuron has been extended in order to get a better performance. It has been successfully applied to the identification problem of a nonlinear dynamical system. It can learn the structure of the dynamical system very well, and once the learning has been completed, the future behavior of the system can be predicted well for a practical use, even though it is quite chaotic.

11.8 ACKNOWLEDGMENTS

Many thanks are due to H. Kusanagi and T. Kubo for their helpful assistance in this study.

11.9 REFERENCES

1. Lee, S. C. and Lee, E. T., "Fuzzy Neurons and Automata." *Proc. of the 4th Princeton Conference on Information Science and Systems*, pp. 381-385, 1970.

2. McCullough, W. S. and Pitts, W. "A Logical Calculus of the Ideas Immanent in Nueral Nets," *Bulletin of Mathematical Biophysics*, Vol. 5, pp. 115-137, 1943.

3. Yamakawa, T. and Tomoda, S. "A Fuzzy Neuron and Its Application to Pattern Recognition," *Proc. of the Third IFSA Congress*, pp. 30-38, 1989.

4. Yamakawa, T. "Pattern Recognition Hardware System Employing a Fuzzy Neuron," *Proc. of the International Conference on Fuzzy Logic & Neural Networks*, pp. 943-948, 1990.

5. Yamakawa, T. and Furukawa, M. "A Design Algorithm of Membership Functions for a Fuzzy Neuron Using Example-Based Learning." *Proc. of IEEE International Conference on Fuzzy Systems*, pp.75-82, 1992.

6. Driankov, D., Hellendoorn, H. and Reinfrank, M. *An Introduction to Fuzzy Control*, Springer-Verlag, 1993.

7. Zurada, J. M. *Introduction to Artificial Neural Systems*, West Publishing, 1992.

8. Nakano, K. *An Introduction to Neurocomputing*, Corona, 1991 (in Japanese).

9. Yamakawa, T., Kusanagi, H., Uchino, E. and Miki, T. "A New Effective Learning Algorithm for a Neo Fuzzy Neuron Model." *Proc. of the Fifth IFSA World Congress*, pp. 1017-1020, 1993.

10. Korn, G. A. and Korn, T. M. *Mathematical Handbook for Scientists and Engineers*, McGraw-Hill, 1968.

11. Rumelhart, D. E. and McClelland, J. J. PDP Research Group. *Parallel Distributed Processing Vol.* 1. MIT Press, 1986.

12. Kawakami, H., *Chaos CG Collection*, Science, 1990 (in Japanese).

13. Uchino, E., Yamakawa, T. and Kubo, T. "An Extension of a Neo Fuzzy neuron Model and Its Applications to Signal Prediction." *Proc. of the Second European Congress on Intelligent Techniques and Soft Computing*, pp. 484-488, 1994.

14. Uchino, E. and Yamakawa, T. "High Speed Fuzzy Learning Machine with Guarantee of Global Minimum and Its Applications to Chaotic System Identification and Medical Image Processing." *Proc. of the 7th IEEE International Conference on Tools with Artificial Intelligence*, to appear, November 1995.

12

SLIP CONTROL AND TORQUE OPTIMIZATION USING FUZZY LOGIC

R. Palm and K. Storjohann
Siemens AG Corporate Research and Development
GERMANY

A crucial point for modern electrically driven locomotives is the optimal force transmission between wheel and rail. By means of computerized locomotive control a more or less optimal adjustment of the slip and, with this, the friction force between wheel and rail becomes possible. In this chapter a torque maximizing algorithm for optimal adjustment of the slip based on fuzzy logic is presented. The method has been tested by means of a model of the vehicle including the nonlinear friction force between wheel and rail. In addition to this, the slip control proper has been performed using a fuzzy controller. The simulations show that, compared with conventional techniques, both the fuzzy torque maximizer and the fuzzy slip controller included cope better with disturbances, parameter fluctuations, and the problem of limit cycles.

12.1 INTRODUCTION

For modern electrically driven locomotives a force transmission between wheel and rail takes place because of the slip defined as the deviation between the tangential velocity of the wheel at the point of contact between wheel and rail and the velocity of the vehicle. Thus, the actual driving force of the vehicle is the friction force acting between wheel and rail. The friction force is in balance with the forces at the wheel and to the vehicle. Further, there is an equilibrium between the forces at the wheel and the motor shaft in which the driver introduces the motor torque. The friction force is determined by the weight of the vehicle and the friction μ which is dependent on the slip *s*. Generally, the function μ(s) is nonlinear and mainly determined by the parameters *velocity of the locomotive* and *quality of the wheel-rail conditions*. In the ideal case the slip-friction curve has either a friction maximum or a plateau. However, real measurements showed that such ideal conditions never appear. Instead, the curves are strongly disturbed and, because of the nonhomogeneous conditions on the track, they represent mixtures of the ideal formes [1,2]. These mixtures originate from a drastic change in the conditions on the track e.g. the change from a dry rail in a tunnel to a wet rail outside the tunnel. Therefore, ideal curves represent averages originating from regression analyses with regard to measured μ (s)-curves.

The friction curve can be classified as a stable and an unstable region (see Figure 12.1). For the stable region, the rising branch of the curve, the noncontrolled dynamical system vehicle rail, is always stable if the required motor torque in addition to the forces of inertia, damping forces and spring forces do not exceed the maximum of the friction force. One has to notice that two slip values belong to one specific friction force. Nevertheless, for the stationary case the stable state, appears, namely the slip with the smaller absolute value. However, if the increase of the motor torque disturbs the force balance the system increases the slip producing a friction force which is able to stabilize the system again. However, such an action can lead the system into the unstable region in which the increase of the friction force by rising the slip fails. The opposite is the case: the friction force decreases (friction characteristic with maximum) or keeps constant (characteristic with plateau). Drastic slip changes can be always expected. as the vehicle changes its velocity These drastic changes have to be avoided because of the following reasons:

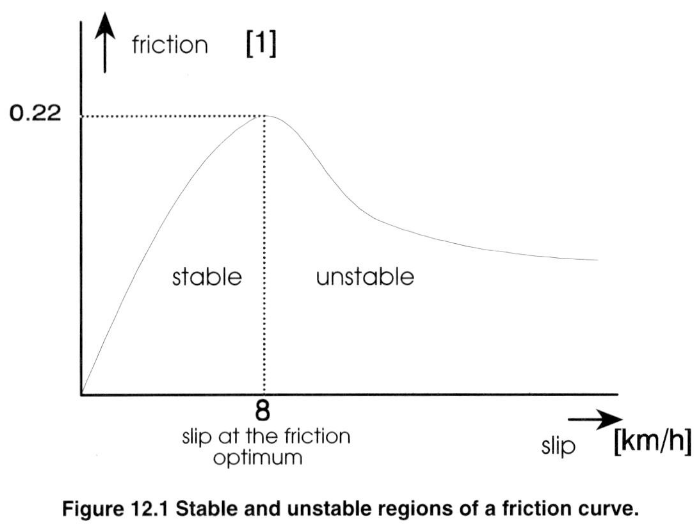

Figure 12.1 Stable and unstable regions of a friction curve.

1. avoidance of high mechanical stress
2. reduction of high energetic stress of the driving motor
3. optimization of the force transmission from wheel to rail.

The last point, the optimal force transmission, is the subject of torque optimization. The aim is to adjust a particular slip for which the force transmission is optimal. This optimum is reached for the maximum of the friction characteristic if such a maximum exists.

The control of the slip includes two tasks:

1. Adjustment and control of the slip according to a reference slip value (desired slip)
2. Search for an optimal reference slip value that makes the optimal force transmission between wheel and rail possible.

The vehicle control is performed in such a way that the noncorrected driving torque F_1^* is given either by the driver or automatically by a higher hierarchy level LZB whose decisions originate from a desired train velocity v_{des} based on a schedule [3]. Within block AFB, v_{des} is compared with the actual velocity v_{act} where a noncorrected driving torque F_1^* is produced by means of a controller in AFB (see Figure 12.2).

The slip control is embedded in the total control hierarchy such that F_1^* is corrected by the output ΔF of the slip controller. The corrected driving torque $F^* = F_1^* + \Delta F$ is fed to the vehicle. Performing the correction of F_1^* one has to ensure that for the acceleration phase the condition $\Delta F \leq 0$ and for deceleration $\Delta F \geq 0$ holds. This means that the absolute value of the driving torque F_1^* must not be exceeded.

Slip optimization or torque maximizing with the help of conventional means has already been described [2,8]. The main disadvantage of the conventional methods described is the constant width of the searching step and, with this, the occurrence of limit

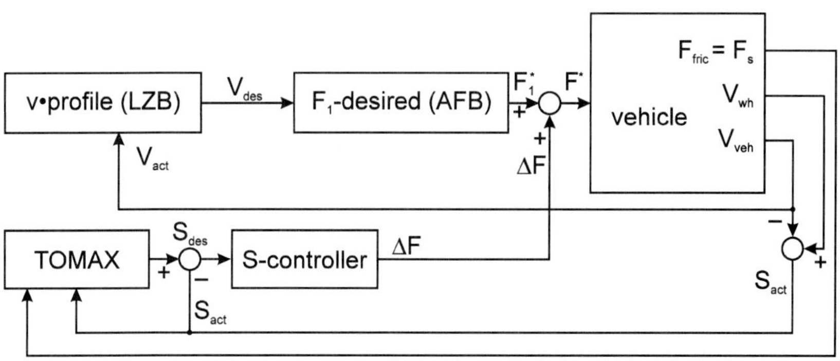

Figure 12. 2 Block scheme of slip control.

cycles near the friction maximum. On the other hand, conventional methods do not cope satisfactorily with disturbances and parameter fluctuations.

Therefore, a method is required which is able to adapt the width of the searching step and which copes better with disturbances and parameter changes.

Such a method has been developed by using an algorithm on the basis of fuzzy rules [5].

The slip control proper has been performed by sliding mode fuzzy control [4,6].

12.2 PROCESS MODELS

12.2.1 Model of the Vehicle Dynamics

The modeling is performed in two steps: First, the vehicle dynamics is modeled in terms of differential equations. Second, the wheel rail process is modeled mainly in terms of the friction curve. In a first model the shaft between motor and wheel has been assumed to be stiff (see Figure. 12.3).

On the other hand, modeling finite torsion of the shaft is more realistic and contributes essentially to the dynamics of the whole process. That is why a second model has been built which takes into account the torsion between motor and wheel. For the model the following mathematical symbols have been used:

U = motor torque
$\dot{\varphi}_a$ = angle velocity of the motor
C_a = torsion spring of the shaft
D_a = damping of motor
J_a = moment of inertia of the shaft
$\dot{\varphi}_r$ = angle velocity of the wheel
D_r = damping of the wheel
J_r = moment of inertia of the wheel

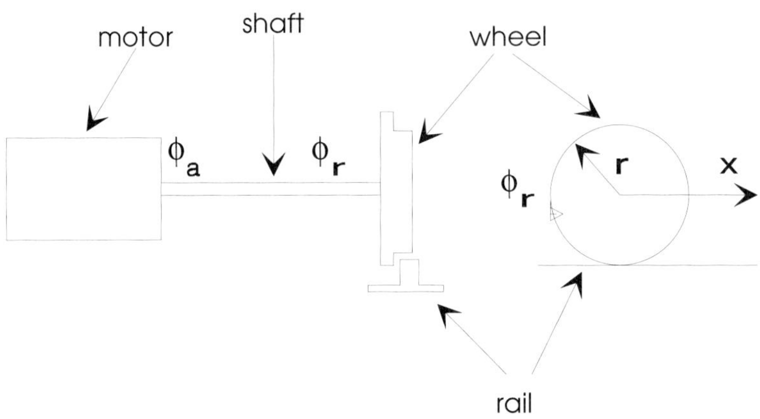

Figure 12.3 Model of the vehicle dynamics.

C_x = wind force parameter
m = mass of the vehicle
g = gravitation constant
$h(x)$ = profile of the track
r = radius of the wheel
μ = nonlinear friction coefficient

12.2.2 Model of the Vehicle Dynamics without Torsion between Motor and Wheel

For a stiff shaft ($\varphi_a = \varphi_r$) the transmission of the torque onto the rail takes place through the slip exclusively. Therefore, the balance equation yields

Motor torque = damping of the motor + damping of the wheel
+ slip force * radius of the wheel

where

Slip force = wind force + track slope force
+ inertia force of the vehicle

The analytical formulation of the force-torque-balance yields

$$\begin{aligned} U &= (D_a + D_r)\cdot\dot{\varphi}_a + (J_a + J_r)\cdot\ddot{\varphi}_a + mgr\mu\cdot(r\varphi_r - \dot{x}) \\ mg\mu\cdot(r\varphi_a - \dot{x}) &= mg\sin(\arctan\frac{dh(x)}{dx}) + C_x\dot{x}^2 + m\ddot{x} \end{aligned} \qquad (12\text{-}1)$$

12.2.3 Model Of The Vehicle Dynamics With Torsion Between Motor And Wheel

In reality a shaft is a compliant mechanical element and can be modeled as a torsion spring. The transmission of the torque onto the rail takes place both through the spring force and the slip force. Then the balance equation yields:

Motor torque = damping of the motor + inertia force of the motor shaft
+ spring force of the motor shaft

Spring force of the motor shaft = damping of the wheel + inertia force of the wheel
+ slip force * radius of the wheel

Slip force = wind force + track slope force + inertia force of the vehicle

$$
\begin{aligned}
&U = D_a \cdot \dot{\varphi}_a + J_a \cdot \ddot{\varphi}_a + C_a \cdot (\varphi_a - \varphi_r) \\
&C_a \cdot (\varphi_a - \varphi_r) = D_r \cdot \dot{\varphi}_r + Jr \cdot \ddot{\varphi}_r + mgr\mu \cdot (r\varphi_r - \dot{x}) \\
&mg\mu \cdot (r\varphi_r - \dot{x}) = mg\sin(\arctan\frac{dh(x)}{dx}) + C_x\dot{x}^2 + m\ddot{x}
\end{aligned}
\tag{12-2}
$$

The torque f_s that drives the vehicle is the result of its weight mg and the slip-dependent friction $\mu(s)$:

$$f_s = mg\mu(s) \tag{12-3}$$

12.2.4 Model of the Friction Curve

The friction curve describes the nonlinear relation between friction and slip where the friction is proportional to the driving force transmitted to the rail (see Figure 12.4).

The friction curve has been modeled by a parabola and an exponential function: the parabola $\mu_r(s)$ describes the curve for $s < s_u$ and the exponential function $\mu_r(s)$ describes the curve for $s \geq s_u$:

$$\mu(s) = \mu_l(s) = a_l \cdot s^2 + b_l \cdot s \tag{12-4}$$

for $s < s_u$

$$\mu(s) = \mu_r(s) = e^{a_r \cdot s + b_r} + c_r \tag{12-5}$$

for $s \geq s_u$.

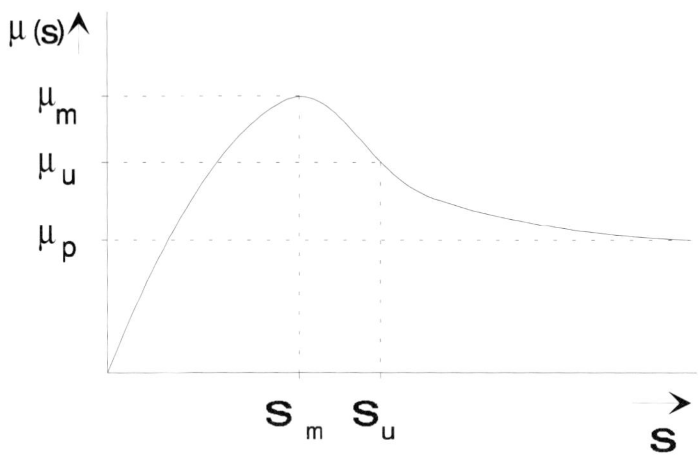

Figure 12.4 Modeling of the friction curve.

The parameters a_r, b_r and c_r are determined by the point $\mu_u = \mu(s = s_u)$, the height of the plateau $\mu_p = \lim_{s\to\infty} \mu(s)$ and the slope of the exponential function $\mu_r(s)$. To obtain a smooth transition in $\mu_u = \mu(s = s_u)$ the slopes of $\mu_r(s)$ and $\mu_l(s)$ are set to be equal.

12.3 SLIP CONTROL

12.3.1 Slip Control Using a Conventional PI Controller

The inputs of the conventional PI slip controller are the actual slip s_{act} and the desired slip s_{des}. The output is a correction $u = \Delta F$ of the driving force . It has to be stressed that any correction $u = \Delta F$ must not exceed the originally applied driving force F_1^* $|F^* + \Delta F \le F_1^*|$ from which follows that the correction $u = \Delta F$ has to be negative or zero for the acceleration phase and positive or zero for the deceleration phase. The PI control law is

$$\Delta F = K_P \cdot e + K_I \cdot \int_0^t e \cdot dt' \tag{12-6}$$

where $e = s_{des} - s_{act}$.

12.3.2 Slip Control Using a Sliding Mode Fuzzy Controller (SMFC)

The principle of a SMFC is similar to that of a Sliding Mode Controller with boundary layer [7,4,6].

The control characteristic is based on the input $s = \lambda \cdot e + \dot{e}$ where λ is the slope of the so-called switching surface in state space (see Figure 12.5) and the control law is

$$\Delta u = -K \cdot sat(s/\Phi) \tag{12-7}$$

with K, positive constant; Φ, width of the boundary layer (BL); and $sat(x) = \begin{cases} x & \text{if } |x| < 1 \\ 1 & \text{if } |x| \ge 1 \end{cases}$.

Figure 12.6 shows the regarding $\Delta u/s$ - characteristic.

In our case, the control result Δu has to be integrated so that the control law (12-7) turns into

$$u = -K \cdot \int_0^t sat(s(t')/\Phi)dt'$$

For the SMFC $e, \dot{e}$ and the control variable Δu are normalized with respect to a common standard interval $e_N, \dot{e}_N, \Delta u_N \in (a,b)$:

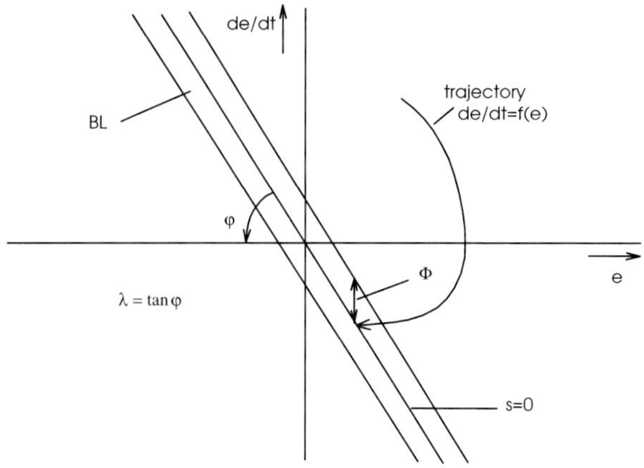

Figure 12.5 Sliding line in state space.

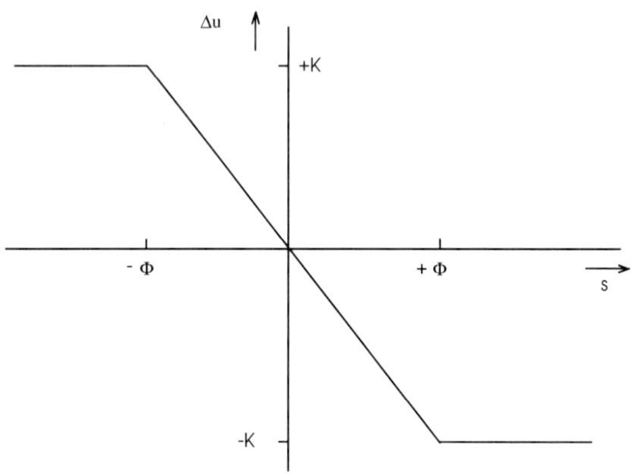

Figure 12.6 SMC with boundary layer.

$$\begin{aligned} e_N &= N_e \cdot e \\ \dot{e}_N &= N_{\dot{e}} \cdot \dot{e} \\ \Delta u_N &= N_u \cdot \Delta u \end{aligned} \tag{12-8}$$

Then, we compute $s_N = e_N + \dot{e}_N = 0$ in the normalized state space using the linguistic variables: VL, Very Low; L, Low; M, Medium; H, High; VH, Very High. We define the following linear fuzzy rules

$$\begin{aligned}
&IF\ |s_N|\ \text{is}\ VL\ \text{THEN}\ K_{FUZZN}\ \text{is}\ VL\\
&IF\ |s_N|\ \text{is}\ L\ \text{THEN}\ K_{FUZZN}\ \text{is}\ L\\
&IF\ |s_N|\ \text{is}\ M\ \text{THEN}\ K_{FUZZN}\ \text{is}\ M\\
&IF\ |s_N|\ \text{is}\ H\ \text{THEN}\ K_{FUZZN}\ \text{is}\ H\\
&IF\ |s_N|\ \text{is}\ VH\ \text{THEN}\ K_{FUZZN}\ \text{is}\ VH
\end{aligned} \tag{12-9}$$

The output of the FC is $K_{FUZZ\,N} \geq 0$ which increases its value with a growing distance $|s_N|$ of the state $\vec{e}_N = (e,\dot{e})_N^T$ from $s_N = e_N + \dot{e}_N = 0$. After that, the output $K_{FUZZ\,N}$ has to be normalized in order to obtain a physical control signal

$$K_{FUZZ} = K_{FUZZ\,N} \cdot N_u \tag{12-10}$$

The sign of the controller output Δu is obtained from the location of $\vec{e}_N = (e,\dot{e})_N^T$ in state space depending on whether $\vec{e}_N$ is laying below or above the switching line $s_N = 0$. To avoid drastic changes of the control variable Δu in the neighborhood of $s_N = 0$ an additional BL with the width $|s_N| = \Phi$ is introduced in which the controller works linearly. In addition to the fuzzy term a compensation term $-\lambda \cdot \dot{e}$ is introduced with $\lambda = N_e / N_{\dot{e}}$.

The final structure of the controller is

$$u = -\int_0^t \left(\lambda \cdot \dot{e} + K_{FUZZ}\, sat\left(s(t')/\Phi\right)\right)dt' \tag{12-11}$$

The fuzzy part K_{FUZZ} of the controller was designed in the normalized interval $|s_N|, K_{FUZZ\,N} \in (0,255)$ independent of the physical domain. The scaling factors $N_e, N_{\dot{e}}, N_u$ serve for the tuning of the controller with respect to a desired behavior of the system to be controlled. Figure 12.7 shows the characteristic of the control part $\Delta u = -K_{FUZZ}\, sat(s/\Phi)$.

12.4 TORQUE MAXIMIZER (TOMAX) ON THE BASIS OF FUZZY RULES

12.4.1 Conventional TOMAX

A conventional torque optimizer (or torque maximizer) checks out on which branch of the friction curve the system is working (see Figure 12.8). If the system reacts with a positive (or negative) change of the friction force Δf_s when a positive (or negative) slip correction Δs is introduced (formally: if $\text{sgn}(\Delta f_s) = \text{sgn}(\Delta s)$, then the system is working on the stable branch of the friction curve. A positive correction of the desired slip value *corrslip* leads the system closer to the friction maximum μ_m. This *corrslip* has to be added to the previous desired slip s_{des}. Otherwise, if the signs of the slip correction *corrslip* and the change of the friction force Δf_s are different, then the system is working on the

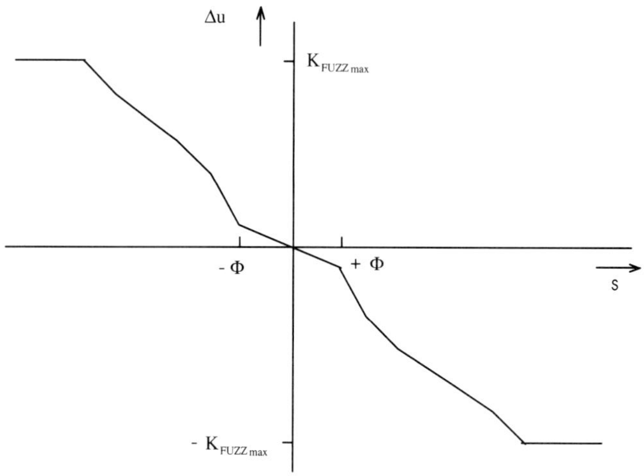

Figure 12.7 Characteristic of the SMFC.

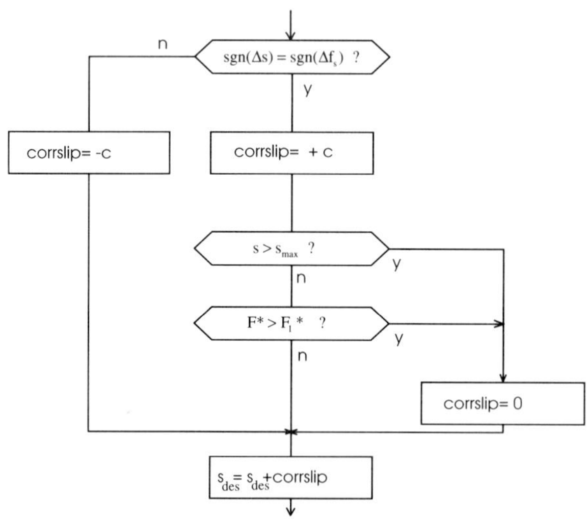

Figure 12.8 Conventional TOMAX.

unstable branch of the friction curve. Thus, in order to bring the system closer to the friction maximum, a negative *corrslip* has to be applied. This is valid for the acceleration phase. For the deceleration phase the sign of *corrslip* has to be changed. If either a predefined slip maximum or a torque maximum is reached, then *corrslip* has to be set to zero. The conventional TOMAX is characterized by the fact that it only evaluates the signs of Δs and Δf_s with a constant change of the desired slip: *corrslip* $= \pm c$. This leads automatically to limiting cycles at the maximum of the friction curve.

12.4.2 Fuzzy TOMAX

Following the principle of the conventional TOMAX a fuzzy TOMAX has been developed in which Δs, Δf_s, and *corrslip* are fuzzy variables. This leads not only to an evaluation of the signs but also to the values of the respective variables. The input variables Δs and Δf_s and the output variable *corrslip* are scaled (normalized) in order to be independent of the real system in the design phase.

The normalization is computed by

$$\Delta s_N = slip_{scale} \cdot \Delta s$$
$$\Delta f_{s_N} = f_{s_{scale}} \cdot \Delta f_s \tag{12-12}$$

$$corrslip_N = corr_{scale} \cdot corrslip \tag{12-13}$$

For Δs_N, Δf_{s_N}, and $corrslip_N$ the following attributes have been chosen: PB, Positive Big; PS, Positive Small; Z, Zero; NS, Negative Small; NB, Negative Big.

The fuzzy TOMAX has been designed with the help of an appropriate rule base (see Figure 12.9) following from two simple design rules:

1. The sign of *corrslip* has been chosen as for the conventional TOMAX.

2. The absolute value of $corrslip_N$ should increase for increasing values of $|\Delta s_N|$ and $|\Delta f_{s_N}|$, respectively.

Δs \ Δfs	NB	NS	Z	PS	PB
PB	NB	NS	Z	PS	PB
PS	NS	NS	Z	PS	PS
Z	Z	Z	Z	Z	Z
NS	PS	PS	Z	NS	NS
NB	PB	PS	Z	NS	NB

corrslip

Figure 12.9 Fuzzy TOMAX

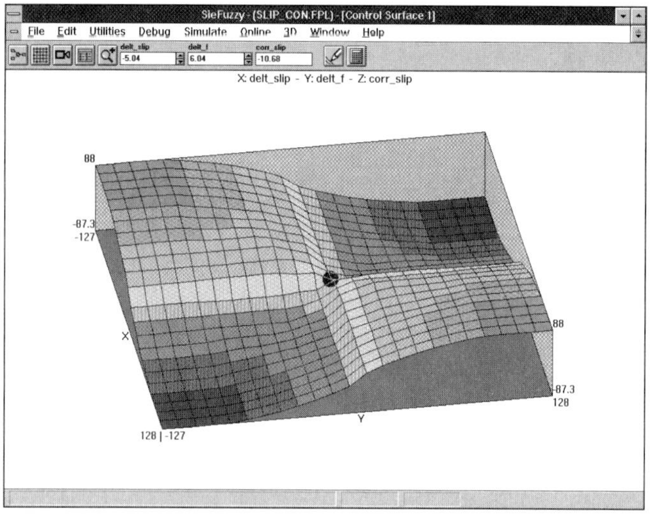

Figure 12.10 TOMAX characteristics.

The result is a rule base with 25 rules. The membership functions are chosen such that the resulting three dimensional control surface $corrslip_N = f(\Delta s_N,\ \Delta fs_N)$ is steep near the boundaries of the universe of discourse and flat in the middle of it (see Figure 12.10).

12.4.3 Structure of System

Figure 12.11 shows the block scheme of the train simulator INTERSIM. The train simulator runs on a PC whereas the slip controller and torque maximizer (TOMAX) run on a different control unit SIBAS. The train simulator is based on the dynamic Equation (12-1).

The train model, i.e., the state equations (procedure MOST1) and the friction curve (function RWKL), represent the process to be controlled. Both of them have to be processed in a quasi analog way which means a high sample rate (sample time: 1ms).

The display output is updated every 10 ms (procedure OUT).

12.4.4 Data Transmission

The SIBAS control algorithm is updated every 20 ms by the procedure SAMPLE1. The data transmission between INTERSIM and SIBAS is done by polling (see Figure 12.12). INTERSIM provides SIBAS the actual slip s_{act} and the driving torque f_s. SIBAS returns the required change of the driving torque ΔF to reach the desired slip s_{des}. Normally, the torque maximizer has to be activated at a slower rate than the slip controller. Therefore, the slip controller runs every 20 ms whereas TOMAX is activated only every 100 ms.

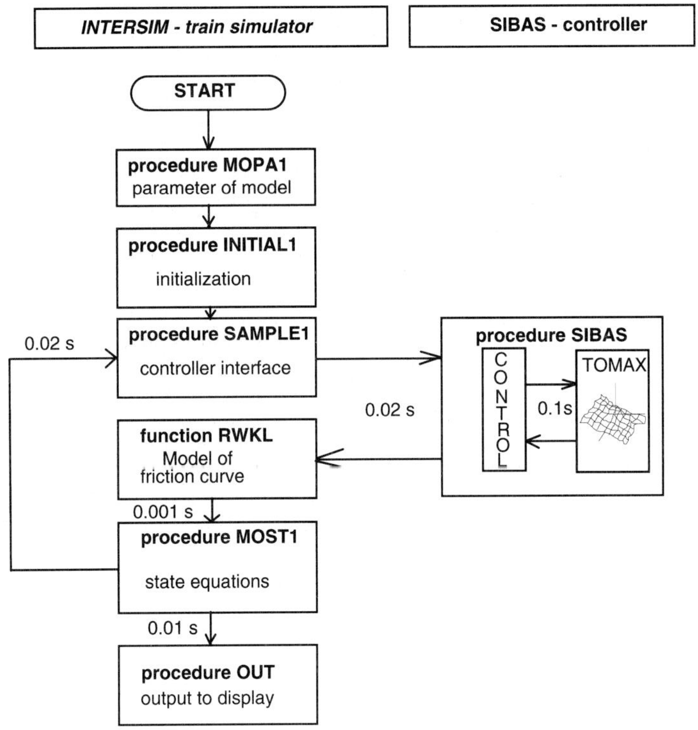

Figure 12.11 Block scheme of the train simulator.

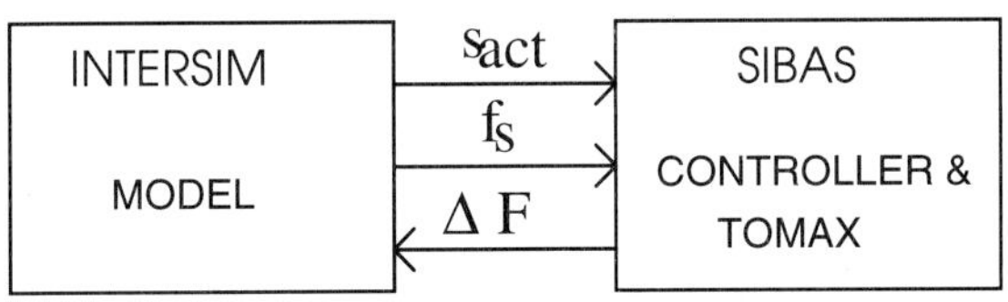

Fig12-12 Data transmission between INTERSIM and SIBAS.

12.5 ADJUSTMENT OF SLIP CONTROLLER AND TOMAX

12.5.1 Generation of Lookup Tables for the Slip Controller and TOMAX

In order to save memory and computing time both the slip controller and TOMAX have been changed from programs into corresponding look-up tables.

The generation of look-up tables requires the following steps (see Figure 12.13):

- Generation of the files **slip_con.til** for the controller and **search.til** for TOMAX by means of the fuzzy tool TILShell (SIEFUZZY)
- Translation of the two files into **slip_con.c** and **search .c**, respectively.

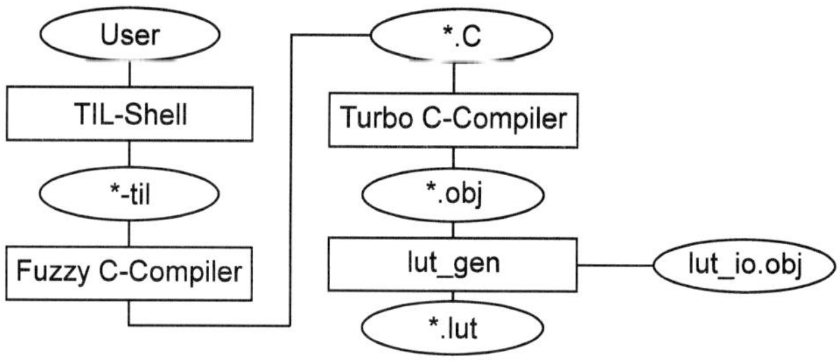

Figure 12.13 Generation of the look-up tables.

- Generation of the regarding object codes **slip_con.obj** and **search.obj.**
- From these procedure **lut_gen.exe** provides the one-dimensional look-up table **slip_con.lut** for the slip controller and the two-dimensional look-up table **search.lut** for TOMAX.

12.5.2 Adjustment of the Slip Controller

The adjustment of the conventional and fuzzy slip controller has been carried out by applying a step function to the input of the control loop.

PI controller. Neglecting the integral term for the time being, the proportional parameter K_P of the PI controller was adjusted such that the step response showed no overshoot. The remaining error accepted was e = 0.3 *m/s*. The corresponding result was

$$K_P = -5000 \;\; kg/s$$

Then, the integral parameter K_I was increased until the actual slip s_{act} reached the set point with a maximum overshoot of 10%. The corresponding result was

$$K_I = -100 \;\; kg/s$$

SMFC. For the adjustment of the slip controller four parameters are of importance: the scaling factors $N_e, N_{\dot{e}}, N_u$, and the width Φ of the boundary layer.

Scaling factor N_e. The choice of N_e depends on the normalized s-domain. In our case, *e* has to match the domain [0, 255]. On the other hand, a frequent absolute value for the nonnormalized error *e* was lying between 0 and 0.2 m/s. Then, error *e* was normalized such that, in the case of $\dot{e} = 0$, an error *e* > 0.2 m/s would have driven the controller to its limits. From this followed a scaling factor N_e = 1500 s/m.

Scaling factor $N_{\dot{e}}$. The scaling of the error velocity $\dot{e}$ is closely connected with the scaling of *e* via the slope λ of the switching line. The corresponding relation is

$$\lambda = N_e / N_{\dot{e}} \tag{12-14}$$

We chose $\lambda = 25\ 1/s$ so that all disturbing frequencies greater than 25 1/s are to be filtered out.

From this we obtained

$$N_{\dot{e}} = N_e / \lambda = 1500\frac{s}{m} / 25\frac{1}{s} = 60\frac{s^2}{m}$$

Scaling factor N_u. The parameter N_u has great influence on the overall gain of the system to be controlled. On the other hand, the control loop involves a large number of nonlinearities which cannot simply be modeled analytically. For example, a control output $u = \Delta F$ may not lead to an overall driving force F^* being greater than the driving force F_1^* applied by the driver. Therefore, the sum of control outputs $u = \Delta F$ has to be bounded. Therefore, we tried to avoid an extensive analysis of this sort of problem. Instead, we observed the step response of the system and changed the parameter N_u heuristically as long as the control result led to an acceptable rise time together with a minimum overshoot. With the scaling factor

$$N_u = 0.3N$$

we obtained good simulation results.

Boundary layer Φ. The width Φ of the boundary layer was chosen approximately as 1/5 of the normalized domain [0, 255] in order to obtain a linear control behavior in the close vicinity of the origin of the control characteristic. Hence, we chose Φ = 40.

12.5.3 Adjustment of TOMAX

Conventional TOMAX. By experiment it has been found that a constant **corrslip = 0.01 m/s** is an acceptable choice for the searching process. Greater values lead to a faster adaptation but to large amplitudes of the limit cycles. On the other hand, for smaller values of **corrslip** the searching process was too slow.

Fuzzy TOMAX

Scaling factors $slip_{scale}$ and $f_{S\,scale}$. The scaling factors $slip_{scale}$ and $f_{S\,scale}$ are determined in similarity to the way as scaling factors for the slip error are determined. The change of slip Δs and the change of force Δf_s are in the interval [−128, 127]. It has been found that slip changes Δs in the vicinity of the maximum of the friction curve are in the interval [−0.075, +0.075]. For Δf_s we found the corresponding interval [−20, +20]. From this we obtained the scaling factors

$$slip_{scale} = 1700 \ \ \mathrm{s/m}$$

$$f_{S\,scale} = 6 \ \ 1/N$$

The scaling factor for the slip correction *corrslip* works as gain in the outer slip correction loop overlying the inner slip control loop. Using a $corr_{scale} \in [0.1, 0.4]$ an acceptable searching process on the friction curve could be found. Finally, we chose the scaling factor

$$corr_{scale} = 0.14 \ \ \mathrm{m/s}$$

Filtering. In order to smooth out disturbances, low-pass filters were used for the actual slip s_{act} and the force changes Δf_s. The regarding filter equations are

$$s_{act\,filt}(t_{k+1}) = s_{act\,filt}(t_k)\cdot(1 - filt_s) + s_{act}(t_k)\cdot filt_s$$

and

$$\Delta f_{S\,filt}(t_{k+1}) = \Delta f_{S\,filt}(t_k)\cdot(1 - filt_f) + \Delta f_S(t_k)\cdot filt_f$$

with

$$filt_s = 0.6,$$

$$filt_f = 0.9$$

12.6 SIMULATION RESULTS

We used the following parameters for simulation:

F_1^* = motor torque (2250 N)
D_r = damping of the wheel (4 kg m/s)
J_r = moment of inertia of the wheel (120 kg/m)
C_x = wind force parameter (6 kg/m)
m = mass of the vehicle (11100 kg)
g = gravitation constant (9.81 m/s^2)
r = radius of the wheel (1 m)

To evaluate the fuzzy method both the fuzzy slip controller and the fuzzy TOMAX were compared with corresponding conventional ones. In addition to this, different fuzzy TOMAX characteristics were produced by changing the input membership functions.

Several tests were performed using the following features:

1. Constant friction curve with a maximum.

2. Change of the slip optimum with constant friction value.

3. Stochastic change of slip optimum and friction value.

4. Change from acceleration phase to deceleration phase.

5. Friction curve with plateau.

For a fair comparison of the methods the following criteria were chosen:

SLIP CONTROLLER:

1. Sum of error squares sum_s

2. Sum of control squares sum_u

3. Rise time t_1

4. Settling time t_2

5. Overshoot a

TOMAX:

1. Time period t_1 until the optimal friction force is reached first

2. Settling time t_2

3. Transition time $t_2 - t_s$ from one friction optimum to another

4. Overshoot a

5. Amplitude b of limit cycles.

Figure 12.14 shows the comparison of an optimized PI controller and a SMFC. In both cases the control variable ΔF is restricted to the interval $[0, -2000\ N]$. From $t = 0$ until $t = t_0$ the slip increases its values up to $s_{act} = 0.6$ m/s without any control. At $t = t_0$ a desired step function $s_{des} = 0.2$ m/s is introduced. At the same time the control algorithm starts working. The result is that the *fuzzy slip controller* performs better than the *conventional PI-controller* especially with regard to the parameters *rise time* t_1, *settling time* t_2, and *overshoot* a.

Figure 12.15 shows the comparison of a conventional TOMAX and a fuzzy TOMAX.

Generally, the conventional TOMAX leads to limit cycles at the friction optimum. A fuzzy TOMAX with a flat characteristic avoids limit cycles at the optimum. A fuzzy TOMAX with a steeper characteristic leads to limit cycles but with a lower amplitude as with the conventional method.

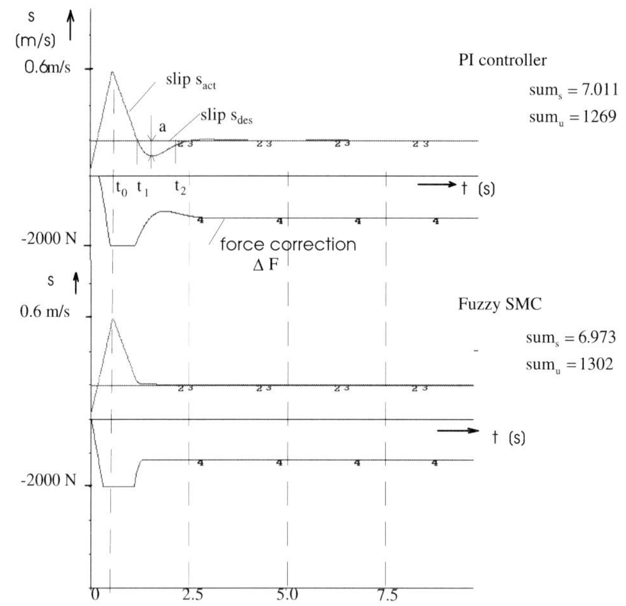

Figure 12.14 PI control and SMFC of slip.

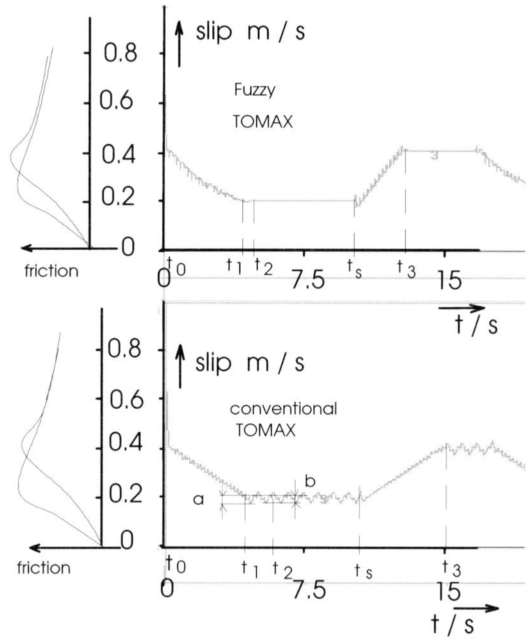

Figure 12.15 Fuzzy and conventional TOMAX.

In the case of a change of the friction optimum, the fuzzy TOMAX performs better regarding the occurrence of limit cycles and the time $t_3 - t_s$ to reach a new optimum. Here, the friction curve changes its optimum (maximum) drastically from $s_{m_1} = 0.2$ m/s to $s_{m_2} = 0.4$ m/s.

Further tests have been done with regard to a stochastically changing optimum. In this case, after reaching the optimum the fuzzy TOMAX works smoother than the conventional one. However, in order to cope with high rates of change of friction force, a low-pass filter had to be installed.

A last experiment dealt with an additional torsion of the motor shaft incorporating in the process model which led to high eigenfrequencies of the system. Therefore, higher frequencies have been filtered out by means of a built-in low-pass filter. In summary; the evaluation shows that both the fuzzy slip controller and the fuzzy TOMAX work better than the corresponding conventional ones especially regarding a smoother behavior at the slip optimum.

12.7 REFERENCES

1. Harprecht, W. "Anfahrverhalten, Leistung und Zuverlässigkeit der Lokomotiven der Baureihe 120 der deutschen Bundesbahn." *eb - Elektrische Bahnen* 82 (1984) 2, pp. 30-54

2. Lang, W. "Gleit - und Schleuderschutz - Grundsätzliche Überlegungen und deren Realisierung beim ICE." Siemens AG VT, internal report, 1991.

3. Roth, G. "Ein neues Gleit- und Schleuderschutzsystem für Hochleistung-striebfahrzeuge." Siemens AG VT; internal report, 1991.

4. Palm, R. "Sliding Mode Fuzzy Control." *Proceedings of IEEE International Conf. on Fuzzy Systems FUZZ-IEEE '92*, pp. 519-526, San Diego, CA, 1992.

5. Palm, R. and Storjohann, K. "Torque Optimization for a Locomotive using Fuzzy Logic." *Proceedings of Symposium on Applied Computing 1994*, Phoenix, AZ, pp.105-109, 1994.

6. Palm, R. "Robust Control by Fuzzy Sliding Mode." *Automatica*. Vol. 30, No. 9, 1994, pp. 1429-1437.

7. Slotine J.-J., Li, W. *Applied Nonlinear Control.* Prentice Hall, 1991.

8. Vogel, U. "Untersuchung eines Verfahrens zur Hochausnutzung des Rad-Schiene-Kraftschlussesbei Triebfahrzeugen." *eb - Elektrische Bahnen* Vol. 89, No. 10, pp. 285-292, 1991.

13

ON MINIMAL MODELS AND AUTOMATED RULE GENERATION WITH THE FUZZY BEHAVIORIST APPROACH FOR VEHICLE NAVIGATION

F. G. Pin
Oak Ridge National Laboratory
USA

Sensor-based operation of autonomous robots in unstructured and/or outdoor environments has revealed an extremely challenging problem, mainly because of the difficulties encountered when attempting to represent the many uncertainties which are always present in the real world. These uncertainties are primarily due to sensor imprecisions and unpredictability of the environment, i.e., lack of full knowledge of the environment characteristics and dynamics. Two basic principles, or philosophies, and their associated methodologies are proposed in an attempt to remedy some of these difficulties. The first principle is based on the concept of "minimal model" for accomplishing given tasks and proposes to utilize only the minimum level of information and precision necessary to accomplish elemental functions of complex tasks. This approach diverges completely from the direction taken by most artificial vision studies which conventionally call for crisp and detailed analysis of every available component in the perception data. The second principle which is proposed is based on the representation of the system's uncertainties using fuzzy-set-theory-based approximations and on the representation of reasoning and control schemes as sets of elemental behaviors. A framework merging these principles is presented, which we have named the "Fuzzy Behaviorist Approach" (FBA). Using the FBA and its associated formalism for rule-base development, an automated generator of fuzzy rules has been developed. This automated system can automatically construct the set of membership functions corresponding to fuzzy behaviors that have been expressed in qualitative terms by the user. The system also checks for completeness of the rule base and for non-redundancy of the rules. Two major conceptual features, the suppression and inhibition mechanisms which allow an expression of dominance between behaviors, are discussed in detail. Some experimental results obtained with the automated fuzzy rule generator applied to the domain of sensor-based navigation in a priori unknown environments, using one of our autonomous test-bed robots as well as a real car in outdoor environments, are then reviewed and discussed to illustrate the feasibility of large-scale automatic fuzzy rule generation using the combination of the "minimal model" and "fuzzy behaviorist" concepts.

13.1 INTRODUCTION

Significant research activity has taken place over the past two decades in the area of autonomous mobile robot navigation. Several successful applications have been reported in which robots showed robust sensor-based navigation features in nonlaboratory, although fairly well structured, corridor-type environments such as hospitals, energy producing and manufacturing plants, and/or warehouses (e.g., see [1], [2], [3], [4]). One of the greatest challenges now facing the research community is to extend the domains of application of autonomous sensor-based navigation to the general class of unstructured environments, i.e., environments which are generally dynamic, not fully known a priori, and typically unpredictable. The paramount complexity of the sensor-based navigation problem in unstructured environments arises mainly from the uncertainties which exist and become pervasive in the overall system (which includes the robot and its surrounding domain). In addition to the typical sensor inaccuracies (there are no such things as "perfect" sensors), the dynamics and unpredictability of the environment generate very large uncertainties in the perception and reasoning systems, i.e., it becomes impossible to generate complete or exact models of the system (robot and environment) and/or of its behavior. These uncertainties, in turn, typically propagate through the control systems and lead to further inaccuracies or errors (e.g., in the robot position, in the sensor orientation) which compound the problem by increasing the uncertainties on the perception. In these conditions, the overall cost (time, computational needs, computing resources, etc.) of achieving the type of precision which was common in structured and static environments jumps by orders of magnitude, either from a requirement for much more refined sensors and perception data, or from the need for very time- and computation-expensive methods such as uncertainty analysis and propagation techniques. The impact of this cost increase is particularly important for real time systems (where real time is defined as the guarantee of producing a response within a prescribed amount of time), which become much more difficult to design, if not impractical to implement in realistic situations.

Humans on the other hand, seem to handle/cope very well with uncertain and unpredictable environments, often relying on approximate or qualitative data and reasoning to make decisions and to successfully accomplish their objectives. Furthermore, humans seem to gather information in what we will refer to as the "approximate first" fashion: they look for and/or perceive first some "general-type" information, of a symbolic, iconic, approximate, or even "blurry" nature, and then progressively focus their perception on details, particular regions, or further precision as they judge necessary to supplement the "general" information. This is quite opposed to the approach which conventional artificial vision techniques seem to pursue, i.e., precision-aimed analysis of every piece of data acquired is performed first [e.g., every pixel in every region of every image produced by a Coupled Charged Devices (CCD) camera], followed by progressive extraction of more and more "general" information through successive applications of "crisp" mathematical models (e.g., finders of edges, regions, geometric properties), with each extractive pass generating more "general" but also "approximate" information because of the cumulative propagation of the basic uncertainties existing on the original data. In other words, humans seem to perceive and reason using information cycles progressing, on an "as needed" basis, from "general and blurry" to "precise and detailed," while conventional artificial perception systems seem to operate using the opposite direction, from "detailed analysis of every data" to "general" information extraction.

Although we concede that full duplication, simulation, or even emulation of the characteristics of human perception and reasoning are very far from being accomplished, and their underlying processes are equally far from being fully understood, we believe that the observations above may point to two directions of research for sensor-based "intelligent" machines which seem to fall in horizons diametrically opposed to most of the directions conventionally followed today. These two research directions revolve around (1) the need for, and use of, *minimal models* in accomplishing tasks, i.e., models based on the minimal levels of information (e.g., with respect to quantity, precision) that are required to reach a goal, and (2) the need for, and almost exclusive use of, approximate reasoning in perception-based decision making. Again, we believe that solutions may be far away for these problems which are currently in their infancy and, therefore, the objective of this chapter is limited to only presenting initial steps in the research directions pointed out by the observations above, while illustrating promising results based on our recent experiments in sensor-based vehicle navigation.

13.2 NEED FOR MINIMAL MODELS

One of the significant features which humans exhibit when they reason is their capability to rapidly make *tradeoffs* among various aspects of their inferencing. In particular, humans are very apt at trading the quantity and precision of the information which they could acquire or derive (but chose not to), for some other characteristics of the reasoning process which they deem much more essential in accomplishing their objectives. Speed of achieving a decision, for example, is one of the most intuitively known factors for which humans will tradeoff precision or quantity of information. There are of course many other factors which humans will trade off in their reasoning, but for the sake of analogy with real-time sensor-based robotic systems, speed versus quantity and/or precision will be the focus of the discussions in this chapter.

It is interesting to note that people who are consistently successful in resolving previously unencountered challenges in changing environments are those who accept to deliberately change and/or adapt their reasoning to correspond to the situation at hand, i.e., they consistently look for a reasoning strategy that has a chance of solving the problem, find what the "right amount" of information to implement that reasoning is, and then adapt their information gathering process to focus (and generally, in an exclusive fashion) on acquiring that particular information. The key, therefore, to successfully and consistently accomplish objectives over a variety of tasks in real-time conditions, seems to consist in finding what "the *minimal* amount" of information needed to reach the desired goals is, and what other non essential process can be traded off so that focus or priority can be given to acquiring the essential information first. For robotic systems, this translates into capabilities to (1) select, on a task-dependent basis, the minimal information, be it qualitative or quantitative, that can be assembled (with available resources or from memory) and is needed to reach *any* decision that would satisfy the goal and (2) process that information, with a reasoning scheme *commensurate with the acquired qualitative or quantitative information*, to produce a decision that meets the objective. It is clear that the selected "reasoning scheme" also is a part of the needed information, and therefore becomes an integral part of the "minimal model" necessary to successfully reach the objective.

Note that in this context, the notion of optimality of an "intelligent machine" may diverge significantly from the conventional concept of optimization with respect to time, length, etc., typically studied in robotic systems, and instead would become an indication of the percentage of times for which objectives have been reached, i.e., for which a successful trade off strategy has been found among the multitude of available data and methods. Furthermore, if longevity and survivability are some of the desired characteristics of future "intelligent robots," a 100% successful trade off capability becomes the implementation goal for these systems (as opposed, for example, to the goal of acquiring 100% of possibly available data, for every task and/or for every memorized reasoning scheme).

Clearly, the first capability outlined above with an "optimal" or even "near optimal" ranking, is a very far away goal, and may require decades of research, prior to reaching practical implementation. The concept however brings up a clear direction for future developments: systems that can perform systematic search for the minimal information sets that are needed for given tasks, appear to be essential components of future versatile "intelligent" machines, and basic research toward developing such systems needs to proceed with a fresh look or a fresh approach with regard to what requirements such systems place on artificial perception. To illustrate this statement and provide one simple example of what a minimal model may be for a specific task, consider the problem of a camera-equipped mobile robot having to follow a moving object such as an automobile, while maintaining a safe separation distance. The conventional artificial vision approach would consist in analyzing every pixel of the image and every detail of the background environment to eventually identify the car and try to determine its motion parameters (e.g., through a very computation-intensive optic flow method). This seems quite an inefficient perception procedure since the only desired information for the task is the approximate size in the image plane of the "form" which is ahead and which has "approximately this color." On the other hand, if the system could first acquire an indication of the approximate size in the image plane of the "form" which has this approximate color, it could quickly determine its approximate location and recent size variation (e.g., the size in the image plane increases slowly, decreases rapidly), and take appropriate control actions relative to the task (e.g., speed up, slow down) using a few simple and computationally inexpensive rules, with the consequence that the system could react and operate at a much faster speed than in the previous case. Thus the capability to rapidly extract data that is approximate, but has high information content, directly from perception systems, is one of the items that we argue would put "intelligent machines" on a better footing (with respect to speed, survivability, efficiency/cost, etc.) for operation in changing and unstructured environments by allowing the use of the necessary, albeit *minimal*, models to accomplish their goals.

13.3 NEED FOR APPROXIMATE REASONING

From the discussion of the previous section, it is clear that the information to be processed during the reasoning phase may appear in either quantitative or qualitative form. Furthermore, significant uncertainty ranges may be associated with the data, be it of a quantitative or qualitative nature. Methods for representing, encoding, and processing quantitative data are quite advanced today with the computer technology explosion. This is not the case, however, for qualitative data or information, for which few methods of representation, encoding, or processing methods are formally or firmly established.

Qualitative reasoning (also termed approximate reasoning) refers to a set of methodologies which have been developed to provide decision-making capabilities in systems where all uncertainties cannot be fully engineered away (e.g., there are limits on maximum sensor precision, on the predictability of the environment). These methodologies attempt to capture some aspects of the reasoning methods typically exhibited by humans when controlling systems; i.e., they aim at implicitly incorporating uncertainties in the information gathering and reasoning processes, rather than explicitly determining and propagating them through numerical calculations or representations. Several approximate reasoning theories and associated mathematical algebra have been developed over the past two decades (e.g., see methods and references in [5]), the most commonly used today for applications to control systems being Zadeh's theory of fuzzy sets [6–8]. This theory is at the basis of very successful implementations varying from control of subway cars, elevators, cement kilns, washing machines, cameras and camcorders, inverted pendulums, to painting processes and color image reconstruction, to even Ping-Pong playing robots [9–15].

One of the important factors which has prevented the wide-spread utilization of approximate reasoning methodologies in real-time systems has been the lack of computer hardware allowing processing and inferencing directly in terms of approximate or linguistic, or "fuzzy" variables (e.g., far, fast, slow, left, faster) and approximate rules (e.g., if obstacle is close, then go slower; if temperature is high and pressure is increasing, then decrease power a lot). Prospective implementations thus had to rely on simulations of the approximate reasoning schemes on conventional computers based on "crisp" processing. The result was a significant penalty in speed of operation, typically prohibiting applications in most "hard real-time" systems. Over the last half decade, however, several innovations have allowed some bridging of this gap; in particular, unique computer boards have recently been developed which use custom-designed VLSI fuzzy inferencing chips (e.g., see [16], [17], and [18]) on VME-bus compatible boards. These systems can be directly programmed in terms of qualitative variables and rules and, when incorporated in a control system, can directly communicate and interface with robotic hardware (e.g., with motors, actuators). Such computer hardware developments have proven extremely useful in supporting the developments needed in the area of approximate reasoning for real-time "intelligent" machines, and have been a strong basis [18,19] for the activities reported here.

13.4 FUZZY BEHAVIORIST APPROACH AND RULE GENERATION FOR VEHICLE NAVIGATION

The problem of autonomous mobile robots navigating in a priori unknown and unpredictable environments was selected as a paradigm for research on qualitative reasoning systems with minimal models because its characteristics rank very high on the list of criteria that typically indicate suitability of a reasoning problem for representation using qualitative logic and resolution using trade offs:

- The input to the control system, particularly when provided by sonar range finders and odometric wheel encoders, is inaccurate, sparse, uncertain and/or unreliable.
- No complete mathematical representation exists of the process termed "navigation" although, as demonstrated by humans, a logic for this process exists which can typically be represented and successfully processed in terms of linguistic variables.
- The approximations involved in the numerical representation of the system and its environment (e.g., geometric representations, map discretization in grid) are significant.
- By its given nature the behavior of an outdoor environment is unpredictable, leading to large uncertainties in its representation and frequent need for trade off of speed vs. precision.

Several research groups have studied approximate reasoning techniques, in particular fuzzy logic, to mimic human reasoning capabilities in navigation tasks (e.g., see References [18] to [22]). In all these applications, the sensor-based decision-making process has been implemented as a set of fuzzy rules which, together, express the desired navigation decisions of the robot for various combinations of the input data. Very successful results have been achieved when the number and complexity of the rules were small. When these increased, however, and/or the perception system grew more sophisticated (i.e., more sensory input data are provided), the typical difficulties encountered with large rule base systems emerged: the lack of established formalism for the development of rule bases—in particular with respect to completeness, interaction, and redundancy of the rules—made the actual coding of the fuzzy rules an iterative empirical process, requiring lengthy trial-and-error experiments.

13.4.1 Fuzzy Behaviorist Approach

In an attempt to alleviate this general shortcoming of rule-base system development, we recently proposed a "Fuzzy Behaviorist Approach" (FBA) which provides a formalism for the development of fuzzy rule systems for control of autonomous robots. The basic premises underlying our proposed FBA are as follows:

- Each action of the robot results from the concatenation of elemental behaviors.
- Each elemental behavior is a direct mapping from a single stimulus mode to a single output control.
- Each behavior is represented by one or a set of fuzzy rules which are defined by the membership functions of the rule's antecedent (stimuli) and consequence (output controls).
- Each mode of stimulus corresponds to a single dimension of the input space and is independent, in a *possibilistic* framework, of other stimulus modes.
- For behaviors effecting *the same output control dimension*, the possible conflicts between behaviors with stimuli that overlap in the multidimensional input space must be resolved through the expression of the respective dominance between the various behaviors.
- Each type of input data provided by the sensors is fuzzified with a membership function expressing, as a *possibility* distribution, the uncertainty associated with the specific measurement or calculation (see [19] or [23] for a detailed discussion of this process).
- Triggering of any behavior takes place when the current input data and the antecedents of the behavior's fuzzy rules have nonempty fuzzy intersection. The triggering strength transferred to the output control is dictated by the fuzzy intersections of the input data and the rule antecedents.
- In the current system discussed below, a typical "center of area" defuzzification scheme is used to generate "crisp" control set points for those output variables that represent direct actuator commands.

Cast in the framework of fuzzy set theory, these principles can be expressed as follows: the inferencing, I, of the (robot) reasoning system provides the relationship between the input universe of discourse U and the output universe of discourse V. The input universe of discourse U is multidimensional, with each dimension representing a type of input data on which the inferencing can act (e.g., distance to obstacle to the right of route, direction to the goal, distance to the goal), i.e., each input dimension is a mode of stimulus, s_i, that can excite the inferencing. Similarly, the output universe of discourse V, is multidimensional, with each dimension representing a type of output data, i.e., a type of control, c_j, (e.g., turn control, motor #1 speed) that can be implemented. Thus we have:

$$I{:}U(s_1, s_2, \ldots, s_i, \ldots, s_n) \rightarrow V(c_1, c_2, \ldots, c_j, \ldots, c_m). \tag{13-1}$$

The total numbers of possible stimulus modes, n, and of control modes, m, are of course dependent on the sensory and actuation capabilities implemented on the robot. Each dimension s_i or c_j is a one-dimensional space on which fuzzy sets can be defined using membership functions in the conventional manner (e.g., see [6], [7], [8], and [9]). An elemental behavior, B_{ij}, is thus defined as a direct mapping from s_i to c_j:

$$B_{ij}: s_i \rightarrow c_j \tag{13-2}$$

which is represented by one or several fuzzy rules relating fuzzy subsets of s_i to fuzzy subsets of c_j. As an example, assume s_3 represents the "direction to the goal" input dimension and c_2 represents the "turn control" output dimension, then the behavior B_{32}, "turn control as a function of the goal direction," would include fuzzy rules of the type: IF (direction to the goal is *left*) THEN (turn value is *positive*), where *left* and *positive* are fuzzy subsets defined by their membership functions on s_3 and c_2, respectively.

Note that the fourth requirement of the approach specifies that the input space be designed such that the stimulus modes (input dimensions), s_i, are independent of each other, i.e., such that the possibility for the ith input to be any fuzzy subset in s_i is completely independent of the possibility for any other input to be any fuzzy subset on their stimulus mode. In other words, the possibilities for any and all stimuli to occur are unrelated and independent of each other. This allows a behavior B_{ij} to be extended to a mapping B^*_{ij} from U to V as:

$$B^*_{ij}: U(\#,\#,\#,\ldots,s_i,\#,\ldots,\#) \rightarrow V(\phi,\phi,\ldots c_j,\phi,\ldots,\phi) \tag{13-3}$$

where the # and ϕ signs represent "nonsignificant" input and output dimensions, respectively. By definition, s_i and c_j are the "significant" input and output dimensions of behavior B^*_{ij}. For example, the behavior B_{32} discussed above has 3 (for input) and 2 (for output) as significant dimensions and would be extended to a behavior B^*_{32} on the multidimensional input and output spaces with fuzzy rules now expressed as:

> IF (input 1 is *anything* and input 2 is *anything* and direction to the goal is *left* and input 4 is *anything* and …)
> THEN (output 1 is *do nothing* and turn value is *positive* and output 3 is *do nothing* and …)

and the fuzzy subsets *anything* and *do nothing* have membership functions uniformly equal to 1 and to 0 over the entire range of their respective input and output dimensions.

With these features, rule bases embodying sets of elemental fuzzy behaviors can be very easily generated (e.g., see [18] and [19]). They can also be readily augmented with additional behaviors, either to handle situations of increasing complexity (e.g., see [19] and [23]) or if additional perception or control capabilities are added to the system, requiring larger dimensions of the input or output spaces. In these cases, however, a very important aspect of the formalism which needs to be emphasized here is the requirement for independence and non-conflict of the stimuli of the behaviors effecting the same output controls. This requirement simply expresses that only one action command can be sent to a single output control for any given stimulus (a single point in the input space). This leads to newly added (or acquired) behaviors having to "dominate," or "be dominated by," some of the existing behaviors in one or more regions of the input space. This concept of dominance between behaviors, which exhibits itself in almost every action of our everyday life, is illustrated here using a simple example: consider two behaviors acting on the *same* output control, say the two object-holding fingers of a child or, for analogy, the gripper of a robot. Initially, the child or the robot only has one behavior which is a "don't get burned" behavior and which could consist of one fuzzy rule expressing that IF (the object being held is hot) THEN (release the object). The input dimension for this behavior is "temperature of object being held," and the fuzzy set "hot" can be represented with a typical membership function. Suppose that the child is now being taught the value of things or assume that the robot is being given a new perception device so that it can recognize the value of objects. A new behavior could be given to the robot or taught to the child, stating that IF (the object being held is expensive) THEN (don't release the object). The input dimension for this behavior is "value of the object being held," and the fuzzy set "expensive" can be represented with a typical membership function. Taken separately, the two behaviors are fully independent and each has a different one-dimensional input space. They could therefore be developed and exhibited independently of each other. When these two behaviors are merged to create a more complex reasoning system, then the new input space for the system becomes two-dimensional. Within that new system, the two behaviors are still independent since they trigger from stimuli on different input dimensions that do not affect each other. However, every time the child or the robot will handle one of great grandmother's priceless china cups filled with very hot tea, the implicit dominance of one behavior over the other in their overlapping region of the input space will be exhibited: the cup will, or will not, be dropped, signifying respectively that the "don't get burned" behavior dominates, or is dominated by, the "don't drop valuable objects" behavior at that particular point in the input space.

It is clear that for both the child and the robot, some instruction, reinforcement learning, experience, etc., may cause the dominance process and its resulting outcome to vary over time. However, if some "mistakes" and/or "trial and error experiments" are not allowed to the child or to the robot, then an indication of what behavior dominates the other in what region of their overlapping areas in the input space *must be given to the child or the robot at the time the second behavior is added and merged with the other in the system's reasoning module*. In the automated system, this is accomplished through the "suppression" mechanism, in which the output membership function of the dominant behavior is modified, so that its weight will appropriately overpower that of the other behavior in the center of gravity calculation. If due to previous dominances and/or suppression requirements with respect to other behaviors, the membership function cannot be made adequately overpowering, then the system uses the concept of "inhibition" to express the behavior's dominance. Applying inhibition within the overlapping region of input space consists in partially truncating the input membership function of the "weaker" behavior so that the dominant behavior always triggers with a greater strength. Although both suppression and inhibition mechanisms result in expressing the desired dominance, their concepts are quite different: one basically operates on the relative weight of the behaviors in the output space while the other modifies the triggering conditions of the behaviors in the input space. This difference, which is transparent in the automated system, becomes significant when one seeks to couple the system with learning modules for refinement of the behaviors and/or of the rule membership functions through reinforcement learning.

13.4.2 Automated System

Just as different people may use different strategies, different rules, and different qualitative variables to express their navigation process and still navigate efficiently "in their own way," several strategies may be used to embody a particular process in a rule base, i.e., there is not a single or unique rule base representation of a given process. For example, a rule base for obstacle avoidance may be built on the basis of a distance-to-obstacle strategy, as was done in [10], with rules organized and developed for input conditions in which obstacles are very near, near, far, very far, etc.; or on the basis of the direction of obstacles, with rules organized and developed for obstacles located on the right, center, or left of the travel direction. Because of the requirement of our FBA for each behavior to trigger from a single input dimension, the expression into rules of various possible strategies may appear quite different, even though the overall process and resulting actions of the robot may be similar. In the automated system, the user inputs the strategy for the rules in a "qualitative" form using the format shown in Figure 13.1.

The five-line format in Figure 13.1 describes one rule, with the first line giving the "reference name" of the rule, the second line listing the names of the rules or behaviors which are suppressed (or inhibited) by this rule, the third line giving the suppression parameter E (which will be described in the following paragraphs), and the last two lines specifying what the desired input and output of the rule on the "significant" (see Section 13.4.1) input and output dimensions of the behavior are. In the current version of the automated system, each rule is assumed to be of the form:

$$\text{If } (A \text{ is } A_1 \text{ and } B \text{ is } B_1 \text{ and } C \text{ is } C_1 \text{ and } D \text{ is } D_1) \quad \text{THEN } (E \text{ is } E_1 \text{ and } F \text{ is } F_1) \tag{13-4}$$

therefore operating on four input and two output channels. Although extension to any

name:	RN	or	name:	LF
suppressing list:	G? LF		M:	0.3
E:	50		Outputs:	- 10
Outputs:	20 -		Inputs:	-- Far -
Inputs:	- Near --			

Figure 13.1 Input format for the automated rule generation system.

number of input and output channels is possible, this configuration was chosen in the initial version of the automated system because it corresponds to what is available on the custom-designed VLSI fuzzy inferencing chips and boards (see [16], [17], and [18]) which we utilize in our experimental work. Use of the character "?" in the suppression list of line 2, such as "G?" in the example of Figure 13.1, indicates that the rule suppresses all other rules whose name includes the other character, in this case all rules whose name begins with the letter "G." If the subject rule does not suppress other rules, then the first character of line 2 is the letter M followed by a number expressing the desired weight m of the output (explained in the following paragraphs), and the line giving the suppression parameter E is omitted. The line labeled "outputs" in Figure 13.1 indicates which of the two outputs is effected by the behavior, with the number, expressed in percent of the scaled output range, specifying the desired center of mass of the membership function of the output [E_1 or F_1 in Equation (13-4)]. The line labeled "inputs" specifies which one of the four input channels is the behavior's "significant" input dimension, and the qualitative name indicates which fuzzy set constitutes the antecedent [A_1, B_1, C_1, or D_1 in Equation (13-4)] of the rule on the input dimension. The membership functions defining the input related fuzzy sets are defined by the user on a behavior-by-behavior basis (e.g. see [17] or [18]), and can be stored according to their "name" in a "membership function library."

When the user has listed all the rules of the desired behaviors in the format of Figure 13.1, the automated system can generate a "skeleton" of the rule base and check if it verifies the input-related requirements of the approach. In particular, the system constructs the four-dimensional input spaces for each of the two output dimensions, so that it can evaluate completeness of, and redundancy in, the rule base and report all instances to the user. For any region of incompleteness, i.e., regions of the input space not covered by any of the behaviors stimuli, the user decides on either the addition of a behavior to cover these possible stimuli, extension of the current behaviors (through extension of their input membership function) to include these input regions, or no modification if input data within these uncovered regions or "blind spots" are never expected to occur (for example if these regions correspond to values outside the operating range of the sensors). For the regions of redundancy, i.e., areas where stimuli from two or more behaviors are overlapping, the system reports every rule for which a dominance has not been specified but may be required because of the input overlapping. The user can then interactively add to or modify the dominance specifications in lines 2 and 3 of each rule, until all requirements of the approach are verified and all desired dominances are expressed in the rule base. The actual generation of the rule base, including the suppression and/or inhibition mechanism, can then proceed as follows: initially, all rules are given a "standard" output membership function equal to 1 over a width of one bit, centered at the bit value expressed in line 4 of the "qualitative" expression of the rule (see Figure 13.1). The system checks the sets of rules that are effecting the *same* output dimension. If no suppression mechanism has been expressed between the rules because dominance is not necessary, then the output membership functions are unconstrained and they remain at their "standard" value. If a dominance has been expressed between two or more rules, then the dominant rule is the one that is modified if suppression is possible; otherwise the dominated rules are modified using inhibition, as explained in the following paragraphs.

Given a rule i, and its output function $\mu_i(y)$ on the y output dimension, the center of mass y_i and weight m_i of its output are given by:

$$y_i = \frac{1}{m_i} \int y \mu_i(y)\, dy \tag{13-5}$$

$$m_i = \int \mu_i(y)\, dy \tag{13-6}$$

and, by definition, $m_i > 0$ and $y_{\min} \leq y_i \leq y_{\max}$, where $y_{\min}$ and $y_{\max}$ are the extremum values of the y output dimension. Assume that two rules, which control the same output y, have respective output membership functions $\mu_i(y)$ and $\mu_j(y)$, with corresponding centers of mass and weights given by Equations (13-5) and (13-6). When both rules trigger under the same input conditions, the resulting output membership function is the union (or max operation) of $\mu_i(y)$ and $\mu_j(y)$ and, since a "center of gravity" defuzzification scheme is used on the VLSI chips [16,17,18], the overall output y_0 is:

$$y_o = \frac{m_i y_i + m_j y_j}{m_i + m_j} \tag{13-7}$$

From Equation (13-7), it is clear that the *relative* value of the output weight of the rules can change the output y_o. In particular, if the weight of one rule is strong enough, the output y_o can be "attracted" *within* the output membership function of this rule, *even though the other rule still "works" and contributes to the overall output.* This constitutes the basis for implementing the suppression mechanism.

13.4.2.1 Suppression Mechanism

In the general case, a rule, labeled rule h, may be required to suppress a set S of other rules i, $i \in S$. The suppression mechanism can be expressed as the requirement that, when all the rules h and i, $i \in S$, trigger, their combined output y_o must fall within an "allowable error," ε_h, of the center of mass y_h of rule h. The suppression condition thus can be written as:

$$| y_o - y_h | \leq \varepsilon_h \tag{13-8}$$

This allowable error, ε_h, can take any positive value between 0 and 63 (the maximum bit number of the output membership function range), and is the user-specified suppression parameter E appearing in line 3 of the qualitative rule description shown in Figure 13.1. Typically, a small value ε_h indicates a "strong" dominance of rule h over the rules in S, while a large ε_h represents a "weak" dominance or what we will refer to as a "preference." Since:

$$y_o = \frac{m_h y_h + \sum_{i \in S} m_i y_i}{m_h + \sum_{i \in S} m_i}, \tag{13-9}$$

we have:

$$\frac{\sum_{i \in S} |y_h - y_i| m_i}{m_h + \sum_{i \in S} m_i} \leq \varepsilon_h \tag{13-10}$$

from which a minimum value for m_h can be calculated. Since the distances between the centers of mass $|y_i - y_h|$ are always less than or equal to $Y_r = y_{\max} - y_{\min}$, selection of m_h as:

$$\left(\frac{Y_r}{\varepsilon_h} - 1\right) \sum_{i \in S} m_i \leq m_h \tag{13-11}$$

guarantees that rule h suppresses the other rules i, $i \in S$. From m_h, and the selected shape (rectangular in the current system) of the output membership function, the width of, and/or the full $\mu_h(y)$, can be easily determined.

13.4.2.2 Inhibition Mechanism

In some cases, the suppression mechanism described above may not be implementable because of the discretization of the membership functions in 64 bits. This will typically occur when ε_h is specified to be very small, or if several dominance mechanisms need to be implemented within the rule base, resulting in progressively large weights calculated from Equation (13-11). In this case, the inhibition mechanism is used instead of the suppression mechanism, and the dominance of rule h over rules i, $i \in S$, is forced by appropriately truncating the *input* membership functions of the nonsignificant dimensions of rules i so that these rules do not trigger when rule h does. IF (A_1, B_1, C_1, D_1, E_1, F_1), with significant input dimension B, defines the rule [see Equation (13-4)] which dominates rule (A_2, B_2, C_2, D_2, E_2, F_2), then B_2 is the overlapping input to be truncated, and the truncated condition B_2^* is expressed as $B_2^* = B_2 \cap \overline{B_1}$, where $\overline{B_1}$ represents the complement of condition B_1. Note that this effectively removes the overlap of the rules' input in the multidimensional input space (and therefore the conflict between the rules), and that no truncation is necessary in the other input dimensions (that are non-significant for the dominant rule).

Figure 13.2 schematically shows these very intuitive suppression and inhibition mechanisms on a simple example in which two rules, each with two input and one output, act on the speed control of a robot. The first rule, which may be part of a behavior for speed control as a function of frontal obstacle proximity, states that if the obstacle is very close, then the speed should be very small. The second rule states that if the goal is

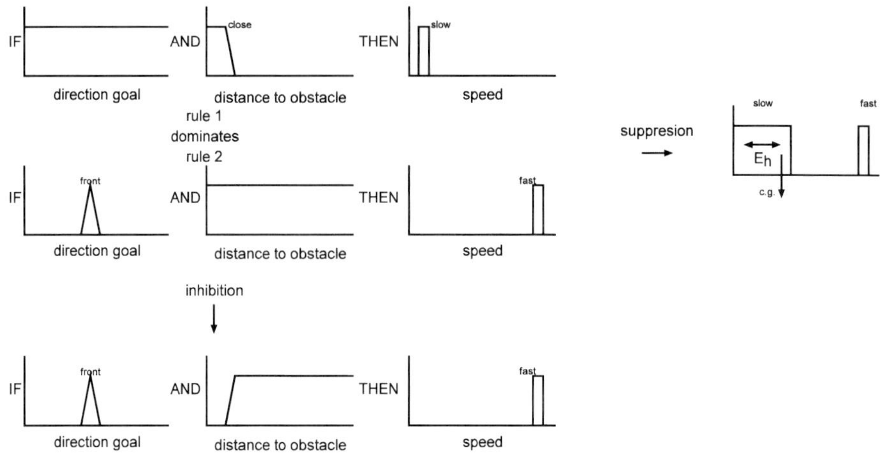

Figure 13.2 Schematic of the suppression and inhibition mechanisms.

straight ahead then the speed is fast. The membership functions uniformly equal to 1 over their entire range represent the non-significant input dimensions of each behavior.

13.5 SAMPLE EXPERIMENTAL RESULTS

Using the automated fuzzy rule generation system, a variety of rule bases were generated (e.g., see [19], [23]) for the sensor-based navigation of mobile robots in a priori unknown environments. The experiments involved both indoor and outdoor environments, and progressed from simple goal tracking navigation in laboratory-type situations to sensor-based driving of a car in parking lot conditions, and from platforms with omnidirectional capabilities to robots with complex car-like kinematics including non-holonomic constraints. In what follows, sample results from some of these experiments are reviewed to illustrate the automatic rule generation process including the suppression and inhibition mechanisms. As discussed in detail in [18] and [19], the four sensory input to the fuzzy inferencing system are the goal direction (or target direction), which is updated at loop rate using the odometry system; and the minimum distances to obstacles obtained using groups of acoustic range finders in three 75° wide sectors at the left, center, and right of the robot's travel direction. The two output of the inferencing system are the commands for turn increments and speed of the robot. Thus, navigation behaviors with these data can involve Goal Orientation (GO) and "front," "left," or "right" Obstacle Proximity (OP) as input, and Turn Control (TC) and Speed Control (SC) as output.

13.5.1 Example of Basic Navigation Behaviors

As mentioned previously, several rule bases, representing various "strategies," may be developed to solve a complex problem or to embody a complex behavioral process. Figure 13.3 shows the rules for one of the rule bases we investigated for sensor-based navigation, in which front obstacle proximity is only used for speed control, while side obstacle proximity is only used for turn control. The behaviors shown in Figure 13.3 are thus organized as follows:

$$\begin{aligned} \text{GO} &\rightarrow \text{TC (3 rules)} \\ \text{GO} &\rightarrow \text{SC (1 rule)} \\ \text{"front" OP} &\rightarrow \text{SC (4 rules)} \\ \text{"left" OP} &\rightarrow \text{TC (4 rules)} \\ \text{"right" OP} &\rightarrow \text{TC (4 rules)} \end{aligned}$$

The input file that was used to generate the rules through the automated system is shown on the left of the figure. The "qualitative" rule descriptions (in the format of Figure 13.1) of the input file are presented next to the automatically produced rule membership functions graphically displayed on the right side of the figure. Each of the 16 rules in the figure is displayed as a vertical arrangement of six graphs of membership function. The top four graphs in each rule show the membership functions corresponding to the input variables A_1, B_1, C_1, D_1 in Equation (13-4) (the direction angle to the goal and the distances to obstacles in the left, center, and right sectors) while the bottom two graphs correspond to the output variables E_1 and F_1 (turn command and speed command). The three rules of the GO → TC behavior simply express that if the goal is to the center, left, or right, respectively, then the robot should make a zero, positive (i.e., left), or negative (i.e., right) increment of turn, respectively, and there is no command for speed change. The GO → SC rule states that if the goal is toward the back sector of the robot, the robot should slow down (i.e., apply a negative contribution of speed). The four rules of the "front" OP → SC behavior express that if the obstacles in front are very far, then a speed close to the maximum possible can be applied, and that the closer the frontal obstacles, the slower the robot should go, eventually stopping when the obstacles are dangerously close. Note that the weight of the velocity command (i.e., the output membership function) increases with "increasing danger." Similarly, the eight rules of the OP → TC behaviors express that the closer the obstacles on a side, the greater the increment of turn in the opposite direction and the "heavier" the turn command should count in the output control calculation. Note the large weight of the output membership function of the "very near" OP → TC rules, which results from this behavior having been selected as suppressing the GO → TC behavior, i.e., expressing that when obstacles are very close, their avoidance is always of greater importance than tracking the goal. It is clear that without this expression of dominance, the GO → TC and OP → TC behaviors would often result in dead-lock or oscillatory situations in which the robot would not turn at all or would oscillate between two orientations. This type of situation constitutes one of the very serious drawbacks of the reactive navigation methods which use potential field techniques, and has been alleviated here by using the suppression mechanism.

```
Behave : GF
M     : 1.0
Outputs : 50 -
Inputs  : Front - - -

Behave : GL
M     : 1.0
Outputs : 10 -
Inputs  : Left - - -

Behave : GR
M     : 1.0
Outputs : 90 -
Inputs  : Right - - -

Behave : GB
M     : 0.5
Outputs : - 0
Inputs  : Back - - -
```

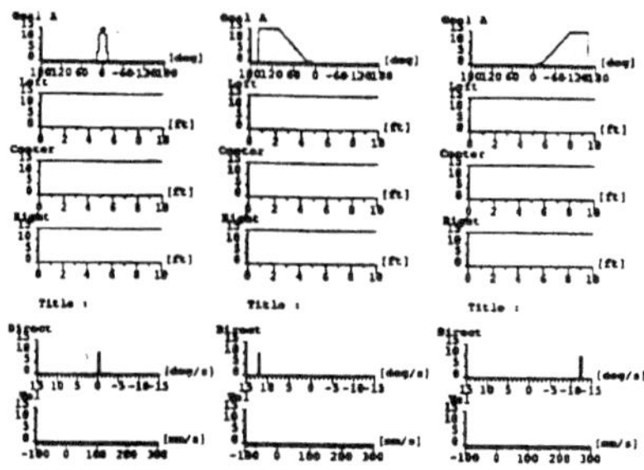

GO ⟶ TC

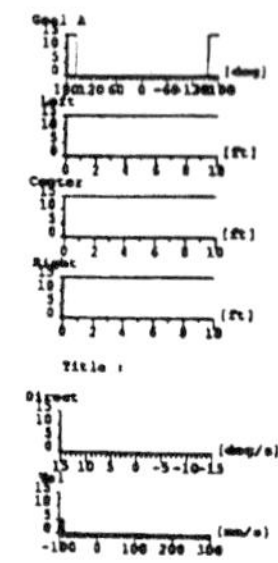

GO ⟶ SC

```
/***   Front    ***/
Behave : FC
Sup   : G?
E     : 10
Outputs : - 25
Inputs  : - - Close -

Behave : FN
Sup   : G?
E     : 50
Outputs : - 50
Inputs  : - - Near -

Behave : FF
M     : 0.5
Outputs : - 80
Inputs : - - Far -

Behave : FNo
M     : 0.3
Outputs : - 100
Inputs : - - No -
```

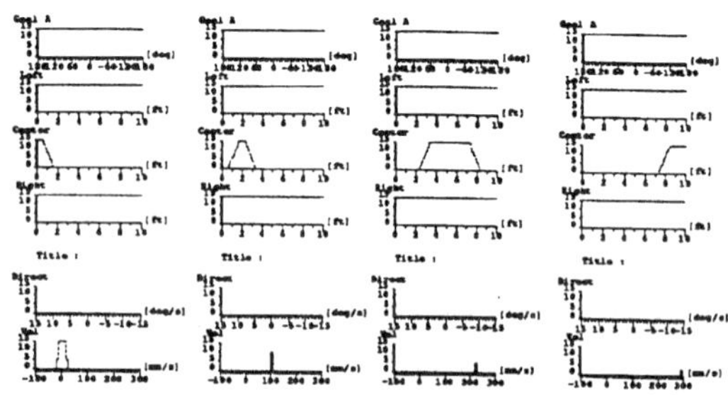

front OP ⟶ SC

```
/***   Left    ***/
Behave : LC
Sup   : RN
E     : 20
Outputs : 90 -
Inputs  : - Close - -

Behave : LN
Sup   : G?
E     : 20
Outputs : 80 -
Inputs  : - Near - -

Behave : LF
Sup   : G?
E     : 100
Outputs : 70 -
Inputs : - Far - -

Behave : LNo
M     : 0.3
Outputs : 50 -
Inputs : - No - -

/***   Right    ***/
Behave : RC
Sup   : LN
E     : 20
Outputs : 10 -
Inputs  : - - - Close

Behave : RN
Sup   : G?
E     : 20
Outputs : 20 -
Inputs  : - - - Near

Behave : RF
Sup   : G?
E     : 100
Outputs : 30 -
Inputs  : - - - Far

Behave : RNo
M     : 0.3
Outputs : 50 -
Inputs  : - - - No
```

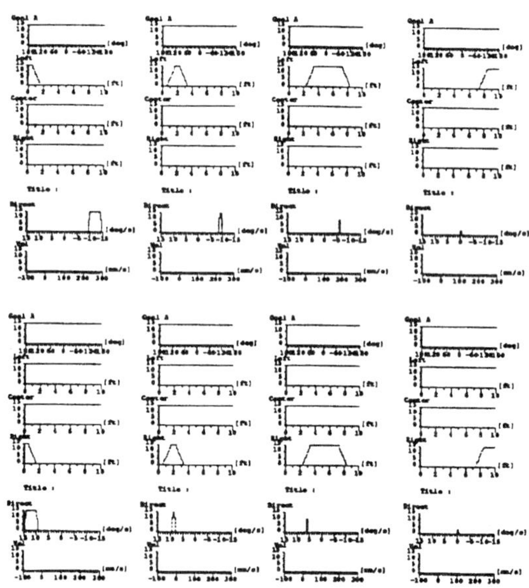

side OP ⟶ TC

Figure 13.3. Sample rule base for sensor-based navigation.

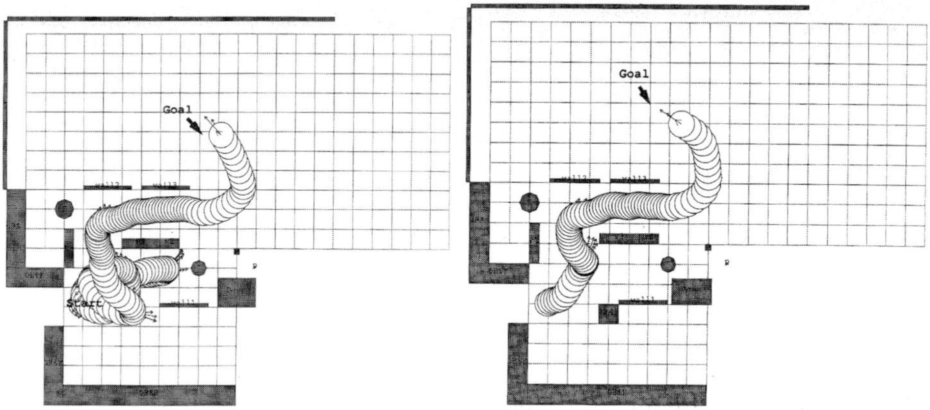

(a) (b)

Figure 13.4 Actual runs of the robot using automatically generated fuzzy rule bases with a) a strong behavioral dominance of obstacle avoidance over goal tracking, b) a lesser behavioral dominance of obstacle avoidance over goal tracking.

Figure 13.4 shows plots of actual runs made with one of our small omnidirectional robots [25] to illustrate the overall reactive navigation obtained with the automatic generation of fuzzy rules. These plots are also given here to provide an example of the effect on the navigation behaviors which a dominance mechanism (suppression or inhibition) can produce. In the figures, the shaded areas represent the obstacles which were placed in the room, while the path of the robot is illustrated using the succession of circles showing the position of the robot every 20 loop rates. In Figure 13.4(a), the rules shown in Figure 13.3 were used, which embody a very strong dominance of the obstacle avoidance (OP $\rightarrow$ TC) rules over the goal tracking (GO $\rightarrow$ TC) rules. Consequently, due to the almost constant proximity of the corridor walls, the suppression mechanism is quite effective in the early part of the run and the robot wanders around for quite a long time, guided principally by obstacle avoidance. It eventually gets positioned ideally to enter the corridor and then turns right, in a direction closest to the goal direction. It follows the corridor, and when reaching the end of the wall, turns left toward the goal. Clearly, the dominance of the obstacle avoidance rules over the "move to the goal" behavior may be too strong in this environment. For the sample run shown in Figure 13.4(b), this dominance has been decreased, i.e., the suppression parameter [ε_h explained in Equations (13-8) to (13-11)] has been doubled from 20 to 40 in the obstacle avoidance rules named LC, LN, RC, RN (see input file on the left-hand side of Figure 13.3), and a corresponding rule base has been generated using the automated system. The robot is now seen to negotiate the entrance of the corridor much more rapidly because of the greater effect of the goal tracking behavior, resulting in a much shorter run to the goal. From an overall behavioral point-of-view, the simple change of the dominance of the "obstacle avoidance" over "move to the goal" behaviors in the rule base, has transformed the "shy" robot, driven mainly by obstacle avoidance in Figure 13.4(a), into a "much braver" robot proceeding more rapidly toward its goal in Figure 13.4(b). Note that this "transformation" has been accomplished without direct modification of the logic embodied in the elemental behaviors.

13.5.2 Extension to Robots with Car-Like Kinematics

One of the expected strengths of our proposed fuzzy behaviorist approach using elemental "human-like" behaviors is that the *linguistic logic* embodied in the behaviors should be invariant across systems of similar characteristics. In other words, for robots with similar perceptive abilities and motion capabilities, the linguistic expression of given behaviors, and therefore their representation in the fuzzy framework, should be the same for compatible input and output. For example, a "goal tracking" behavior connecting the perceived goal direction to a rate of turn [e.g., IF (goal is to the right) THEN (apply increment of turn to the right)] should be invariant for any robot which has a means to perceive the goal direction and to perform the required turn. Using this property (and realizing that the rate of turn of a car is proportional to the steering angle of the wheels) [2,23] all navigation behaviors developed for the laboratory omnidirectional platform appear directly applicable to the driving of a car of similar size, except for those behaviors which require a rate of turn too large for the car to perform because of its limited steering angle. The "very near" OP → TC rules (see Figure 13.3), which require the platform to perform high rates of turn (using its omnidirectional capability) when obstacles are detected at dangerously close ("very near") distances on the left or right of the travel direction, are the only ones that would therefore require attention when changing from the omnidirectional platform to a car with a small maximum steering angle.

As a demonstration of this transportability of invariant behaviors from one system to another, the same behaviors (except for the "very near" OP → TC behavior) and the same fuzzy rules that were utilized for the omnidirectional platform were used to implement the autonomous control of a car on the basis of the same three 75° "wide and blurry sonar eyes" and goal direction input. To complete the navigation rule base for the driving of the car, a behavior had to be included to handle situations where "very near" obstacles are detected and the car does not have enough space to complete a turn away from obstacles because of its limited steering angle, therefore requiring some maneuvers using reverse gear. Another strength of our proposed fuzzy behaviorist approach is its capability for superposition of elemental behaviors, allowing for progressive addition of behaviors to the system to resolve situations of increasing complexity. Thus, by observing human reactions in such driving situations, a behavior was created that can be expressed as follows: IF [obstacle is very near on right (alternatively, left)] THEN [steer right (alternatively, left) AND (back up)]. This response was further divided into a steer control behavior: "very near" OP → TC, and a speed control (back up) behavior: "very near" OP → SC, to respect our approach's requirement for independence and single input dimensions of behaviors. Both simulation and outdoor experiments were performed with the maneuvering behavior augmented rule base and results can be found in [19] and [23]. Because of the unavailability of a car with automated actuation, no autonomous navigation tests were performed outdoors. However, the system was investigated for use as a "driver's aid" using one of the company cars. In the driver's aid mode, the same rule base, sensors, and inferencing system as in the previous experiments are used, however the commands produced by the rule base are displayed to the driver to guide him/her in driving the car. The driver conventionally uses the gas and brake pedals and steering wheel to implement the commands that are displayed on a portable computer screen located next to him/her in the cabin.

For the experiments, the sensor platform was placed on the rear trunk of the car and the driving was done mainly in reverse gear so that the driver was effectively prohibited from seeing the environment while driving. In this "driver's aid" mode, the commands are not displayed to the operator as crisp control values but as bars of variable lengths over the generic speed and steering scales, effectively providing only the direction of the command (left or right, forward or back) and the *relative strength* (i.e., more steering, faster, slower, etc.) the driver should apply on the controls between the maximum steering and speed values. Had a speech synthesizer been available on the portable computer, these commands could have been given in linguistic form using a few simple words. The types of environments in which the tests were performed were the diversely occupied parking lots of Oak Ridge National Laboratory (ORNL), as can be seen in the background of the photograph in Figure 13.5. During the experiments, it was interesting to observe each operator develop his/her own interpretation of, and response to, these relative

Figure 13.5 Photograph taken during one of the outdoor sensor-based navigation experiments with "driver's aid" mode in one of the ORNL parking lots.

commands, leading to different routes and maneuvering situations for the same start and goal positions. From the system's development point of view, this inclusion of the human in the control chain effectively consisted in including a source of unpredictable noise and delays in the actuation system of the autonomous operation mode. The successful operation of the rule base in this mode of driving thus provided a stringent robustness test of the qualitative inferencing scheme and navigation system.

13.6 CONCLUDING REMARKS

An automated system to generate fuzzy rules from the qualitative description of a reasoning process has been developed. The automated system is built on the basis of the minimal models and fuzzy behaviorist approaches which we proposed for the development of fuzzy rule bases embodying "human-like" behaviors in sensor-based decision-making systems. The concepts of suppression and inhibition of behaviors and the inclusion of corresponding mechanisms in the automated system have been described. Examples of the use of the automated system to generate fuzzy rule bases for the sensor-based navigation of autonomous robots have been discussed. Sample runs of the robots have been presented to illustrate the navigation behaviors obtained with the automatically generated fuzzy rule bases as well as the effect of a change in the inter-behavior dominance expressed through the suppression and/or inhibition mechanism. Experiments with a real car have also been discussed to illustrate the capability of readily adding behaviors to the fuzzy rule base to resolve situations of increasing complexity and, as shown in the driver's aid feasibility study, the straightforward "linguistic" interfacing capability of the fuzzy behavior-based system.

A variety of lessons and observations can be drawn from these experiments. Several of these, relevant to the topics dealt with here, are listed and/or discussed below.

- The entire navigation code or scheme consists of less than 20 fuzzy rules. Compared to the 30,000+ lines of "crisp" coding which were previously utilized to accomplish the same task, the efficiency and gain in code development time, code and data storage space, etc., of the approach seems to be indisputable in this respect.
- Our observations showed the approximate reasoning scheme to be much more robust and reliable than the previously used "crisp" codes when faced with sensor inaccuracies and environmental uncertainties. This seems to clearly support the claim that, for situations where precision is not the primary goal, robotic tasks can be very efficiently accomplished using an approximate reasoning scheme, but also can be made more robust with respect to uncertainties through an implicit "folding" of these uncertainties within the approximate variables of the reasoning scheme.
- As illustrated through the various experiments involving small omnidirectional and indoor robots and/or an outdoor car, approximate reasoning schemes can be built to embody very generic functions, i.e., neither the sensory data nor the code itself need to be specifically developed for a particular robotic platform. This seems to indicate that, contrary to what is typically obtained with "crisp" logic codes, approximate reasoning schemes need not be system-specific but rather can be *function-specific*,

i.e., they can embody a reasoning *strategy* rather than a specific instantiation of the strategy.

- Furthermore, not only does this genericness of the approximate reasoning schemes provide straightforward transportability of codes among various systems with *similar* perception and motion *means*, but it also points out that for such sets of robots, trade offs in reasoning strategies depend and can be done primarily on the basis of the task to be performed. Additionally, the trade offs can take place rather independently of which robot system will perform the task.
- A significant consequence of this genericness of approximate reasoning schemes is of course their scale up capability. As shown in the experiments, basic navigation schemes could be augmented and/or enhanced with additional behaviors *without* rewriting the previously tested behaviors. The gain in development time resulting from this property of approximate reasoning schemes is expected to be substantial.
- With respect to the minimal model concept, it should be noted that the sparsity of the data typically utilized with approximate reasoning schemes may be somewhat misleading. This sparsity only exists at the time scale of one sampling period. If accumulated over time, however, a very large part of the environment is sampled since the robot, and consequently the sensors, are translating and rotating. This obviously brings the question of trade off between the type of information which is minimally needed at the sampling rate versus that which is minimally needed at the overall task rate. It is clear that in making such trade offs, consideration must be given to the amount of information that will result from accumulation over time of the data, rather than simply snapshot type of information. We believe that the answer to this question is the key to a method for systematically estimating task-specific or function-specific minimal models, and developments in this area are the focus of our on-going research.

13.7 REFERENCES

1. Skewis, T., Evans, J., Lumelsky, V., Krishnamurthy, B. and Barrows, B. "Motion Planning for a Hospital Transport Robot." *Proceedings of the 1991 IEEE International Conference on Robotics and Automation*, pp. 58–63, Sacramento, CA, April 1991.

2. Vasseur, H. A., Pin, F. G. and Taylor, J. R. "Navigation of Car-Like Mobile Robots in Obstructed Environments Using Convex Polygonal Cells." *Robotics and Autonomous Systems*. Vol. 10, No. 2-3, 1992, pp. 133–146.

3. Brooks, R. A. "Elephants Don't Play Chess." *Robotics and Autonomous Systems*. Vol. 6, No. 1-2, 1990, pp. 3–15.

4. Andersen, C. et al. "Navigation Using Range Images on a Mobile Robot." *Robotics and Autonomous Systems*. Vol. 10, No. 2-3, 1993, pp. 147–160.

5. *Uncertainty in Artificial Intelligence.* (Kanal, L. N. and Lemmer, J. F., eds.) North-Holland, 1988.

6. Zadeh, L. A. "Fuzzy Set." *Information and Control.* Vol. 8, 1965, pp. 338–353.

7. Zadeh, L. A. "Outline of a New Approach to the Analysis of Complex Systems and Decision Processes." *IEEE Transactions on Systems, Man, and Cybernetics* Vol. 3, No. 1, 1973, pp. 28–45.

8. Zadeh, L. A. "Fuzzy Logic." *IEEE Computer.* Vol. 21, No. 4, April 1988, pp. 83–93.

9. *Fuzzy Sets and Their Applications to Cognitive and Decision Processes.* (Zadeh, L. A., Fu, K. S., Tanaka, K. and Shinamura, M., eds.). Academic Press, 1975.

10. Yasunobu, S., Miyamoto, S. Takaoka, T. and Ohsihima, H. "Application of Predictive Fuzzy Control to Automatic Train Operation Controller." *Proceedings of the IECON '84*, pp. 657–662, 1984.

11. Holmblad, L. P. and Ostergaard, J. J. "Control of a Cement Kiln by Fuzzy Logic." *Fuzzy Information and Decision Processes.* Elsevier Publishing, (Gupta, M. M., and Sanchez, E., eds.). 1982, pp. 389–399.

12. Larkin, L. I., "A Fuzzy Logic Controller for Aircraft Flight Control." *Industrial Applications of Fuzzy Control.* Elsevier Publishing, (M Sugeno, ed.). 1985, pp. 87–103.

13. Ono, H., Ohnishi, T. and Terada, Y. "Combustion Control of Refuse Incineration Plant by Fuzzy Logic." *Proceedings of the 2nd International Fuzzy Systems Association Congress*, pp. 345–348, July 1987.

14. Yamakawa, T. "Stabilization of an Inverted Pendulum by a High-Speed Fuzzy Logic Controller Hardware System." *Fuzzy Sets and Systems.* Vol. 32, 1989, pp. 161–180.

15. Hirota, K. A., Arari, Y. and Hachisu, S. "Fuzzy Controlled Robot Arm Playing Two Dimensional Ping-Pong Game." *Fuzzy Sets and Systems*, Vol. 32, 1989, pp. 149–159.

16. Watanabe, H., Dettloff, W. and Yount, E. "A VLSI Fuzzy Logic Inference Engine for Real-Time Process Control." *IEEE Journal of Solid State Circuits* Vol. 25, No. 2, 1990, pp. 376–382.

17. Symon, J. R. and Watanabe, H. "Single Board System for Fuzzy Inference," *Proceedings of the Workshop on Software Tools for Distributed Intelligent Control Systems*, pp. 253–261, September 1990.

18. Pin, F. G., Watanabe, H. Symon, J. R. and Pattay, R. S. "Autonomous Navigation of a Mobile Robot Using Custom-Designed Qualitative Reasoning VLSI Chips and Boards." *Proceedings of the 1992 IEEE International Conference on Robotics and Automation*, Nice, France, pp. 123–128, May, 1992.

19. Pin, F. G. and Watanabe, Y. "Navigation of Mobile Robots Using a Fuzzy Behaviorist Approach and Custom-Designed Fuzzy Inferencing Boards." *Robotica.* Vol. 12, No. 6, 1994, pp. 491–503.

20. Yen, J. and Pfluger, N. "A Fuzzy Logic Based Robot Navigation System." *Proceedings of the AAAI Symposium on Applications of Artificial Intelligence to Real-World Autonomous Mobile Robots*, Cambridge, pp. 195–199, MA, October, 1992.

21. Takeuchi, T. et al. "Fuzzy Control of a Mobile Robot for Obstacle Avoidance." *Information Science*. Vol. 45, 1988, pp. 231–239.

22. Sugeno, M. et al. "Fuzzy Algorithmic Control of Model Car by Oral Instructions." *Fuzzy Sets and Systems*. Vol. 32, 1989, pp. 207–219.

23. Pin, F. G. and Watanabe, Y. "Steps Toward Sensor-Based Vehicle Navigation in Outdoor Environments Using a Fuzzy Behaviorist Approach." *International Journal of Intelligent and Fuzzy Systems*. Vol. 1, No. 2, 1993, pp. 95–107.

24. Pin, F. G. and Watanabe, Y. "Automatic Generation of Fuzzy Rules Using the Fuzzy Behaviorist Approach: The Case of Sensor-Based Robot Navigation." *Intelligent Automation and Soft Computing*. Vol. 1, No. 2, 1995, pp. 161–178.

25. Pin and Killough, S. M. "A New Family of Omnidirectional and Holonomic Wheeled Platforms for Mobile Robots." *IEEE Transactions on Robotics and Automation*. Vol. 10, No. 4, 1994, pp. 480–489.

14

A FUZZY CONTRACT NET APPROACH TO MANUFACTURING CONTROL

N. Vojdani
University of Dortmund
GERMANY

The manufacturing industry actually adopts the most intensive and comprehensive steps toward rationalization. The development toward intelligent control of manufacturing and logistics processes and quick reaction on rapid changes of production parameters require more efficient and intelligent manufacturing control systems in order to achieve maximum results.

A fuzzy contract-net-based system devoted to solving the scheduling problem in a variety of manufacturing environments has been developed at the University of Dortmund providing workers with current manufacturing schedules and resource availability reports.

It features an architecture inspired by the contract net described by Smith and Davis and a hierarchical fuzzy model. In general the contract net approach models transfer of control in a distributed system with the metaphor of negotiation among autonomous intelligent beings.

14.1 MANUFACTURING CONTROL

Scheduling in manufacturing control is a field of investigation where advances in theory and solution techniques have not produced expected results at the industrial level due to specific features of case studies which prevented the application of general approaches. The scheduling problem in a workshop can be stated as follows [1]: find a feasible sequencing of processing operations in the various facilities which satisfies the production requirements given a set of production facilities and technological constraints and given production requirements expressed in terms of quantity, product quality, and time constraints.

The factory floor has a stochastic behavior. The time required for individual operations is often very different. There is a complex dependency between the current load of a machine and the resources required for an operation at that machine. Machines may fail completely, and the tasks may have to be taken over by other machines less suited to the work. Static models of the capabilities of individual machines are unreliable. We need to monitor the status of resources continually, so as to adapt to changes.

The Production Planning and Control Systems (PPCS) do not satisfy the increasing requirements in short-term manufacturing control such as quick reaction of changing production parameters. In general they support middle-term planning.

As a result, it is necessary to put some tasks of production planning and control in the decentralized operation levels, in which the knowledge exists to take advantage of know-how, experience, and creativity of shop-floor managers and to react adequately to manufacturing situations.

Distributed artificial intelligence and specially the contract-net model described by Smith and Davis [2,3] combined with Fuzzy logic methodology look attractive to combine general-purpose knowledge with dedicated information provided by shop-floor managers. Moreover, the scheduling strategy may explicitly depend upon the production objectives and account for changes in the workshop configuration.

Artificial intelligence scheduling systems have already been developed based on heuristic, constraint-guided search, frame-based knowledge representation [4,5,6], and constraint propagation [1]. CRONOS-II [7] and CRONOS-III [8] are knowledge-based tools devoted to solve scheduling problem in a variety of production environments. YAMS [9] is implemented as a factory control system that partitions tasks using a negotiation protocol.

14.2 THE THEORETICAL BACKGROUND OF CONTRACT NET MODEL

A contract net consists of a set of nodes (agents) which communicate with one another by exchanging messages. Agents represent the distributed computing resources to be managed. There are three classes of agents to be identified:

- The task manager is the agent that identifies tasks to be done and assigns them to other agents for execution.
- The bidders are agents that offer to perform a task.

- The contractor is a successful bidder, one whose bid has been accepted by the manager.

Agents communicate by means of different classes of messages:

- A manager issues a *Task Announcement* describing the task to be done and the criteria for bids.
- Bidders send *Bids* to announce their willingness and ability to perform a task.
- The *Award* message from the manager to the successful bidder establishes that bidder as the contractor for the task.
- the contractor sends an *Acknowledgment* of the award, either accepting or rejecting it.
- The contractor send *Reports* to the manager announcing status or termination of a task.
- The manager may send a *Termination* to a contractor to interrupt its performance of a contract prematurely.
- Idle agents may broadcast their availability with an *Agent Availability Announcement.*

Applications of negotiation for control offer for example:

- Vehicle tracking in a distributed sensor system [3].
- allocating Tasks in a network of CPU's [10,11].
- Modeling a manufacturing factory with YAMS [9].

14.3 WIKOLEIT

WIKOLEIT is the German acronym for a cooperative fuzzy contract net for manufacturing control. WIKOLEIT models functional groups of machinery and workcells which corresponds closely to the traditional manufacturing view of a manufacturing enterprise as Semi Autonomous Control Groups (SACG). Figure 14.1 shows some SACG in a shop floor environment.

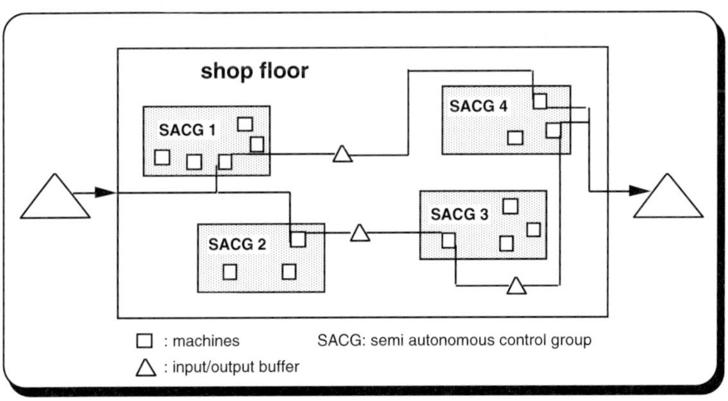

Figure 14.1 SACG in a shop floor environment.

Each SACG include a contract net which is modified as below:

- The task (job) manager is responsible for the management of the manufacturing tasks. It identifies tasks and assigns them to a special agent called dispatcher.
- The dispatcher coordinates the tasks between manager and different bidders e.g., machine agents, personal agent, and resource agent. It starts a negotiation process and sends task announcements to the bidders.
- The bidders are agents that offer to perform a task. We distinguish between three classes of bidders:
    - Machine agents manage the working capacity of an associated machine or workcell.
    - Staff agent is responsible for the service staff (workers).
    - Resource agent manages further resources which are necessary for the execution of process plan (tools, drawings, CNC-programs etc.).
- The contractor is a successful bidder, one whose bid has been accepted by the dispatcher.

Figure 14.2 illustrates a contract net model in a SACG.

The agents are implemented as independent processes with their own data and control segments. They communicate through a set of messages specified by an underlying

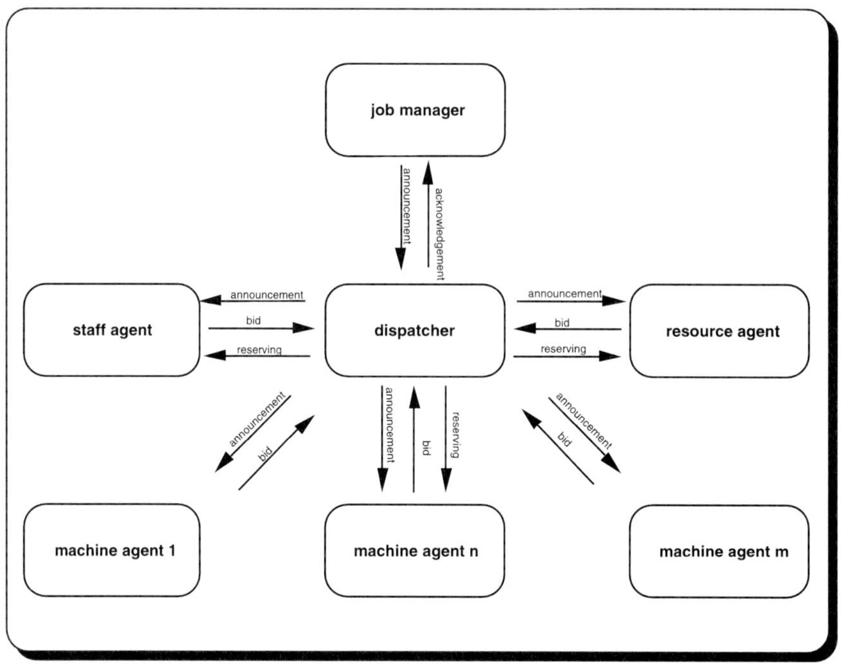

Figure 14.2 A contract net model in a SACG.

communication protocol which describes the way of negotiation between the agents.

Each job/task manager becomes a suitable process plan from the PPC system including the processes (operations) it knows how to perform. There are also references to other processes that are not known to the manager hosting the main process.

The existence of process plans shows that the problem decomposition phase of distributed problem solving is trivial in this case. The process plan constitutes an *a priori* division of the problem, available to the entire system at the outset.

The announcements of the task manager describe the task to be done and the criteria for bids. They will be send to the dispatcher, who chooses the most important one depending on the actual situation in the shop and transmits it forward to the machine agent as well as to the staff agent and resource agent.

Bidders are now able to determine and announce their willingness and ability to perform a task. They check the specification of the announcement and compare it with their own specification and technical abilities. At least bidders send back their willingness and abilities to perform the task to the dispatcher. From the set of incoming bids the dispatcher chooses the best fitting one, conclude an execution contract between the task manager and the successful bidders and inform the task manager as well as the contractors about these contract. A contractor is a successful bidder, whose bid has been accepted by the dispatcher. In the case of the machine agents follows the allocation of the machine capacity while the staff agent reserves the service staff. Further necessary resources are occupied by the resource agent.

The negotiation process of the dispatchers described above continues until all announcements of the task manager are satisfied. During the decision process a set of different alternatives have to be considered. Its necessary to include the knowledge of the responsible manufacturing experts in this decision process. The advices of the experts are rather formulated verbal and vague than with mathematical exactness. Handling this information and the determination of the next task announcement to broadcast and the best fitting bid can be well supported by fuzzy logic.

14.4 FUZZY LOGIC COMPONENT OF WIKOLEIT

To model the dispatcher in the contract net a hierarchical fuzzy system is used. Figure 14.3 shows the fuzzy supported negotiation of the dispatcher. The negotiation process of a dispatcher works in following steps:

Step 1—receiving of announcement data. The dispatcher receives the announcement information from job manager. It consists of the parameter in the process plan such as processing time, start time, delivery due date, local job priority. They are represented as linguistic variables and their linguistic values. Figure 14.4 depicts the linguistic variable announcement priority.

Step 2—determination of announcement priority. In this step the received data will be fuzzyfied. It follows the fuzzy inference process with the aid of a fuzzy rule base. The fuzzy rule base includes the dedicated knowledge of the manufacturing experts. After

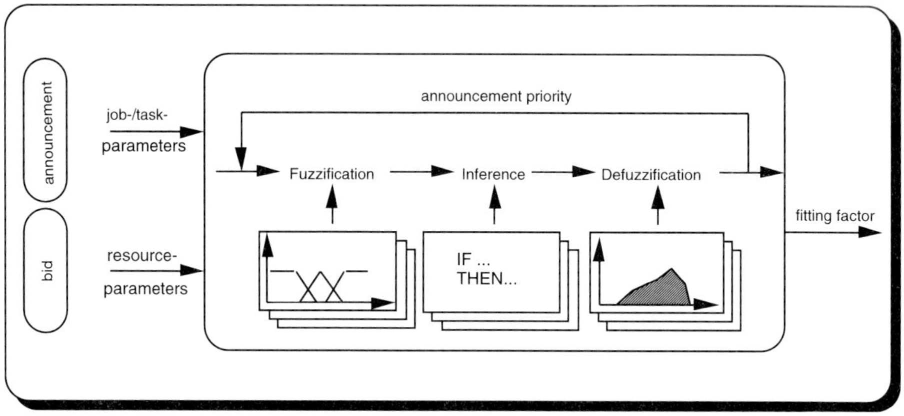

Figure 14.3 Fuzzy supported negotiation of the dispatcher.

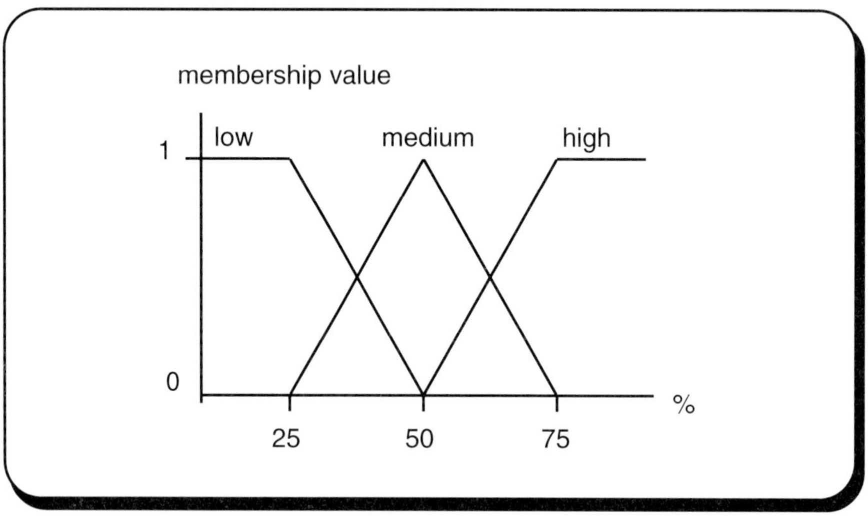

Figure 14.4 Fuzzy membership function for linguistic variable "announcement priority".

defuzzification the announcement priority is determined as a crisp value. Figure 14.5 shows the fuzzy inference process for determination of announcement priority.

Step 3—broadcasting for a bid. After determining the announcement with high priority, it is announced (broadcast) between different agents. The agents try to generate bids considering their specification and the requirements of the announcement. The bids contain resource parameters which are modeled as linguistic variables; their values are crisp input data of the next inference process.

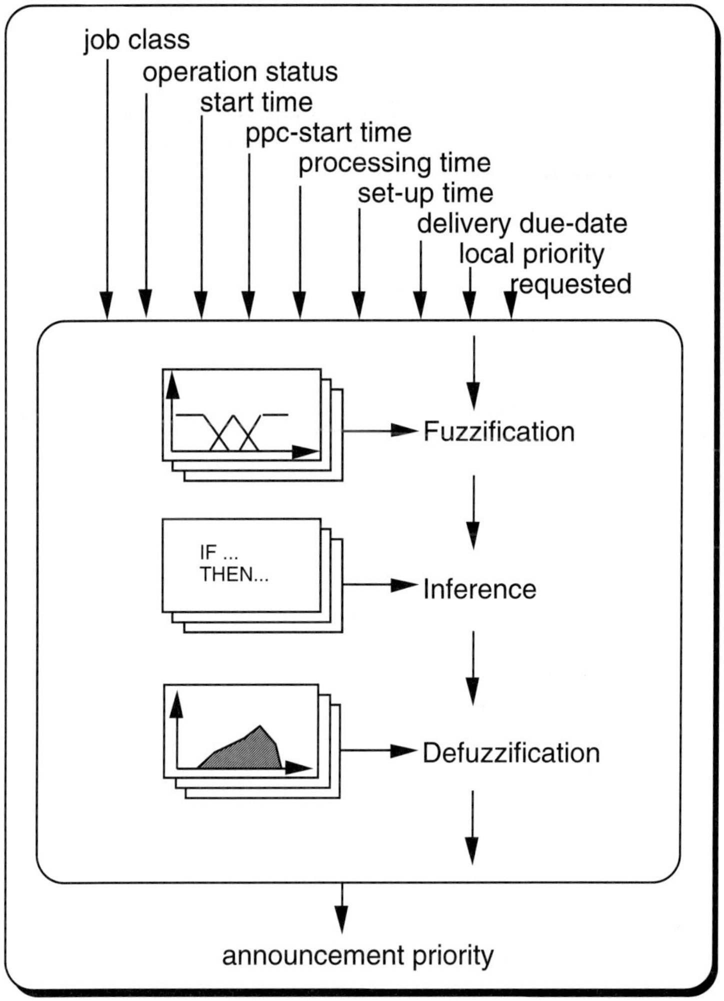

Figure 14.5 Fuzzy inference process for determination of announcement priority.

step 4—matching announcement/bid. To find the possible combination all bids would be matched to the announcement.

Step 5—determination of the fitting factor. For each possible announcement/bid combination the bid data will be fuzzyfied. Together with the fuzzyfied announcement priority, they build the input data for the next inference process to determine the fitting factor. Figure 14.6 illustrates the fuzzy inference process for determination of fitting factor.

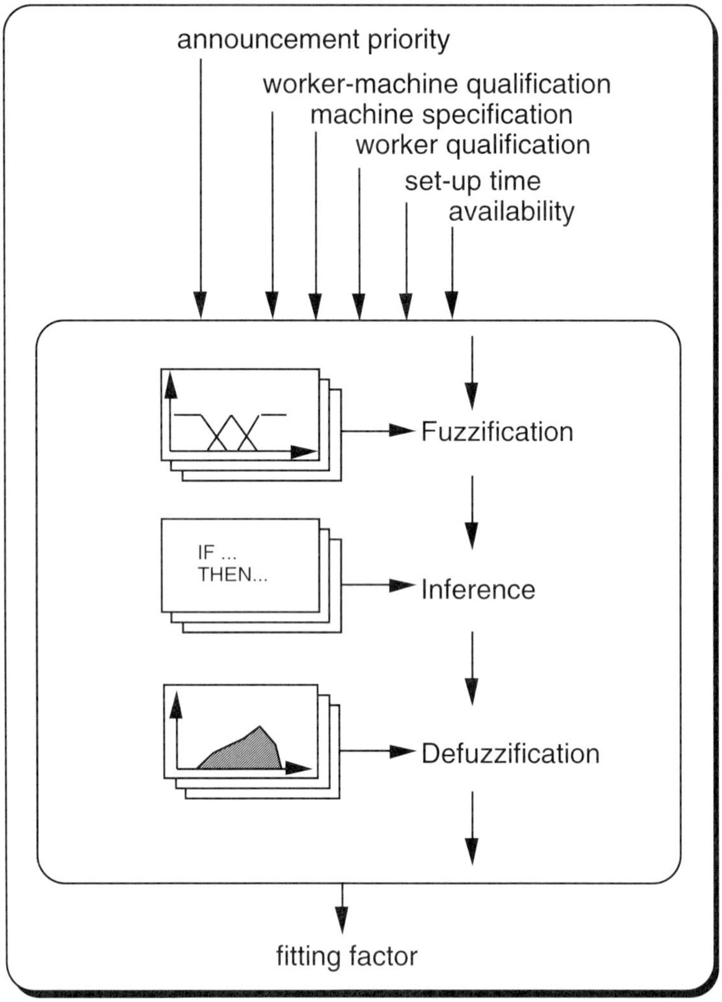

Figure 14.6 Fuzzy inference process for determination of fitting factor.

step 6—determination of the fittest combination. The announcement/bid combinations would be compared with the aid of the fitting factor to determine the fittest one, for which the required resources will be reserved. The successful machine agent, staff agent, and resource agent will become an award message from the dispatcher to establish them as contractor for the task.

14.5 RESULTS

With this prototype we are able to model a discrete manufacturing environment with machines, workcells, service stuff and further resources (tools, drawings, CNC-programs, etc.) associated with different workshops (sawing, lathing, milling, etc.). The system was fed with manufacturing tasks of a superior medium-term production planning and control system. In the scope of a knowledge acquisition several production experts have been interviewed. With their knowledge about connections and interactions in decision situations the fuzzy rule base of the prototype has been configured.

The cooperating fuzzy contract nets take care of a smooth order running through the different shops. With the release of a manufacturing order the attached working instructions are transmitted to the workshops and taken over by the task manager. Through the transmission of the instructions to all involved control areas informations for local planning and global optimization is available.

Following the release of the manufacturing order through the production planning and control system the manufacturing order is taken over from the control area, in which workshop the first process of the working instruction has to be executed. The task manager generates the corresponding announcement and transmits it to the dispatcher, who starts the bidding process. If a contract between the task manager and successful bidders is achieved, all the following workshops are informed about the disposition of the working process in the actual workshop.

In this way the actual data of all preceding workshops are available to the following ones and could be considered by internal planning within the fuzzy contract nets. After the successful execution of a working process a confirmation message is send to the superior planning and control system and the manufacturing order is given over to the responsibility of the subsequent workshop. The procedure goes on until all processes described in the working instruction are executed.

Both job/order and machine schedules could be displayed. For example an actual machine schedule can be visualized through the screen shot in Figure 14.7. In the list box

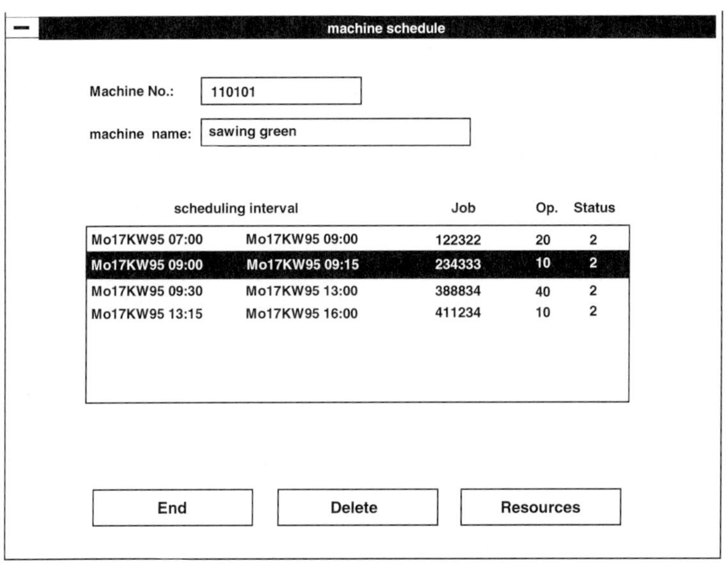

Figure 14.7 A machine schedule.

different scheduling and availability intervals for the machine placed in the upper box are listed.

Following interval status are possible:

1. break
2. idle (available)
3. allocated
4. idle (interruption interval)
5. in process
6. idle (processing interval)
7. terminated

By pushing the "delete" bottom one can cancel a schedule. Pushing the "resource" bottom shows the actual status of required resources.

14.6 CONCLUSION

The fuzzy contract net pursues a solution strategy which is similar to human planning and control and is able to formalize and process uncertain manufacturing knowledge using Zadeh's fuzzy set theory which provides a conceptual framework for handling this kind of knowledge.

In review to the decentralization of the workshops in a production plant, fuzzy contract nets support a manufacturing topology with different control areas (sawing, lathing, milling etc.). They are represented by one fuzzy contract net, which is responsible for planning and controlling the associated machines, service staff, and resources. The interdependencies among the workshops require the cooperation and coordination of the fuzzy contract nets.

WIKOLEIT provides a reactive scheduling of plants and support the user with interfaces to see (and update) the status of all orders and resources. For example, if a machine breaks down, then all affected orders could be rescheduled. It can also be extended to connect to an online data collection system on the plant floor, providing real-time updating of plant status.

14. 7 REFERENCES

1. Bensana, E., Bel, G. and Dubois, D. "OPAL: A Multi-Knowledge-Based System for Industrial Job-Shop Scheduling." *International Journal of Production Research.* Vol. 26, No. 5, 1988, pp. 795-819.

2. Smith, R. G. and Davis, R. "Frameworks for Cooperation in Distributed Problem Solving." *IEEE Transactions on System, Man and Cybernetics*, Vol SMC-II, No. 1, 1981, pp. 61-70.

3. Smith, R. G. and Davis, R. "Negotiation as a Metaphor for Distributed Problem Solving". *Artificial Intelligence*, Vol. 20, 1983, pp. 63-109.

4. Fox, M. S., Allen, B. and Strohm, G. "Job-Shop Scheduling: an Investigation in Constraint-Directed Reasoning." *Proc. 2nd American AI Conference*, Pittsburgh, PA, pp. 155-158, 1982.

5. Le Pape, C. "Soja: A Daily Workshop Scheduling System". *Proc. 5th Technical Conf. of the British Computer Society—Specialist Group on Expert Systems*, Warwick, UK, 1985.

6. Smith, S. F., Fox, M. S. and Peng, St. Ow. "Constructing and Maintaining Detailed Production Plans: Investigation into the Development of Knowledge Based Factory Scheduling Systems." *AI Magazine*, Fall 1986, pp. 45-61.

7. Canzi, U., Guida, G., Poloni, W. and Pozzi, S. "CRONOS-II: A Knowledge-Based Scheduler for Complex Manufacturing Environment." *Proc. 2nd ACM IEEE Int. Conf. on Data and Knowledge Systems for Manufacturing and Engineering*. pp. 76-82, Gaithersburg, MD, 1989.

8. Gottlob, G. and Nejdi, W. "Expert Systems in Engineering." In *Lecture Notes in Artificial Intelligence*. (Sickmann, J. ed.). Springer, 1990.

9. Parunak, H. V. D. "Manufacturing Experience with the Contract Net". In *Distributed Artificial Intelligence. Research Notes in Artificial Intelligence*. (Huhns, M. N. ed.). Pitman, 1987.

10. Ramamitham, K. and Stankovic, J. A. "Dynamic Task Scheduling in Hard Real-Time Distributed Systems." *IEEE Software*, pp. 65-75, July 1984.

11. Stankovic, J. A. and Sidhu, I. S. "An Adaptive Bidding Algorithm For Processes, Clusters and Distributed Groups". *Proceedings of the Forth International Conference on Distributed Computing Systems*, San Francisco, pp. 49-59, 1984.

15

FUZZY DISTRIBUTED INTELLIGENT SYSTEM FOR CONTINUOUS PRODUCTION

A. A. Aliev and R. A. Aliev
Azerbaijan Oil Academy
AZERBAIJAN

The kernel of the modern industrial technology, particularly, such its constituents as Computed Integrated Manufacturing (CIM) and Flexible Manufacturing Systems(FMS) is Distributed Artificial Intelligence (DAI). In this chapter the concept, construction principles, architecture, and interaction mechanisms of elements of distributed intelligent systems for continuous industrial branches, particularly oil refinery plants are considered.

The environment of such plants is characterized by uncertainty, imprecision, missing and incomplete data and knowledge, and is supposed to be fuzzy and thus, the Fuzzy Distributed Multi-agent Intelligent System (FDIS) is considered. Its coordinating intelligent agents are described, namely, fuzzy planning, fuzzy scheduling, dispatching, and local on-line dynamic and other expert systems for technological processes (units).

Rules and conventions, protocols of interactions between intelligent agents, functional models of transferring and processing distributed data and knowledge in FDIS for an oil refinery plant are considered. Some of the experimental results of the working of the FDIS on an oil refinery plant are given.

15.1 INTRODUCTION

Modern industrial technology, particularly, its constituents such as FMS and CIM, are based on the ideology and technique of Distributed Artificial Intelligence (DAI). This is caused mainly by the fact that many application subjects, for instance, industrial enterprises, particularly oil refining and petrochemical plants, are geographically distributed, their units and warehouses at locations remote from one another. At the same time they must coordinate their efforts effectively.

It would not be effective to connect states of each unit globally, especially because of real-time constraints that characterize continuous production operations. Instead of using a centralized artificial intelligence with a single agent carrying the global knowledge (which is usually implemented on computers with monolith programs), it is preferable to use the DAI supplied multi-agent system with a proper distribution of functions, knowledge, and resources of the "center" among intelligent agents [1,2].

Another reason dictating the necessity of DAI follows. A number of local intelligent systems for planning, manufacturing scheduling, technological processes control, which at first realize the separate control functions, are created and designed. These are "isles of intelligence" in the plant technology.

However, they lack coordination of goals, problems, and models and compatibility and integrity of their software and hardware are absent. The above do not allow achievement of expected system effect from making them intelligent.

One efficient method of solving the indicated problem is the creation of distributed intelligent control system on the base of an Expert Systems (ESs) network, which integrates in itself the autonomously functioning ESs and other systems oriented to knowledge. The main task here is an efficient utilization of the global ("center") and local (agents) knowledge, information, and computing resources and reasoning technique to obtain an appropriate coordination in order to achieve the global goals. It should be noted that intelligent agents often deal with incomplete, contradictory, missing, and inaccurate data and knowledge [8]. Furthermore, the agents have to make decisions in uncertain situations, i. e., DAI functions in the real world with an environment of uncertainty and imprecision. Obviously, an adequate approach to deal with the real application subjects with such properties is fuzzy logic [3]. In this chapter the concepts, construction principles, architecture, elements (entity of agents) coordination of elements (entities of agents), protocols of knowledge exchange and other immediate problems of the Fuzzy Distributed Intelligent System (FDIS) are considered as they related to an oil refinery plant.

Many publications are devoted to problems of creating DAI [4,5]. Just in [6], 573 works are referred to in the area of cooperation, coordination, distributed control, distributed knowledge, and knowledge representation, multi-agent planning, etc. Though there are only a few of works [7,8]. devoted to FDIS as well as the practical aspects of DAI and FDIS.

15.2 FUZZY DISTRIBUTED INTELLIGENT SYSTEMS (FDIS) DEVELOPMENT CONCEPTS [7]

In this chapter the idea of FDIS development on the basis of hierarchical networks of fuzzy (or nonfuzzy) processing nodes is suggested. In the hierarchical structure, intelligent systems (nodes), which perform the functions of coordinator, and a number of local systems (nodes), which implement control functions of the lower levels of hierarchy, are defined.

The coordinator performs dual functions: on the one hand, it is an intelligent element of a high level of control and it solves functional problems on the level of supervisor and main specialists; on the other hand, it provides interactions between local systems. Regarding the end users of the local systems, the coordinator must provide recognition of fuzzy queries, and perform re-addressing to the specific local system.

As a rule, the local system, which sends queries to the coordinator, has no information about the locations and sources of needed knowledge and data.

During the knowledge communication process performed by the coordinator, transformation of queries and responses must be provided. These are done in accordance with accepted formalisms, used by individual intelligent systems. Thus, the coordinator should support the functions of data and knowledge communication protocols in the network.

Users of local intelligent systems should solve the specific problems, associated with plant control and performance of individual control functions.

The information compatibility of different local systems must be provided by the special unified interfaces and communication scheme. The created FDIS system is characterized as a cooperative knowledge-based system, where nodal knowledge-based elements are autonomous heterogeneous agents, and the rational interaction among cooperating agents is supported.

15.3 FDIS DEVELOPMENT PRINCIPLES [7]

To realize the suggested concept of fuzzy distributed intelligent system development, the following main principles are defined [7].

1. Decomposition of intelligence. There is need for reasonable decomposition of the global intelligence into a set of particular intelligences during development of a distributed control system on the basis of an intelligent systems network. Each of those particular intelligences embraces a limited and specific area of problems and tasks. The decomposition can be accomplished either by natural or artificial manner.

2. Possibility of coordinating intelligences. The decomposition of intelligence, distributed nature of knowledge, control, and decision making processes require the

provision of necessary conditions for integrating particular intelligences during separate problem solving processes. The possibility of coordinating and integrating particular intelligences allow successful decisions making under conditions of information insufficiency.

3. Nonadditive nature of the system's intelligence. The intelligence of the system is not additive regarding its components' intelligences. Knowledge communication and new knowledge generation cause the intelligent potential of the system to exceed the total intelligence of its components.

4. Possibility of information communication between nodes. In a hierarchical network knowledge communication is performed both between the levels of hierarchy (vertically) and between nodes of the same level (horizontally).

5. Adaptability. The network is subjected to difficult demands for the reliability and vitality of system. It must preserve its functional capabilities. Also, the accordances between algorithms of control of current situations and objects' states must be provided. AI systems, functioning in real time, forming and realizing control actions directly on control objects, require high system reliability. In case of failure of some node, restoration of functional capabilities of the system must be provided by reconfiguration of structure. It is necessary to provide the mechanism of data-base and knowledge-base updating. Double adaptation—structural and parametric—is the necessary condition of distributed control system creating on the base of intelligent subsystems.

6. Possibility of working in multiple modes. Local systems, included in a distributed control system, solve different content and form problems. Considering the entities of problems and users' requirements, multimode working (batch processing, operative access, real time) is necessary.

7. The open nature of FDIS. As a rule, at the system's creation, the limited problem area for control object is examined. During evolution of object, new problems occur, which must be coordinated with the earlier considered problems solving, i.e., object's development requires an adequate development of control system. Taking this into account, the FDIS must be created as a open system with the modular structure providing the possibility of including new systems.

8. Intelligent nature of nodes. Existing nodes and nodes to be created must have their own intelligences and be interconnected with other nodes through communication network over their powers, competence and implemented functions.

9. Distributed nature of system. It is necessary to provide flexible distribution of intelligent, information, and computation resources among the nodes. In order to rationally use the computational resources of different levels of the hierarchy and increase functional capabilities of system, it is necessary to effectively use the

capabilities of controllers and intelligent systems for information collecting and processing, solving of problems the diagnosis and preparing information for experts on the low level of the hierarchy. The center, using the information about free resources and efficiency of their utilization, should solve the problem of distribution and redistribution of intelligence and other resources.

15.4 ARCHITECTURE OF FDIS (RELATED TO OIL REFINERY PLANT)

The suggested conception and principles were used as the basis of building a FDIS for the oil refinery plant. This fuel plant provides the complete cycle of oil refining into the finished products: automobile petrol, diesel fuel, aviation kerosene, and tar. The simplified technological scheme of the oil refinery plant is shown in Figure 15.1. The primary oil refinery unit is the first section in the technological chain of the oil refinery plant. In this unit as the result of refining the petrol fractions SB 85° C (SB–start of boiling), 85–180° C, aviation kerosene, diesel fuel, vacuum remainder, tar, etc., are produced. Unit ELOU-AVT is closely connected with other units of the plant. Fraction SB 85° C is a component of petrol and fraction 85-180° C is the initial raw material for catalytic reforming process. The vacuum remainder is utilized by the catalytic cracking unit. Petrol fraction 85-180° C enters the catalytic reforming unit in order to produce from it the high-octane component of gasoline. The vacuum remainder of this unit enters the catalytic cracking unit. Outputs of this unit are stable, high-octane gasoline, gas, light and heavy flegma. The efficiency of catalytic cracking unit work essentially depends on raw material quality. The process of gasoline mixing (compounding) is the final state of manufacturing. For preparing different gasolines the following raw materials are used:

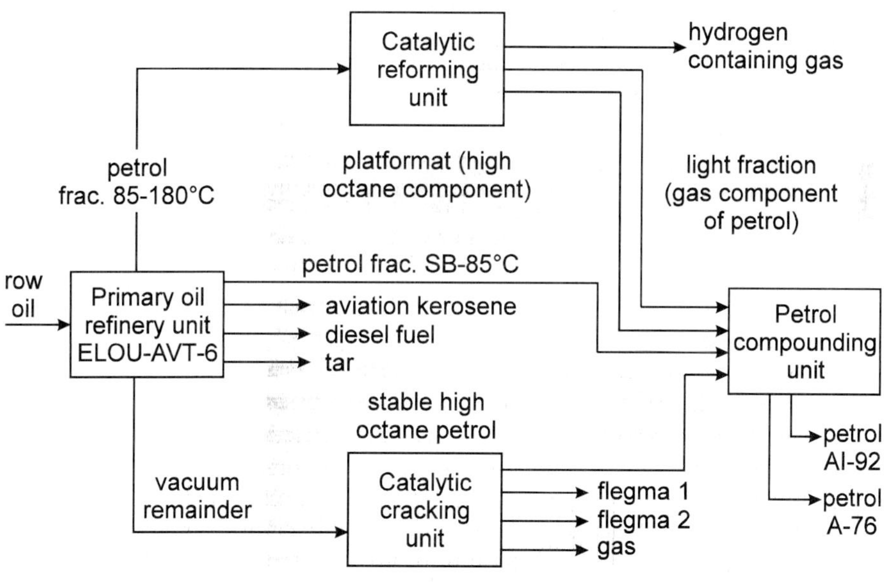

Figure 15.1 The simplified technological scheme of the oil refinery plant.

platformat from catalytic reforming unit, stable high octane petrol from catalytic cracking unit, technical isooctane (imported), petrol fraction SB 85° C from the primary oil refinery unit.

The characteristic features of the object are close interconnections between units where changing of the working mode of one unit influences the others modes that leads to necessity of mutual adjusting of working modes of interconnected units.

There are a large number of external and internal disturbances which influence the plant. The control decisions are made under conditions of a high degree of uncertainty, generated by incompleteness and inaccuracy of initial information and technological processes descriptions, fuzziness of formed goals, limitations, and possible effects of decisions. All of the above limit an application of formal methods for control decisions and requires the use of FDIS.

Proceeding from the indicated specifics of oil refining, the architecture of FDIS is defined in the form of a hierarchical network, the nodes of which are intelligent agents (for example, fuzzy expert systems), and connections between them provide the data and knowledge communication (Figure 15.2).

Coordinating Agents (CAgs) coordinate the work of the bottom-level elements.

Intelligent agents of the first two levels are used for decision making and coordinating the plant's working modes.

Intelligent agents of the third and fourth levels are the intelligent systems of information collecting and processing, diagnosing, and control of mode parameters of

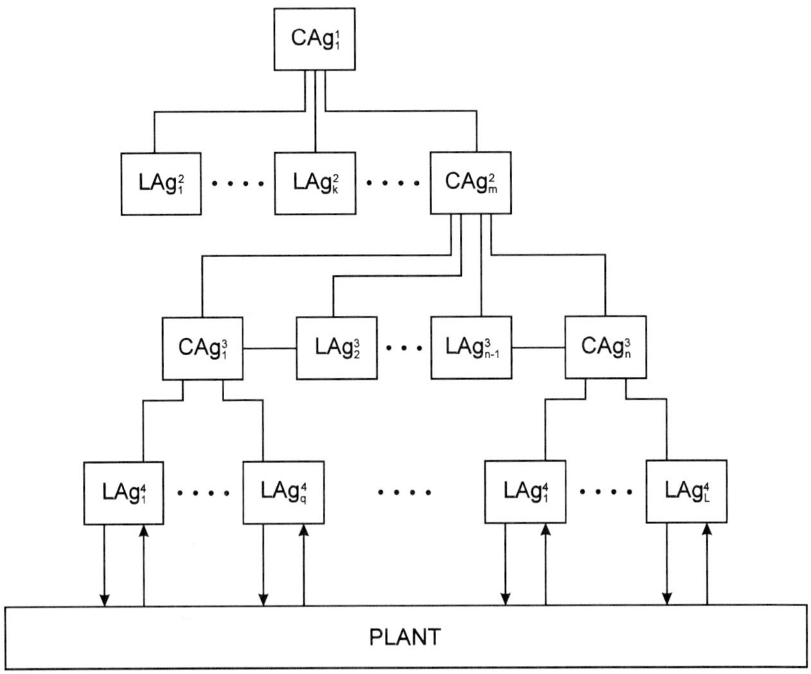

Figure 15.2 The architecture of FDIS.

dynamic objects. Agents of the given level have direct connections with executive mechanisms of local control objects, i.e., they perform direct control of the object's parameters.

The third and fourth levels of this system function in real time. Decision making on selecting rational modes and diagnosing technological processes (third level), automatic control of technological processes in order to support exactly those modes (fourth level) are dynamic problem areas; data and knowledge about them continuously change.

fuzzy expert systems often are used as intelligent agents in FDIS.

Implementation of this architecture requires providing interconnections between agents, i.e., implementing processes of communication between knowledge bases (KBs) which use different knowledge representation; coordinating Ag working modes, i.e., managing knowledge flow and setting priorities of appealing to the concrete agent; removing contradictory nature of knowledge in Ag network; offering to each component of network the necessary computational resources of the system, etc.

Main difficulties occur when different agents (ESs) are connected. To solve this problem the method of complete solutions communication on the basis of unique internode language of information representation is selected.

To support interrelations between agents (ESs) the following modules have been developed: former (F), adapter (A), local library of processes (LLP), global library of processes (GLP), and object glossary (Figure 15.3).

The former is intended for transforming information from the internal representation form into representation form of internode language.

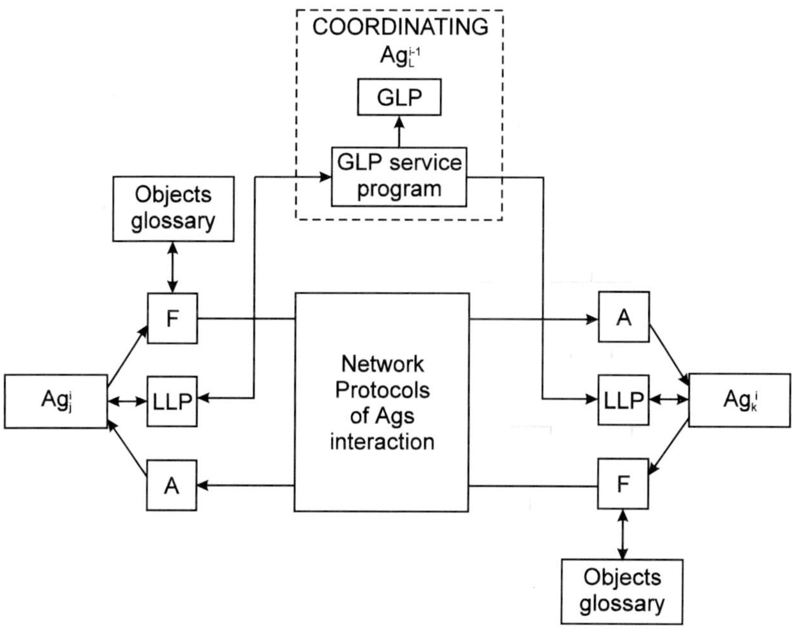

Figure 15.3 The scheme of queries processing.

The adapter is intended for performing reverse functions to former ones.

The local library of processes contains meta-knowledge about knowledge of other nodes of a given level.

The objects glossary gives the node correlations between objects of KBs of network nodes, with which it communicates.

The agent which discovers lack of available knowledge for solving a given problem (during inference) forms a query to another agent of the network. The query is formed in the following way: Ag appeals to its LLP, which provides the required knowledge and necessary list of objects (situations which are needed to solve the problem are defined).

After this Ag activates former which forms objects list and purpose of appealing to KB of another Ag and transforms query in accordance with (requirements of) internode language. Then network software (NS) is activated. By means of NS the formed and transformed query is transferred along data communication network to Ag-receiver in accordance with developed protocols of intelligent systems interaction.

After transferring the query to node (Ag-receiver) adapter of this node performs reverse transformation of the derived query. This query in the form <objects list, purpose> is transferred to Ag. Ag-receiver performs inference and transfers the solution to its former which transforms the solution into internode language representation form and passes it to adapter of Ag-sender, which carries on its inference.

All nodes included into the functional structure of FDIS interact in accordance with the described above interaction mechanism.

Each Ag (node—ES) has its functional purpose and implements some strategy of information communication. Regarding to sender node this strategy determines when the node must transfer information to other nodes what is type of information to be transferred and what the nodes are suitable receivers of this information. Regarding to receiver node the strategy gives conditions of information receiving, types of received information, suitable nodes-senders and methods of local estimation of received information.

15.5 PROTOCOLS OF NODE INTERACTION IN FDIS

The developed protocol of the interaction of intelligent systems (Ags) in FDIS, using generalized service given by levels of open systems interaction, provides transparent communication of knowledge between agents.

The protocol of intelligent systems interaction consists of three protocols [7] (Figure 15.4):

- ARK agent—receiver of knowledge
- ASK agent—sender of knowledge
- protocol of coordination and determination of transferred knowledge semantics, which allows intelligent systems to determine the order and rules of interaction in the cases of used concepts and expressions in semantics coordination;
- protocol of distributed knowledge transference, which realizes the order and rules of knowledge transference, re-reception, and delivery to user;
- protocol of distributed knowledge processing, which determines the order and rules of agents interaction in various situations.

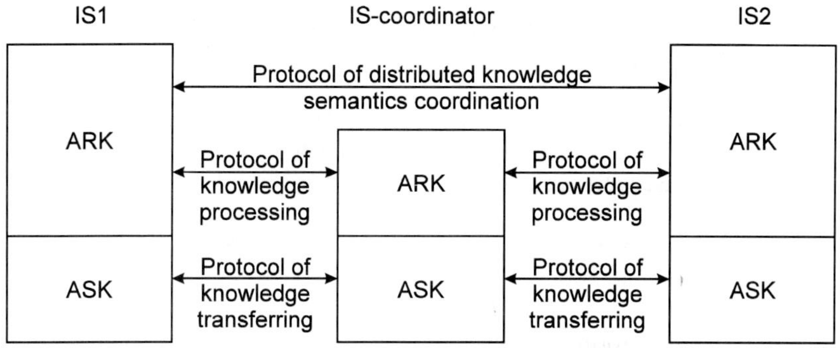

Figure 15.4 The protocol of the interaction of intelligent systems.

15.6 FDIS HARDWARE

FDIS is implemented in the form of a local computer network (LCN) which is an internetwork (Figure 15.5). As one can see from the figure, the internetwork consists of three levels and includes two LCNs integrating hardware and software for solving the following probes.

On the first level problems of planning for the whole plant are solved with the help of the global planner.

The second level integrates computers intended for solving, scheduling and dispatching problems for workshops and with the help of ESs of scheduling and dispatching control.

The third level of LCN integrates computers intended for solving technological unit control problems with the help of compounding ES, catalytic cracking ES, catalytic reforming ES and primary oil refinery ES.

The LCN-1 and LCN-2 are networks with monochannel architecture in the form of bus topology with marker methods of access to data communication media. To LCN-1 and LCN-2 three or four IBM/PC AT computers are connected respectively through network adapters of D-NET type. These networks also include terminators, modules of access, and bridges implemented on IBM/PC AT computers.

For efficient managing the process of distributed knowledge communicating and processing file servers are used on each level of internetwork. They are located in CES computers.

15.7 DESCRIPTION OF INTELLIGENT ELEMENTS (AGENTS) OF FDIS

15.7.1 Planning

The model of the planning problem for the oil refinery plant (Figure 15.1) can be written as follows:

$$\max\ F = C\tilde{\bar{X}};\quad A\tilde{\bar{X}} \leq B;\quad \tilde{\bar{X}} \geq 0 \tag{15-1}$$

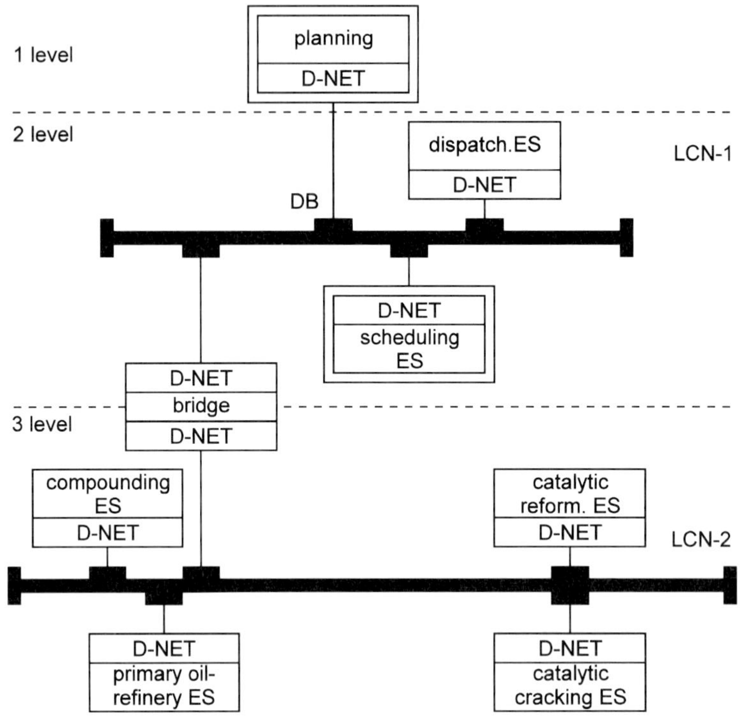

Figure 15.5 Hardware implementation of FDIS.

where F is the goal function; C is the vector of prices; $\tilde{\tilde{X}}$ is the vector of intensity of technological modes of production; A is the matrix of technological coefficients; B is the vector of production resources; ~ means fuzziness.

Let's consider an approach to solving the planning problem in a fuzzy environment based on L-R representation of fuzzy sets that differ from those described in the literature.

By using a definition of the relation between two fuzzy LR-number $\tilde{m}$ and $\tilde{n}$ [9] the crisp equivalent of the system (15-1) will be

$$\max F = \left(CX, C\underline{X}, C\overline{X}\right)$$
$$AX \le B; \quad A(X - \underline{X}) \le B - \underline{B} \tag{15-2}$$
$$A\left(X + \overline{X}\right) \le B + \overline{B}; \quad X,\ \underline{X},\ \overline{X} \ge 0$$

By aggregating the variables

$$x' = x - \underline{x}; \quad x'' = x + \overline{x} \tag{15-3}$$

let's decompose (15-2) to the following systems:

$$\max\ F = CX;\quad AX \le B;\quad X \ge 0 \tag{15-4}$$

$$\max\ F = CX';\quad AX' < B - \underline{B};\quad X' \ge 0;\quad X' \le X^* \tag{15-5}$$

$$\max\ F = CX'';\quad AX'' \le B + \overline{B};\quad X'' \ge 0;\quad X'' \ge X^* \tag{15-6}$$

The systems (15-5) and (15-6) have included the auxiliary constraints reflecting the fact that it is necessary to satisfy inequality:

$$x' = x - \underline{x} \le x^* \le x + \overline{x} = x'' \tag{15-7}$$

where x^* is the optimal solution of the system (15-19).

Having solved the systems (15-4)-(15-6), let's define the values x^*, x'^*, x''^*.

Then the fuzzy solution of the problem (15-7) in LR-representation can be written

$$\widetilde{X}^*_{LR} = \left(x^*, x^* - x'^*, x''^* - x^*\right) \tag{15-8}$$

If lower and upper deviations of vector $\widetilde{B}$ are proportional to the middle value, then the following theorem is true.

Theorem. If for a fuzzy number $\widetilde{B}_{LR} = (B, \underline{B}, \overline{B})$ the condition:

$$\underline{B} = K_1 B;\quad \overline{B} = K_2 B;\quad 0 < K_1,\quad K_2 \le 1 \tag{15-9}$$

is true and the system (15-4) has solution x^*, then (15-5) and (15-6) have solutions:

$$x'^* = \left(1 - K_1\right)x^*;\quad x''^* = \left(1 - K_2\right)x^* \tag{15-10}$$

In this case, to determine the fuzzy solution of system (15-1), it is neecessary to find a solution of system (15-4). The solution of systems(15-5) and (15-6) would be determined by (15-10).

The dimension of the whole model of the problem of planning comprises 208 lines × 212 columns, and the number of elements—902. The software working the algorithm includes the following steps:

1. Technological information A, C to form the model (unit's inputs and outputs, the coefficients of conversion, constraints on quality, prices, etc.) come to Information Base (IB);

2. From the display and IB the values of elements of right-side vectors of constraint and, if needed, other technological information in linguistic form must be given;

3. From IB the information enters the module "Generator" to construct the model for the linear programming task in accordance to requirements to input data formats given in the software complex "LP." At this stage the three models are formed: the model with restrictions on the middle and edge (lower and upper) values of elements of the right-side vector;

4. Models that have formed, come to input "LP";

5. The result of optimization enters IB;

6. Basing on obtained solution and normative and other information of IB, the module "interpreter" forms the full plan which is passed to Scheduling and Dispatching ES to distribute tasks and jobs among units and warehouses for shorter time intervals.

15.7.2 Scheduling Task

The problem of scheduling the oil refinery plant is solved by the fuzzy expert system ESPLAN [1,10,11] that performs calculations and corrections of global plan with division on short periods.

Problems of scheduling are distribution of raw materials and intermediate products among technological units; selection of utilization intensity of these unit work modes; and determination of products' mixing methods.

The knowledge base of the main production current scheduling is built by modular principle. The modules of knowledge base show the partitioning of oil refinery production into stages. The root segment of the knowledge base contains the meta-rules which control the order of subbase operation and have the following form:

```
IF BEGINNING THEN BEG_IIS;
   {in the beginning we put into operation
    the inquiry information subbase (IIS)}
IF BEG_IIS THEN KB(IIS); {IIS activation}
   {Any subbase can activate the object BEG_IIS}
IF END_IIS THEN #KB(IIS) AND KB (INITIAL);
IF END_INITIAL THEN #KB (INITIAL) AND KB (CRACKING);
IF END_CRACKING THEN #KB (CRACKING) AND KB (REFORMING);
IF END_REFORMING THEN #KB (REFORMING) AND KB (COMPOUND);
IF END_COMPOUND THEN #KB (COMPOUND) AND KB (RESULT);
IF END_RESULT THEN #KB (RESULT) AND EXECUTE (GENER);
```

Subbases contain knowledge about various stages of oil refinery production's initial oil refining, cracking, reforming and gasoline compounding. The subbase RESULT forms the output arrays for further generation of output tables. The external program GENER forms the reports (output tables) on the basis of data about current problem solution (output arrays).

Below the fragment of knowledge subbase text is shown:

```
........................................
{Description of objects: short name, full name, minimal,
  value, maximal value, unit of measurement}
OB(S_100, "Plan on local petroleum", 10, 30, "thous.tons");
OB(P_106, "Plan on output of residual oil", 1, 3, "thous.tons");
OB(R_106, "Real output of residual oil", 1, 3, "thous.tons");
OB(E_106, "Excess of residual oil output", 1, 3, "thous.tons");
........................................
OB(E_INIT_106, "Yield efficiency for residual oil at the initial oil refinery unit");
........................................

IF CRACK=Less 0
THEN IN_INIT_100=TP_INIT
AND OUT_INIT_106)=IN_INIT_100*E_INIT_106
AND DISPLAY("Power of the initial oil refinery unit is insufficient")
AND DISPLAY("and demanded plan on residual oil")
BECAUSE "necessary load of the initial oil refinery unit is greater than the top capacity
of the unit"
```

The separate subbase is engaged in the inquiry information maintenance.

The experience in ESPLAN shell application for the creation of a number of scheduling systems shows its high universal abilities, efficiency and comfort of utilization.

The ability of ESPLAN for fuzzy knowledge and data processing is unique among other shells and application expert systems.

The knowledge base for the whole problem contains about 1000 rules which are located in 8 subbases.

15.7.3 Dispatching task

The main task of dispatching ES (DUETS) is monitoring the execution of the unit's tasks and adjusting the plant's balance by generating commands for local agents.

ES DUETS provides materials flow control and accounting; checking availability of raw materials and half-finished products; quantity and quality of finished products; generating and grounding recommendations in concrete situations, etc.

When information for analysis of production situations is prepared, DUETS appeals to ES of technological processes control in order to obtain the operative information about the main technical and economical indices and about states of technological units.

The main properties of the DUETS system are on-line working mode and capability of automatically activating the inference mechanism. To provide that, besides typical blocks such as those for knowledge convertional entry and updating, inference mechanism, explanation and argumentation, monitor, DUETS also contains a block for correspondence with the distributed FDIS data base, served for acquiring actual data about the production process.

The calculation block performs a number of functions for concrete numerical calculations, for example, determining required amounts of different types of raw materials at the input of the technological unit. The DUETS's Knowledge Base (KB) has a three-level structure (Figure 15.6). The lower level of KB consists of the System's World Model (SWM), in which the declarative, extensional and factographic knowledge about the problem area, system's users, and production structure are presented. To represent such knowledge in the SWM the frame formalism is utilized.

The middle level of the knowledge base is formed by the Rule Base (RB), where mainly knowledge of procedural and intentional character is presented. In those production rules, the knowledge of experts and specialists concerning malfunction situations and appropriate actions for their removal, has been accumulated. The rule base has been structured in accordance with peculiarities of the application area and consists of set of rules. Each rule set is an independent production system.

The DUETS inference mechanism works in the following way. After receiving control from the monitor, for example, caused by an inquiry from another agent, the higher level of the knowledge base is activated. Meta-rules define the type of inquiry to the knowledge base and form the goal of the search. Then the set of rules which should be activated is defined. After this, the control is transferred to the rule set, which is the conventional production system.

The inference mechanism is performed in the rule set by a concrete reasoning mechanism and consists of three phases: rule matching, rule selection, and rule execution. Matching means comparing antecedent parts of the rules with the current state of the system's world model. To tell the truth, comparison is carried out among frames-occurrence of the condition elements on left parts of the rules. As a result of accomplishing this phase a conflict set is formed that consists of rules, whose left parts have been matched with the frames-occurrence from the SWM. On the phase of rules selection, the conflict set is cut down to have a single rule in accordance with the priorities of the rules that identifies the strategy of inference.

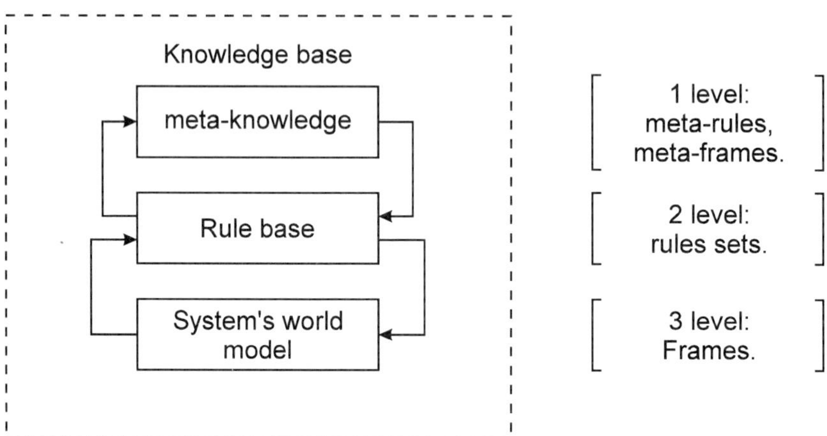

Figure 15.6 Structure of DUETS knowledge base system.

On the phase of execution, execution of the rule is actually performed, and actions from the right part for the rule selected earlier is realized. After this, the comparing phase is reactivated and continued until either goal, stated by the meta-rules, has been reached or the conflict set has no more rules. After one of the conditions is true, control is returned to the high level of the knowledge base. On this level the meta-rules determine whether the goal is reached or not. If not, then a new set of rules is determined which could possibly reach the goal. If there is no such a set or the goal has been reached, the control is returned to the monitor.

An example of interaction of the SWM with the knowledge base during the inference mechanism working time is shown in the Figure 15.7.

15.7.4 Description of Local Intelligent Elements (for Technological Process)

In the result of dispatching problem solving the control tasks to technological units control systems are formed. These systems are implemented in the form of processes expert systems (ES) of primary oil refinery (ESPOR), catalytic cracking (ESCAC), catalytic reforming (ESCAR), compounding (ESCOMP) etc.

The functioning environment of these technological processes are the dynamic domains on which data and knowledge change in time.

These expert systems have a number of features, which are necessary to take into consideration at the design stages [7]: a) large number of the measurable parameters; b) great volume of the knowledge base; c) necessity of the system's high performance speed that provides the real work mode in real-time scale; d) possibility to cope with data varying in time and adaptability to dynamic domain.

The considered expert system includes (Figure 15.8): the information subsystem which consists of the interface with technological process, the dynamic data processing module, and the data base; the diagnostic and decision-making subsystem, including the diagnostic module and decision making module, the results representation and DDC module; relearning subsystem which consists of the results estimation, learning, and interface with knowledge engineer modules.

Close relation between technological process determines necessity of interaction of technological processes ESs between themselves and individual control levels during functioning. For example, primary oil refinery control ES (ESPOR) needs knowledge on raw materials resources and its quality, reagents resources, finished product producing and shipping. This information is given by DUETS. The information about catalytic reforming and cracking unit work is also needed.

ESCAR appeals to DUETS to get the information about row material resources and its quality. To estimate productivity of unit the information about primary oil refinery process states is needed.

Thus, ESCAR functioning requires the interaction of local agents during interrelated control problem solving.

Interaction between agents (ESs) is performed both vertically and horizontally. Conflict situations are settled on the top level, i.e., with the help of CESs of corresponding functional levels.

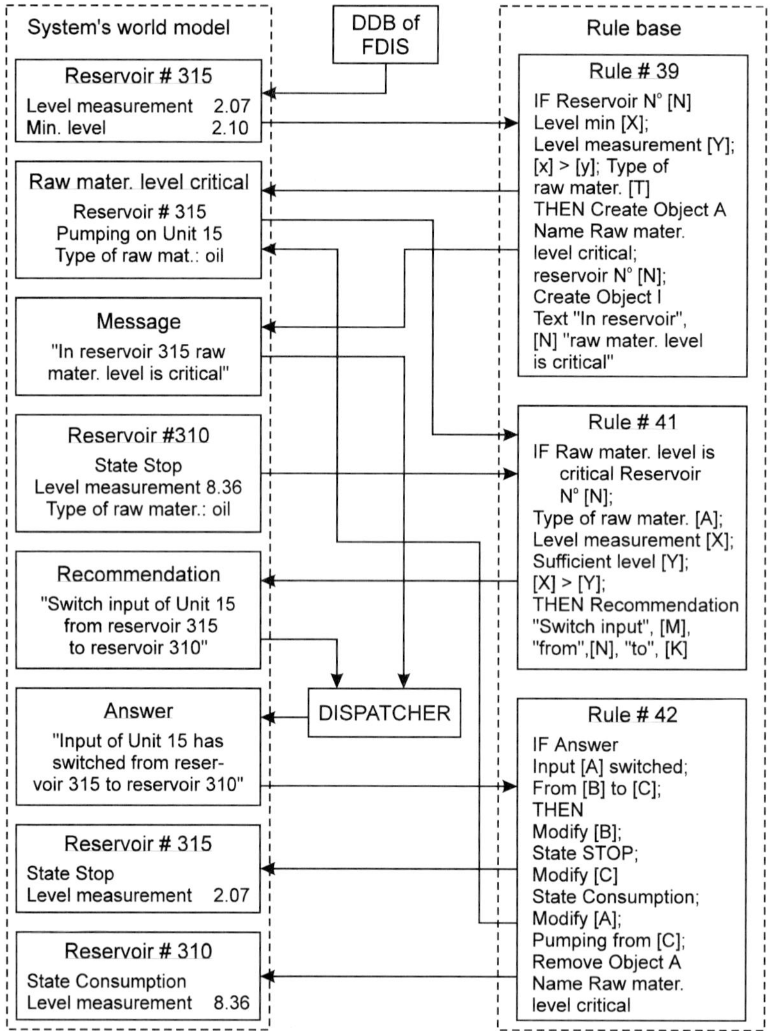

Figure 15.7 Example of working system's inference mechanism.

15.8. EXPERIMENTS ON FDIS FUNCTIONING

Below the process of functioning of FDIS for the oil refinery plant is described on example of the following fragment of two scheduling and compounding systems implemented on the base of ESPLAN shell.

Assume that the production situation is characterized by the facts that platformat is LITTLE and high octane component is ABOUT NORMAL. For successfully solving the problem in this situation one must decrease the plan on AI92 to 15% and correspondingly

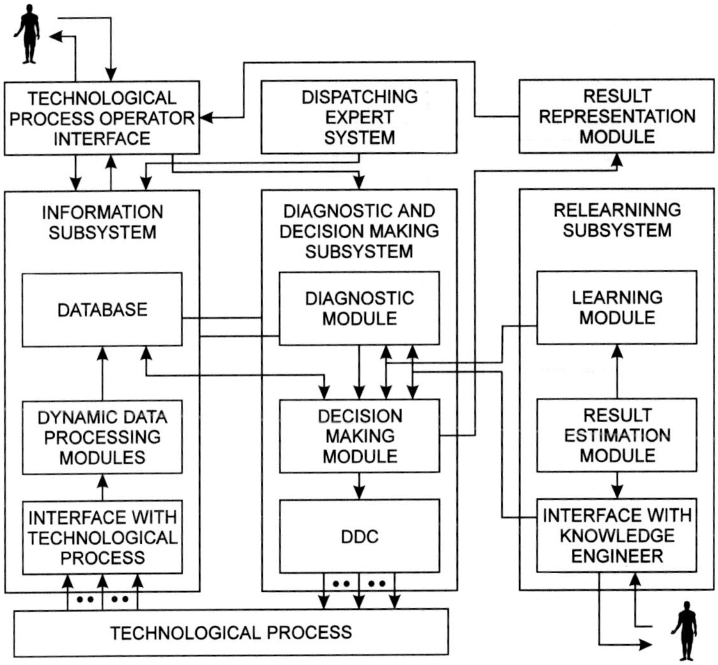

Figure 15.8. The structure of dynamic on-line expert system.

recalculate the recipe of automobile petrol preparation (rule R101 of KB of scheduling agent implemented on the base of ESPLAN shell (Figure 15.9).

ESPLAN forms file query QUERY.WA to ESCOMP to get necessary knowledge (rule R203 of ESPLAN KB). The rule R203 contains special operator-functions of ESPLAN shell: OPEN(file name)—open file, WRITE (object name)—write object name and its value from work area (WA) into file, CLOSE—close file.

Thus, rule R203 must be interpreted as: if calculation of plan on AI92 is needed then appeal to ESCOMP with query QUERY.WA in which current data on quantity of platformat and high octane are written as well as requirement to decrease the plan of AI92 to 15%.

ESCOMP starts its session of work from reading the file-query addressed to it (rules R201 and R108 of ESCOMP KB). With the help of special operator function READ_WA(file name), the data from file QUERY.WA are written into work area of ESCOMP. Then mixing recipe is calculated (rule R113 of ESCOMP KB). After this the file-answer is formed (rule R205 of ESCOMP KB). The file answer query QUERY.WA will contain names of involving components, their quantities and calculated quantity of gasoline AI92. ESPLAN reads the file answer (rule R214 of ESPLAN KB) and continues the work in accordance with its KB.

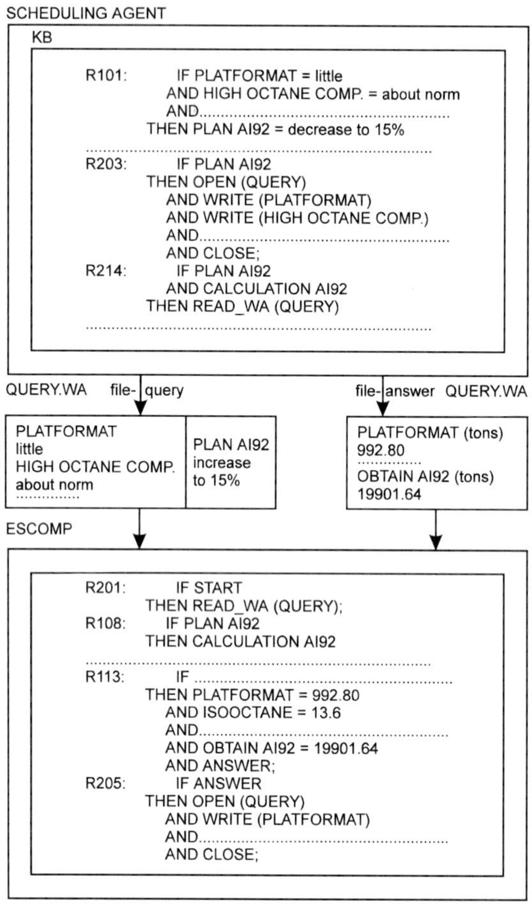

Figure 15.9 An example of intelligent agents interaction.

15.9 CONCLUSION

Originally on the considered oil refinery plant automation of control and production is performed by creating and utilizing separate intelligent systems, including fuzzy ones, for such technological processes as primary oil refining, catalytic cracking, compounding of benzene, dispatching, scheduling, planning etc. Long time these systems were exploiting efficiently on the plants. It took place on the plant that sometimes local systems objectives had priority in respect to the general goal. Moreover, absence of cooperation for solving joint problems concerning many of units, dispatcher, etc., as well as absence of coordination of element's (and intelligent agents) actions didn't allow to get suitable efficiency of the production which was possible to get.

Suggested in this work for FDIS concepts, construction principles, architecture, methods and protocols of interactions of the intelligent agents, models of transferring and processing distributed knowledge and data, including fuzzy ones, were have been put into basis of the integration of the intelligent systems developed in the past. This had provided more consistent work of the plant and achieving better economical indices.

The essence of the developed in the work ideas allows as well come along other way, that is to design original FDIS for the new plants being constructed. Like project has accomplished for a new oil refinery plant.

15.10 REFERENCES

1. Aminzadeh F. and M. Jamshidi (eds.). *Soft Computing: Fuzzy Logic, Neural Networks and Distributed Artificial Intelligence*. Simon & Schuster, 1994, p. 301.

2. Kusiak, A. (ed.). *Artificial Intelligence Implications For CIM*. Springer-Verlag, 1988, p. 527,.

3. Zadeh, L. A. "Fuzzy Logic." *IEEE Computer*. April 1988, pp. 83-93.

4. Stephen, T. and Wong, E. "COSMO: A Communication Scheme for Cooperative Knowledge-based Systems." *IEEE Transaction on Systems, Man and Cybernetics*. Vol. 23, No. 3, 1993, pp. 809-822.

5. Werner, E. "Cooperating agents: A Unified Theory of Communication and Social Structure." *Distributed Artificial Intelligence*. Vol. 2, (Gasser, L. and Huhns, M. N. eds.). Pitman Morgan Kaufman, 1989, pp. 3-36.

6. Bond, A. H. and Gasser, L. "A Subject-Indexed Bibliography of Distributed Artificial Intelligence." *IEEE Transactions on Systems, Man and Cybernetics*. Vol. 22, No. 6, 1992,pp. 1260-1281.

7. Aliev, R. A., Bonfig, K. W., Aliev T. F. and Aliev, R. R. "The Distributed Intelligent Manufacturing System." *Proceeding of First European Congress on Fuzzy and Intelligent Technologies*, Part II, 229-235, Germany, Aachen, 1993.

8. Nakamiti, G., Freitas, R., Predo, J. and Gomide, F. "Fuzzy Distributed Artificial Intelligence Systems." *IEEE*. 1994, pp. 462-467.

9. Aliev, R. A., Aliev, F. T. and Babaev, M. D. *Fuzzy Process Control and Knowledge Engineering Petrochemical and Robotic Manufacturing*. Verlag TUV Rheiland, Köln, 1991, pp. 148.

10. Aliev, R. A., Bonfig, K. W. and Aliev, F. T, *Messen, Stenern und Regeln mit Fuzzy-Logic*. Franzis-Verlag & Co., 1993, p. 310.

11. Aliev, R. A., Mamedova, G. A. and Tserkovny, A. E. "Production Control Under Fuzzy Information." Izd-vo *Energoatomizdat*, M. p. 239, 1991 (in Russian).

16

THE DESIGN OF OPTIMUM FUZZY CONTROL

M. Koch, T. Kuhn, and J. Wernstedt
Technical University of Ilmenau
GERMANY

For about 20 years, fuzzy logic has been used in control applications. The initial goal was to implement human knowledge for the tasks of process control. Today fuzzy logic is applied to model already existing control strategies or to implement new control structures. The large number of applications show the still existing potential of fuzzy control methods. The design procedure for fuzzy control, however, has not changed during the last 20 years. It is still a subjective heuristic strategy. Thus the design of fuzzy control requires much more experience than it is indicated in the literature. The control structures designed cannot be evaluated objectively. By using the methods described in this chapter the objective and optimum design and evaluation of fuzzy control are possible.

16.1 INTRODUCTION

The goal of the first applications of fuzzy logic in control was the imitation of the control activities of a human expert by a machine [1-5]. The only reason for fuzzy control was that classic control methods failed. The use of fuzzy logic for the control of conventional controllable plants was unthinkable because the appropriate industrial computers for control were too expensive.

At the beginning of the 1990s a great boom in fuzzy control caused by the mass production of microcontroller-based fuzzy products in Japan occurred. With the microcontrollers, cheap and efficient hardware for the implementation of fuzzy control was available. The implementation of fuzzy logic was now as simple as the implementation of classical control algorithms. The majority of the application cases were plants that are controllable by conventional methods. Fuzzy logic was implemented to improve control quality, to achieve robustness and a simpler design of control. The imitation of the human expert was not longer the main goal. But the opportunity to formulate and evaluate completely new rule-based control strategies was provided.

Unfortunately the design strategy is still the same as during the 1970s. Pure heuristics while setting up rules as well as "trial and error" while tuning the membership functions are still the most widely used methods for the design of fuzzy control. This procedure is not only subjective, but also dangerous.

In this chapter two objective methods for the design of fuzzy control are presented. First, a method for the primary goal, the optimum imitation of the dynamic control behavior of a human expert is described. Second, the design of optimum and also robust fuzzy controllers, according to the criteria of nonlinear control theory is illustrated.

16.2 FUNDAMENTALS OF OPTIMUM CONTROL STRATEGIES

Optimality is a feature that is defined only by a mathematically formulated criterion. Thus, optimum control is only relevant if the criterion for optimization contains all technical requirements of the control problem. Moreover, the models of the plant and also the exemplary control activities of the human expert are restrictions of the obtainable optimum. When designing a classical optimum state controller, the state space model of the real plant plays the central role. If this model is inexact, the controller is not only the optimum according to the inexact model, but it is also practically useless. Equally, the human expert as an example for control is a criterion of optimality for the fuzzy classifier control. The designed classifier control will not be better than this example.

Thus, it is important to decide which of the design strategies in view of the given control design task has to be used. With the chance to observe and record the activities of a human expert while controlling a very complex plant, the training of a fuzzy classifier controller imitating the expert's behavior is possible. Especially, if the plant is too complex or too complicated for a structured control design, this procedure is appropriate. But if the plant can be modeled and simulated in any way, the control design by criteria of control theory is useful. The conditions and goals of both strategies are summarized in Table 16.1.

16.3 THE DESIGN OF FUZZY CLASSIFIER CONTROL

Besides the basic control tasks the importance of high-level control is increasing at present. The complexity and nonlinearity as well as the incorporation of human decisions characterize the tasks of the high-level control. The control of distilling plants, cement and porcelain kilns, steel production, or energy load management are some technical application examples. But the interest in nontechnical application cases, for instance in

Table 16.1

Method	Fuzzy Classifier Control	Fuzzy Control
Conditions	comprehensive training data, gathered by • interviews • recorded data	a priori knowledge about the plant and the appropriate control structure • static characteristic • dynamic characteristic • simulations • a control strategy • criteria of quality
Goal	optimum of imitation of a human expert	optimum of control quality due to predefined criteria

water economy (barrages, drinking water supply, waste water disposal) or in medicine is still growing.

16.3.1 The Main Principle of Fuzzy Classifier Control

The task of classifiers is the assignment of given objects to predefined classes. When solving the control task, the objects O represent different process situations described by measured or observed data and are stored in the situation vector $\boldsymbol{s}(t)$. The situation vector $\boldsymbol{s}(t)$ is transformed into the feature vector $\boldsymbol{m}(t)$. This feature vector is the input to the fuzzy classifier. The classifier system assigns the decision e by the membership to one of the predefined situation classes. This decision may be a scalar value (a crisp class index) or a vector $\boldsymbol{e}$, consisting of all membership values of the single classes. But the decision $\boldsymbol{e}$ should result in an appropriate control activity $\boldsymbol{u}^*$ by the defined class semantics. The example for the classification, i. e., the appropriate assignment of the situations to the control activities, is the behavior of an experienced human expert.

Based on the available a priori information, the control is structured in the design procedure. By learning from examples, the parameters are estimated. These examples should be prepared as training data LP $\{\boldsymbol{m}, \boldsymbol{u}, q\}$, consisting of different feature vectors $\boldsymbol{m}$, the belonging control actions $\boldsymbol{u}$ of the human expert and a scalar evaluation of the quality of this control action q $(0 \leq q \leq 1)$. This training data should contain all relevant situations of control, because the classifier can only imitate the submitted actions. The interpolation between known situations is possible, but situations outside the trained area could be dangerous. The classifier should detect such situations and respond in a proper manner (keep the last action or alarm).

Being in practical operation, the classifier controller submits a proposal for a control action to the operator or, in automatic operation, it generates a control action directly to the plant. This control actions could be actuating variables or set points for subordinate control loops.

16.3.2 Optimum Design of Fuzzy Classifiers

The goal of the design of fuzzy classifiers is the optimum of imitation of human control behavior and decision making. This method is based on the supposition that a nonlinear plant is driven at well-defined points of its workspace. The characteristic of the plant at such points may be described by linear models.

That gives the opportunity to design linear controllers for these points. For each point and each possible change between these points there exists an exemplary expert behavior. The classifier control should model this behavior [6,7]. Thus the problem of decision making can be divided in two particular problems:

1. recognition of the actual process situation (situation class) and

2. assignment of a situation-dependent control action (class semantics).

The resulting structure is shown in Figure 16.1.

The particular problem (1) is solved by the recognition system, implemented as classifier and having a nonlinear decision behavior. For the solution of the second particular problem (2) appropriate control rules should be formulated. In general, these control rules will have a linear dynamic behavior. The division of the control problem is similar to the known methods in the field of nonlinear identification. When dividing a model into linear, dynamic and nonlinear, static elements, the Wiener or Hammerstein model is estimated. The design of a classifier-based controller should include the available a priori information and also the analysis of the human control actions. This information is shown in Figure 16.3.

The defined structure of the classifier is based on available a priori knowledge. That means:

- the definition of the number of classes, based on knowledge about the relevant set points and the transfer behavior of the plant,
- the selection of relevant information for situation recognition,

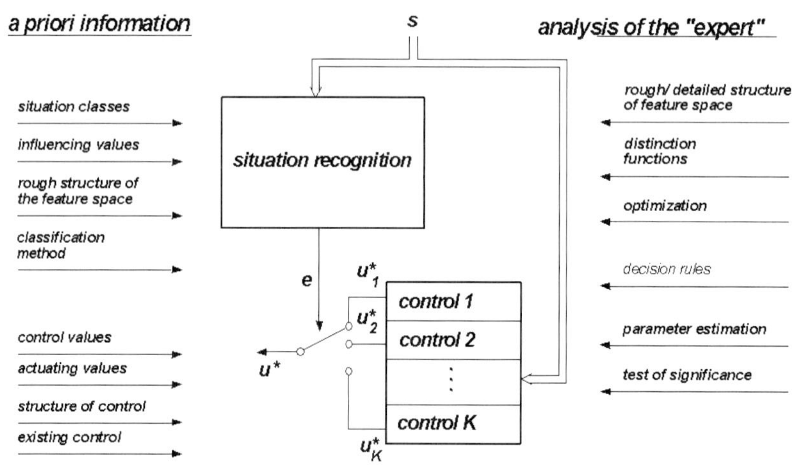

Figure 16.1 Structure and design information of a classifier-based control.

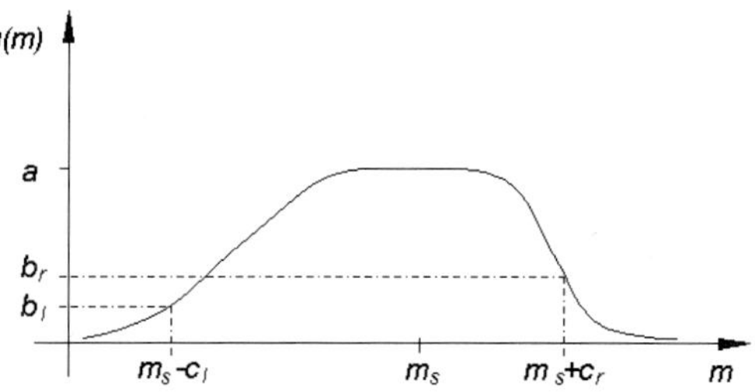

Figure 16.2 Membership function of a fuzzy classifier.

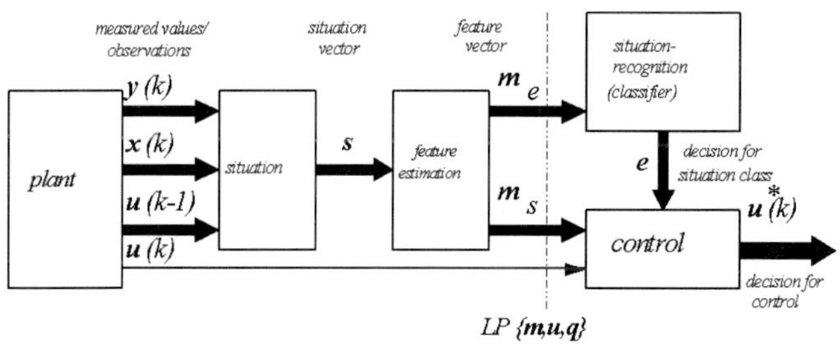

Figure 16.3 Structure and symbolic of a classifier control.

- knowledge about the available and useful action variables,
- the definition of appropriate control rules, and
- the use of existing control.

For the situation recognition a fuzzy classifier with parametric membership functions (Figure 16.2) is used [6,8].

Due to the known sampled controllers the control rules are implemented as polynomial functions $\underline{\mathbf{u}}^* = f(\underline{\mathbf{m}}_s)$. By using the analysis of the control actions, based on the observation of the expert, the parametrization and optimization are carried out. In this step the parameters of the membership functions and the control rules are determined. With the definitions of the structure of the classifier control and the available knowledge the design procedure is given as follows. To illustrate the design process, the cooperation between the plant and the classifier control and moreover the relevant signals are shown in Figure 16.3 [7].

Step I: Definition of the structure of the classifier and the control rules

The design procedure starts with the definition of the features for the situation recognition $\mathbf{m}_e$ and the features for the generation of actuating variables for the control $\mathbf{m}_s$. The features are time-dependent measured values, sampled in an appropriate frequency and preprocessed by data processing algorithms. The vector of features $\mathbf{m}_e$ is the input value for the fuzzy classifier, computing a decision for the actual class of inputs. The membership functions are defined by Equation (16-1).

$$\mu(m) = a\left\{1 + \sum_{i=1}^{r} 0.5 \cdot \left[1 + \operatorname{sgn}(-M_i)\right]\left(\frac{1}{b_l} - 1\right) \cdot \left|\frac{M_i}{b_l}\right|^{d_l} + \sum_{i=1}^{r} 0.5 \cdot \left[1 + \operatorname{sgn}(M_i)\right]\left(\frac{1}{b_r} - 1\right) \cdot \left|\frac{M_i}{c_r}\right|^{d_r}\right\}^{-1} \tag{16-1}$$

The control rules are defined as polynomial functions of the vector $\mathbf{m}_s$ (16-2).

$$u_i^* = \mathbf{f}_s(\mathbf{m}_s)^T \mathbf{a}_i \tag{16-2}$$

where:
$u_i^*(k)$ - computed actuating value for class i
$\mathbf{f}_s(\mathbf{m}_s)$ - transformed feature vector
$\mathbf{a}_i$ - parameter vector for the control of class i.

Step II: Creating a training example
The parametrization of the classifier control is based on real data from the plant, being prepared as a training example. The training example should consist of representative data from the whole workspace. Thus it is a set of feature vectors **m**, containing all relevant information about the system in singular sample steps. This vector **m** consists of all components of $\mathbf{m}_e$ and $\mathbf{m}_s$. The control actions and possibly the evaluation (weights q) are also elements of the training example. The result of this design step is a nonclassified (evaluated) training example LP{**m**, **u**, q}.

Step III: Structuring of the training example and parameter estimation
The division of the control task into a recognition and an actuating part requires the definition of a class structure for the nonclassified training example. It is necessary to find interpretable situation classes which represent typical behavior of the expert. This typical behavior should be approximated by linear model equations. To solve this problem a combination of a hierarchical method of clustering and a modified exchange algorithm is used. The cluster method creates in the first step a rough structure of the training example, based on accumulations of objects. In a second step an optimum class structure, on the basis of the modeling of the given control behavior, is established. Equation (16-3) shows the criteria of quality used.

$$\begin{aligned}
&(1)\ Q &&= \alpha Q_{classification} + (1-\alpha) Q_{control} \\
&(2)\ Q_{classification} &&= \sum_{k=1}^{K} \sum_{O_l \in C_k} p_i \qquad p_i = \begin{cases} 1 \ ; \ k <> e_i \\ 0 \ ; \ \text{else} \end{cases} \\
&(3)\ Q_{control} &&= \sum_{k=1}^{K} \sum_{O_l \in C_k} \left[u_i - f_s^T(m_{s_i})\mathbf{a}_k\right]^2
\end{aligned} \tag{16-3}$$

This criterion evaluates the correct classification of the training situation to the situation class by $Q_{classification}$ and, moreover, the reproduction of the given control action u by $Q_{control}$. By the parameter α a compromise between the contradictory particular criteria can be found. At the same time, the parameters for the recognition system [a, b, c, d, according to Equation (16-1)] and control law **a** [according to Equation (16-2)] are calculated.

Step IV: Structuring of the classifier control
By the means of methods for the selection of relevant situation features the dimension of the feature vector $\mathbf{m}_e$ can be reduced. For testing the significance of the parameters of the control rules, a modified t-test is applied. Nonsignificant parameters also reduce the dimension of the feature vector $\mathbf{m}_s$.

Step V: Implementation and test of the control
In order to avoid an abrupt switchover at the class limits during the running control, it is useful to implement a soft decision rule. This rule should create a compromise in case of a switch between different classes. The combination of the weighted control actions of the neighboring classes with the largest degrees of membership is shown in Equation (16-4).

$$e_1 = \underset{k=1\ldots K}{\arg\max}\, \mu_k(\mathbf{m}_e) \quad e_2 = \underset{\substack{k=1\ldots K \\ k \neq e_1}}{\arg\max}\, \mu_k(\mathbf{m}_e)$$

$$e = \begin{cases} e_1 \rightarrow u^* = u^*_{e_1}; & \mu_{e_1}(\mathbf{m}_e) - \mu_{e_2}(\mathbf{m}_e) > \mu_{\min} \\ (e_1, e_2) \rightarrow u^* = \dfrac{\mu_{e_1}}{\mu_{e_1} + \mu_{e_2}} u^*_{e_1} + \dfrac{\mu_{e_2}}{\mu_{e_1} + \mu_{e_2}} u^*_{e_2}; & else \end{cases} \tag{16-4}$$

16.3.4 AN EXAMPLE OF APPLICATION—THE CONTROL OF THE RIVER WERRA BARRAGE RATSCHER IN SUMMER FLOOD SITUATIONS

To avoid flood disasters the Ratscher and Grimmelshausen barrages were erected in the upper reaches of the river Werra. The Ratscher barrage, discussed in this section, is also used as a recreation area and should guarantee a minimum draining of 10 l/s, according to the ecological requirements. It has a capacity of 5.2 M.m^3, but only 1.2 M.m^3 can be used for control. The classifier-based control mentioned here models the decision making of an experienced human operator during a flood disaster in the summer season.

The task of the control is to catch the peak of the water level by the available capacity of the barrage. Exploiting the control capacity of the barrage, draining should be restricted to a reasonable maximum for the further river basin. As shown in Figure 16.4, the draining of a barrage consists of the base flow, the controllable and the uncontrollable overflow.

The base flow is the actuating value of the system. The controllable overflow passes through the overflow gilled pipes. This is only a safety system serving to avoid an uncontrollable overflow, which may destroy the barrage. The classifier-based control was

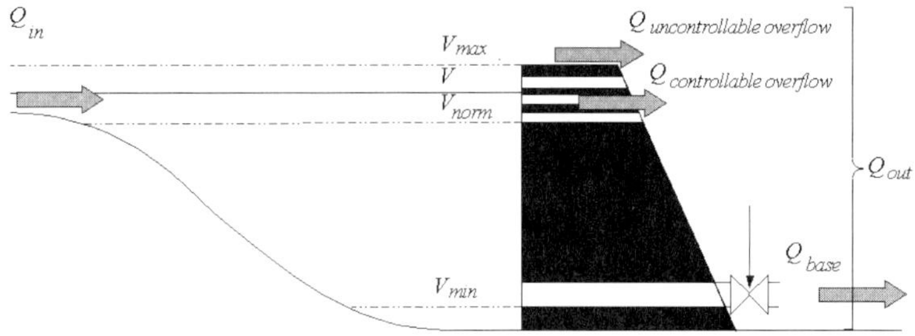

Figure 16.4 Scheme of a barrage.

designed by means of an experienced hydrologist and the historic data of flood disasters. By a simulation model using the available historic data as input the decisions of the human expert were recorded.

Step I: Based on the consultation of the expert, the following structure of the situation and action vectors was defined.
The features describing the actual situation were:

$$\mathbf{m}_e = \begin{pmatrix} N_{gross}(k) \\ N_{eff}(k) \\ Q_{in}(k) \\ Q_{in}(k+1) \\ AWV(k) \end{pmatrix} \begin{matrix} \text{current gross rainfall} \\ \text{current effective rainfall} \\ \text{current inlet} \\ \text{predicted inlet} \\ \text{available capacity of the reservoir} \end{matrix} \qquad (16\text{-}5)$$

- The current brutto rainfall is a measured value.
- The available capacity is the difference between the maximum and the current amount of water in the reservoir.
- The current inlet is estimated by the level of the river at the beginning of the reservoir.
- The actual effective rainfall is estimated by the Lorent-Gevers-Model.
- The prediction of the inlet is done by a rainfall/draining-model, based on the actual measured rainfall.

For adjusting the base outlet, the expert used a linear function of the actual inlet only. There are two situation classes: "normal" and "flood."

Step II: The data of the training examples, based on historic, recorded flood disasters, consists of rainfall and flow measurements of the relevant catchment area and the control decisions of the hydrologist. One training example, the flood disaster of June 26 to July 7, in 1966 is shown in Figure 16.5.

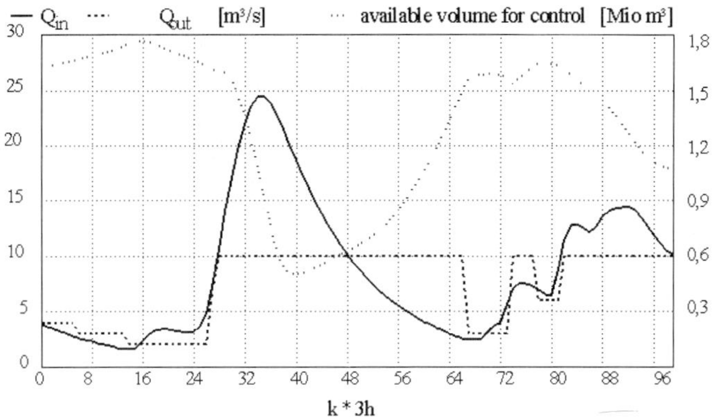

Figure 16.5 The data of the flood disaster from June 26 to July 7, 1966.

Step III: For structuring the training example, the algorithm "complete linkage" was used. The required parameters of the situation-related control were estimated by the cost function, given in Equation (16-3) with $\alpha = 0.5$.

Step IV: The selection method of relevant features using the error of classification as a cost function reduced the optimum feature set to $Q_{in}(k)$ and $AWV(k)$. The significant parameters for control were estimated by an error probability of 5%. The membership functions of classes 1 and 2 are shown in Figures 16.6 and 16.7.

The estimated parameters of control are given as follows:

$$\begin{array}{ccc} \textit{class} & 1 & 2 \\ A = & \multicolumn{2}{c}{\begin{pmatrix} 0.12 & 1.00 \\ 0.87 & 0.00 \end{pmatrix}} \end{array} \tag{16-6}$$

This result shows that the situation classes 1 and 2 may be interpreted as a normal and flood situation, respectively, which is in conformity with the setting up of the problem. The parameters of control harmonize with the statements of the expert. The extracted decision rules are:

IF the situation is normal THEN adjust the same draining as inlet.
IF the situation is flood THEN adjust a draining of 10m³/s.

These control rules are given in Equation (16-7).

$$\begin{array}{cc} \text{class 1 (normal situation)} & \text{class 2 flood situation} \\ Q_{out}(k) = 1.22 + 0.86 Q_{zu}(k) & Q_{out}(k) = 10 \end{array} \tag{16-7}$$

Step V: The designed classifier-based control was completed by a soft decision rule, as given in Equation (16-4) ($\mu_{min} = 0.2$). Using the simulation system, the designed control

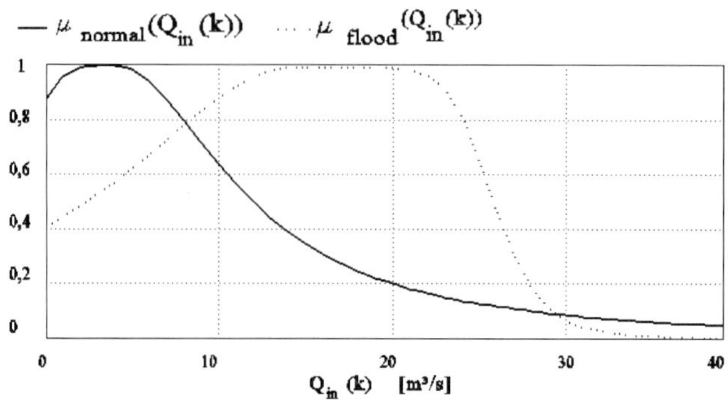

Figure 16.6 The membership functions of the actual inlet, with the capacity amounting to 12 M.m^3 constantly.

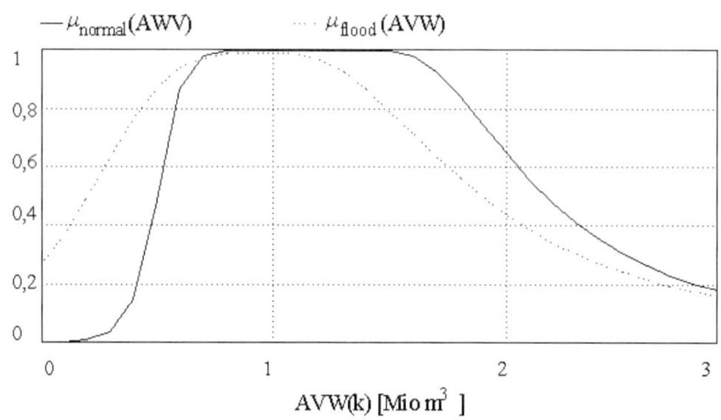

Figure 16.7 The membership functions of barrage capacity, while the inlet is 10 m^3/s constantly.

was verified by the training example and other flood data. The comparison with the expert control shows (Figure 16.8) that the designed classifier works as well or even better. Instead of the soft decision rule, the switches between the situation classes are visible.

16.4 THE DESIGN OF FUZZY CONTROLLERS

16.4.1 Application Cases for Fuzzy Controllers

In the last few years, there were a large number of scientific and industrial contributions published in the field of fuzzy control. The original idea, the application of fuzzy methods to imitate human control for plants being too complicated or too complex for conventional control, expands classic control problems more and more. With the aim to implement human control intelligence in a machine in the 1970s, the recent fuzzy

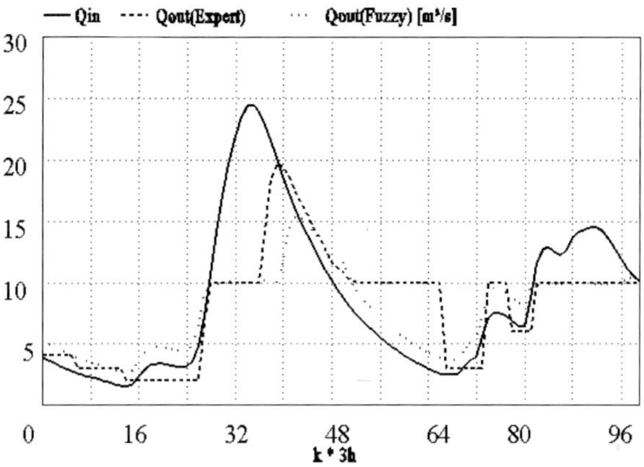

Figure 16.8 The comparison between classifier and expert control (flood disaster from June 26 to July 8, 1966).

control applications try to improve the quality of still existing control solutions.

Fuzzy control is mostly applied in the following cases:

1. Fuzzy data processing and model estimation.

2. Control design for very complex or complicated plants, if conventional methods fail.

3. Simplification of the design of basic control, substitution of conventional control, controller design without control theory.

4. Improve conventional control loops by using fuzzy control.

The 1970s were characterized by the application case (2). But today, having cheap and efficient microcontrollers and PLCs, the case (3) seems to be the favorite one. The danger of this development is clear.

The design method, described here can be assigned to case (4). It is a combination of a heuristic search of structure and an exact parameter optimization. If the parameter optimum of conventional controllers is not satisfactory and the structure optimization fails because the state space model cannot be estimated, this new method can be applied.

16.4.2 Optimum Design Of Fuzzy Control

Using nonlinear criteria of quality, the following strategy allows the controller design directly at the plant or at a model. The main advantage of this strategy is in independence of the existence and quality of training data. Moreover it is possible to implement available linguistic knowledge in fuzzy rules. A disadvantage of this strategy is that the plant should be uncritical, because unstable states during the optimization may destroy it. In such cases a sufficiently exact model of the process is needed.

The strategy is supported by three main ideas [9,10]: on the one hand the modification of known criteria of quality of control for the nonlinear case. On the other hand the parametrization of the fuzzy logic that means a reduction of the large number of degrees of freedom to get an optimizable structure, and last but not least, the selection of appropriate search strategies for parameter optimization. These three particular problems are discussed in the following sections.

Problem I: The modification of known criteria of quality of control To evaluate the quality of nonlinear control loops, various criteria can be used, not only the known integral criteria, but also direct methods, evaluating the characteristic control values that may be applied to force the desired control behavior. Three possible criteria are discussed here.

Integral criteria

The controller design for linear systems by parameter optimization generalizes the control behavior in the whole workspace by a single standard change of set point. Thus the integral criteria could be formulated by a single change of set point in the linearized area. These integral criteria will be used as the core for the nonlinear case.

To evaluate the quality of a single change of set point, the integral of the quadratic control error is commonly used. The criterion to be minimized is given by Equation (16-8).

$$Q_R = \int_{t=0}^{\infty} e^2(t)\,dt \tag{16-8}$$

If the time should be considered too, the criterion is modified to (16-9):

$$Q_R = \int_{t=0}^{\infty} te^2(t)\,dt \tag{16-9}$$

The consideration of the expense of control is often important. Having an integral plant, then the actuating variable reaches the zero value at the end of the control action (16-10):

$$Q_S = \int_{t=0}^{\infty} u^2(t)\,dt \tag{16-10}$$

Having a compensatory plant, then the actuating variable for keeping the set point must be different from zero. In this case the criterion for the expense of control is given by (16-11):

$$Q_S = \int_{t=0}^{\infty} \left(\frac{du(\tau)}{d\tau}\right) dt \tag{16-11}$$

This criterion evaluates only the change in actuating value. In most cases, a compromise between quality and expense of control is desired. This compromise may be achieved by a weighted addition of both particular criteria. In the equation:

$$Q = \alpha Q_R + (1-\alpha)Q_S \quad 0 < \alpha < 1 \tag{16-12}$$

the coefficient α is the weight of compromise between the quality and the expense of control. This criterion may be minimized by analytic methods for simple linear control loops. Having only a simulation model of the plant, the integration can be done only for a finite time limit. Thus the control procedure must be treated as finished at the time limit T_{max}. Moreover the general simulation, containing the fuzzy controller, must be treated as nonlinear. Thus the quality of control can only be defined according to a predefined set point and disturbance trajectory. These trajectories must be representative for the whole workspace of the control loop.

Example

A given fuzzy control loop is sampled at a sampling time T. All relevant signals of the control loop are measured at the discrete sample steps. The procedure of set point change can be treated as finished after K_{max} sample periods. The control loop should be able to change between $n = 3$ set points (W_1, W_2, W_3). Every change between these set points is possible, but the quality of control in W_2 is especially important. The goal is to achieve a compromise between quality and expense of control.

The integral criteria are converted into a time discrete form and combined to a comprehensive combination of set point changes. Using the forward rectangle conversion, the sampling time is a constant factor and may be ignored. For the different weighting of the set points the weight vector is established. The whole criterion and the vectors are given in Equations (16-13) and (16-14), the set point trajectory is shown in Figure 16.9.

$$W = \begin{pmatrix} W_1 \\ W_2 \\ W_1 \\ W_3 \\ W_2 \\ W_3 \\ W_1 \end{pmatrix} \quad \lambda = \begin{pmatrix} \lambda_1 \\ \lambda_2 \\ \lambda_1 \\ \lambda_3 \\ \lambda_2 \\ \lambda_3 \\ \lambda_1 \end{pmatrix} \quad \begin{array}{l} \lambda_1 = 0{,}1 \\ \lambda_2 = 0{,}8 \\ \lambda_3 = 0{,}1 \\ \sum_{v=1}^{n} \lambda_v = 1 \end{array} \tag{16-13}$$

$$Q_{ges} = \sum_{i=2}^{n(n-1)+1} \lambda_i \cdot \left(\alpha \cdot \sum_{k=0}^{K_{max}} k\big(W_i - y(k)^2 + (1-\alpha)\big) \cdot \sum_{k=1}^{K_{max}} k\big(u(k) - u(k-1)\big)^2 \right) \tag{16-14}$$

$$0 < \alpha \le 1$$

Various penalty terms may be added to this criterion, for instance the avoidance of overshoot. However, the use of the characteristic value criteria is much better.

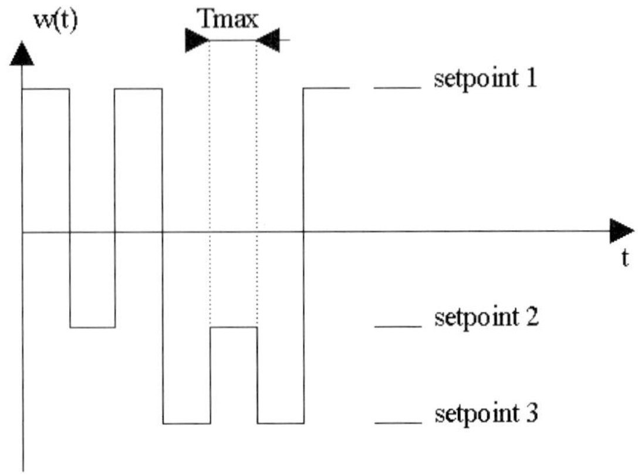

Figure 16.9 Set point sequence.

Characteristic value criteria

There are often special requirements to control parameters in industrial practice. Thus it is useful to formulate special quality functions, considering these parameters such as:

- rate of overshoot Δh
- overshoot time T_m
- final time T_e
- permanent error of control e_B

The criteria may minimize these parameters or adjust them to desired values. Some of these criteria that may be combined are given as follows:

goal	criterion
minimal permanent error	$Q = e_B$
pre - defined rate of overshoot	$Q = \lvert \Delta h - \Delta h^* \rvert$
reduced final time and overshoot	$Q = \alpha \cdot T_\epsilon + \beta \cdot \Delta h$

(16-15)

Trajectory criteria

Criteria of this kind try to force the real trajectories of control into ideal ones (Example I) or into predefined constraints (Example II). Thus the divergences are formulated as penalty terms by an integral criterion. The ideal trajectory or also the constraints must be defined as functions in the time horizon of evaluation.

Example I

The trajectory of the controlled value $y(t)$ should be similar to the ideal trajectory $y^*(t)$ (Figure 16.10). The mathematical criterion is the integral of absolute divergence or quadratic divergence [Equation (16-16)].

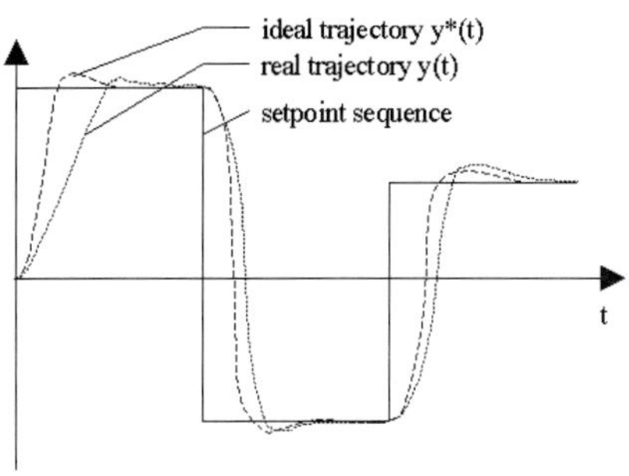

Figure 16.10 Ideal trajectory criterion.

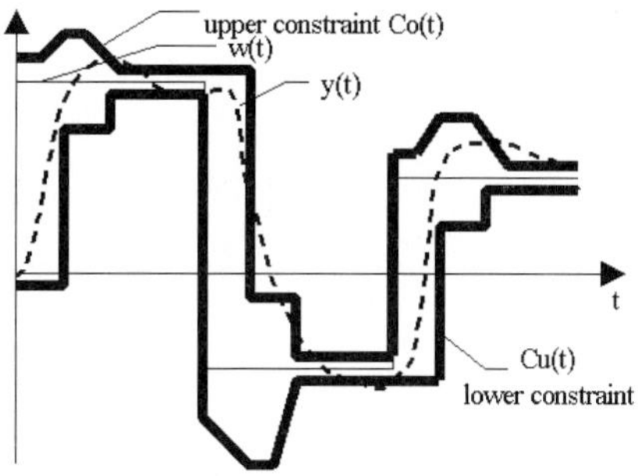

Figure 16.11 Constraints criterion.

$$Q = \int_{t=0}^{T_{\max}} \left| y(t) - y^*(t) \right| dt \quad \text{absolute penalty term}$$

or (16-16)

$$Q = \int_{t=0}^{T_{\max}} \left(y(t) - y^*(t) \right)^2 dt \quad \text{quadratic penalty term}$$

Example II
The controlled value $y(t)$ must not violate the constraints $c_O(t)$ and $c_U(t)$ (Figure 16.11). Thus an integral criterion punishing the violations and consisting of two penalty terms is formulated in Equation (16-17):

$$Q = \int_{t=0}^{T_{\max}} \eta\left(c_O(t) - y(t)\right) + \eta\left(y(t) - c_U(t)\right) dt$$

$$\eta(x) = \begin{cases} 0; & x \geq 0 \\ -x; & x < 0 \end{cases} \tag{16-17}$$

The application of such criteria is very advantageous, because the characteristic control parameters, such as the overshoot rate, can be adjusted in a very transparent manner.

Guarantee of robustness
The robustness is the ability of a controller to tolerate variable parameters of the plant. Robust control is very important in industrial applications. To formulate a mathematical criterion for nonlinear robustness, a new level of hierarchy in optimization is generated. The main idea is the use of more than one models in the design procedure. This models should be different and will reproduce the different states of the variant plant. The goal is now the design of a controller which is able to control all of these models in a satisfying quality. By using the weighted addition of single criteria the criterion of compromise is found by Equation (16-18):

$$Q_{ges} = \sum_{\nu=1}^{\Omega} \xi_\nu \cdot Q_\nu$$

$$\sum_{\nu=1}^{\Omega} \xi_\nu = 1$$

Q_ν quality with model $_\nu$ (16-18)

ξ_ν weight with model $_\nu$

Ω number of considered models.

The weights may be used to consider the importance of the single models.

Problem II: The parametric fuzzy concept. As mentioned above, the problem of parametrization of the fuzzy concept forces the reduction of the large number of degrees of freedom of the common fuzzy logic concept. This can be achieved if the following requirements are fulfilled:

- The rule set is treated as invariant.
- Singletons are used as membership functions for the output terms.
- MIN or PROD is used as an AND-operator and MAX– as OR– and aggregation operator.
- Linear, parametric membership functions for the fuzzy inputs are defined.

The membership functions of the fuzzy inputs

Definition: The membershipfunction of a term of the fuzzy input A is a trapezium (Figure 16.12). All points of the membership functions of the input A are stored in the vector η_A, *where*:

$$\sum_{i=1}^{n} \mu_{A_i}(A) = 1 \tag{16-19}$$

The effect of this definition is that for *a* membership functions of the fuzzy input *A* only $2n$–2 numbers must be stored. Thus the vector η_A is given by:

$$\eta_A^T = (\eta_{A1}\ \eta_{A2} \cdots \eta_{A2n-2}) \tag{16-20}$$

The elements of this vector could be treated as start and endpoints of the interval of full membership of a fuzzy set μ_{Ai}. By setting two of these points to the same value, the definition of triangular and rectangular membership functions is possible. Starting at $-\infty$ or also ending at ∞ the marginal membership functions need only one parameter, the end- or the startpoint of full membership, respectively. In control tasks, there are often centered values to deal with. Thus the number of parameters can be reduced by using symmetric parameterization. That means, only the positive parameters are concentrated in the parameter vector.

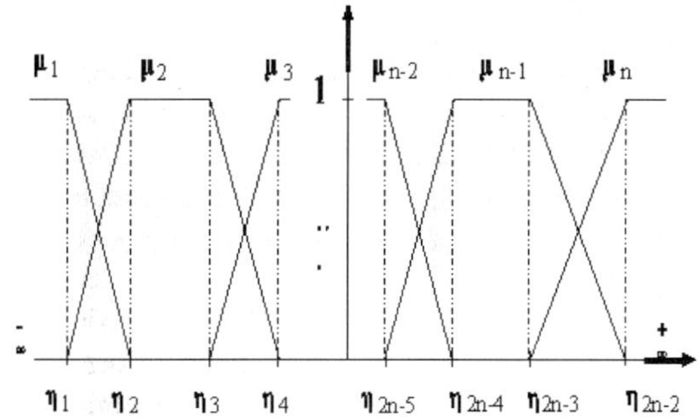

Figure 16.12 Membership functions of input A.

In addition to the parameter vector a restriction vector has to be defined. The restrictions of the parameters keep the membership functions in a correct form. The parameter for the beginning of full membership of linguistic POSITIVE should be greater than the endpoint of full membership of linguistic ZERO.

The rule set

The rule set contains the control strategy. It can be assumed that the control engineer is able to formulate appropriate control rules for the concrete application case. On the basis of our examinations a library of rule sets for typical industrial plants was compiled. These rule sets represent the human decision behavior, characterized by contradictory and superposing rules. An example is given in Section 16.4.3.

When setting up the rule set, it should be considered that this design step is equal to the heuristic structure search. This main advantage of fuzzy control, the implementation of human knowledge and experience, is not given up in this new design strategy. Experience shows that the setting up of the rule set (knowledge base) is much easier than the definition of appropriate membership functions (data base). But the evaluation of the subjective rule set will be objective, because the optimum of membership functions, belonging to this rule set is estimated. Thus the rule set and its operators should be treated as invariant by the parametric concept.

The membership functions of the fuzzy outputs

Definition: The membership functions of the terms of the fuzzy output are singletons, stored in the vector λ_Z.

The theoretical investigations of the transfer behavior of fuzzy structures and a lot of industrial control applications show that the use of singletons is not only practicable and easy to implement. It also shows that complicate membership functions for the outputs and expensive algorithms for defuzzyfication do not cause a better behavior of the controller. Therefore, it is only necessary to store z points for the fuzzy output Z. Thus the vector λ_Z is defined by:

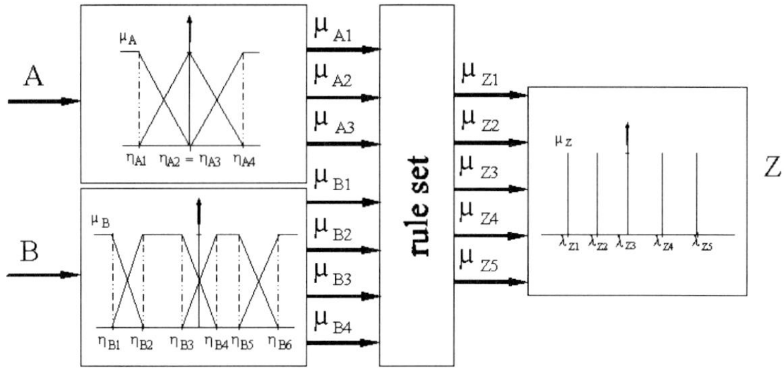

Figure 16.13 Example system.

$$\lambda_Z^T = (\lambda_{Z1}\ \lambda_{Z2} \ldots \lambda_{Zz}) \tag{16-21}$$

In many cases this parameter vector should also be restricted. The use of symmetric parametrization is useful, too.

An example for parametric fuzzy logic

Summarizing the vectors of inputs and outputs the parameter vector $\mathbf{x}$ can be established. Its dimension is defined by n. The parameters should be restricted by the restriction m-dimensional vector $\mathbf{C}(\mathbf{x})$ to save the appropriate design of membership functions. The establishment of the parameter and the restriction vector is shown in the following example (Figure 16.13).

The exemplary fuzzy system has two inputs, A and B, having three $a = 3$, or four $b = 4$ terms and one output Z, having five $z = 5$ terms. The simple summarization of the single vectors creates a 12-dimensional parameter vector. Using a priori knowledge, this dimension can be reduced. The following requirements are made:

1. The membership functions should be symmetric to zero.

2. The marginal terms of Z are equal to the bounds of the actuating variable; thus it is useless to move them.

3. The input A is the control error. To avoid a permanent control error, it is necessary to keep the start- and endpoint of the centered membership function equal to zero.

The resulting parametervector is defined by:

$$x^T = (\eta_{A4}\ \eta_{B4}\ \eta_{B5}\ \eta_{B6}\ \lambda_4) \qquad (16\text{-}22)$$
$$n = 5$$

And the restriction vector is given by:

$$c = \begin{pmatrix} \eta_{A4} \\ \eta_{B5} - \eta_{B4} \\ \eta_{B6} - \eta_{B5} \\ \lambda_4 \\ U_{\max} - \lambda_4 \end{pmatrix} = \begin{pmatrix} x_1 \\ x_2 - x_1 \\ x_4 - x_3 \\ x_5 \\ U_{\max} - x_5 \end{pmatrix}$$
$$m = 5 \qquad (16\text{-}23)$$
$$c_i \geq 0$$

Problem III: Appropriate search strategies. to solve the established optimization problem

$$\min\{Q(\underline{x})\} \quad ; \quad \underline{c_j}(\underline{x}) > 0 \qquad (16\text{-}24)$$

appropriate search strategies should be selected. While larger fuzzy systems have high dimensional parameter vectors, efficient algorithms are needed. We have gained good experiences by using:

- evolutionary and genetic strategies [12],
- sequential quadratic programming (SQP) [11].

The SQP algorithm is very fast and efficient, but the fuzzy control quality functions are particularly nonsmooth and have local optima. To increase the chance finding the global optimum the evolution and genetic strategies should be preferred but they take much more computing time.

A further problem is the consideration of the restrictions. They cannot only be punished by penalty terms, because a violation causes a nonsense membership function. Thus the SQP strategy deals with exact penalty functions for the restriction violations. The evolution strategy treats the parameter vectors violating the restrictions, as stillborn individuals. This saves much computing time, because the simulation for the computation of the cost function is not done if the parameter vector violates the restrictions.

16.4.3. Example of Application—Fuzzy Adapted PID Control of an Unstable Mechatronic Plant

In many industrial control loops linear PID controllers are applied. The industrial PID algorithms contain the knowledge and practical experience of the last decades of industrial control. They are very reliable and therefore control engineers often prefer them to state or fuzzy controllers.

Especially control loops in robotics [13], controlled by such industrial PID controllers, could be improved by the application of fuzzy logic-based adaptors. The PID controller is still used, but its parameters are adapted by a fuzzy system, according to the current situation (Figure 16.14). Such structures can be applied to nonlinear plants and also for critical dynamic control loops.

In this chapter the fuzzy adapted PID control of the unstable mechatronic education system "Arthur" by a low-cost PLC is described.

To demonstrate nontrivial control tasks, the unstable mechatronic system Arthur was developed at the Technical University of Ilmenau. Figure 16.15 shows the scheme of the system.

The task of the controller is to balance of the propel arm in the range from –50° to 50° and the fast and oscillation- free change of the set points. The difference between the speeds of both propels is used as actuating variable. This value is scaled to a range from –100% to 100%, that means:

$u = -100\%$; motor 2, full speed, motor 1, stop
$u = 100\%$; motor 1 full speed, motor 2 stop

The angle (controlled value) is measured by the voltage over an variable resistor. The estimation of the model was done by the analytic mathematical method. But the model parameters were determined by experiments at the real plant.

The analytic model contains a double integrator and two nonlinear feedback loops. The angle determining the force caused by the weight of the arm is modified by a sine function inside the feedback. The second feedback, the friction is a function of the velocity. This model was implemented by a block-oriented simulation in Matlab/Simulink™.

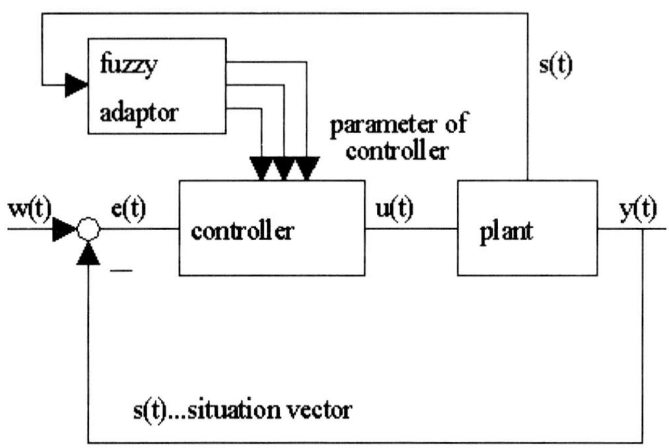

Figure 16.14 The structure of a fuzzy adaptor.

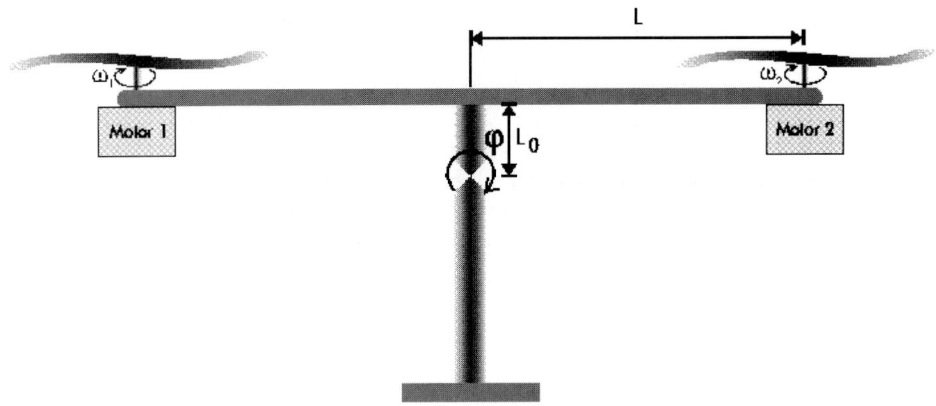

Figure 16.15 The principal scheme of the system.

The cost function is a multicriterial, hierarchical integral criterion, defined by Equation 16-14. This criterion provides the opportunity to consider the quality and the expense of control. Due to the nonlinearity of the plant, the sequence of set points consists of three different set point levels. It is the complete sequence of set point changes between the different levels in the whole workspace of the control loop. The relevant set point angles are –40°, 0° and 20°. The weights for these three set points are equal. Quality and expense of control are also weighted equally.

To achieve the best quality of control, an adaptive fuzzy PID controller is used. The adaptor is implemented by the parametric fuzzy concept in a two-level structure. The first level contains the situation recognition and the second one contains the tuning rules for the PID controller.

The main idea of the adaption algorithm is shown in Figure 16.16.

All situations during a set point change are detected by very simple and transparent rules. Only the control error and its derivative are used, to formulate the fuzzy rules for the situation recognition.

These rules are given by:

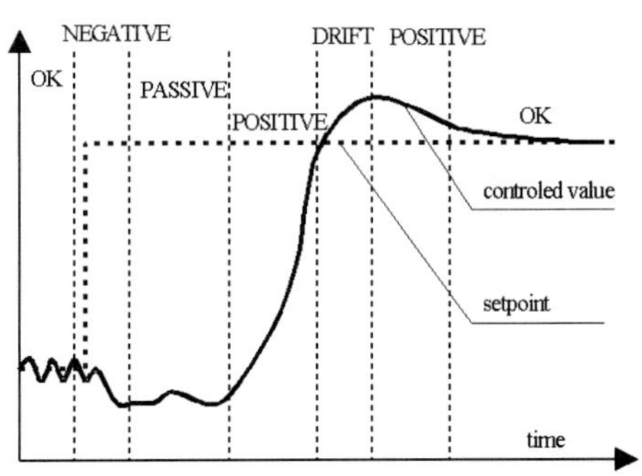

Figure 16.16 The idea of dynamic situation recognition.

IF (E =N) AND (DE =N) THEN SITU :=NEGATIVE;
IF (E =P) AND (DE =P) THEN SITU :=NEGATIVE;
IF (E =Z) AND (DE =Z) THEN SITU :=OK ;
IF (E =Z) AND (DE =P) THEN SITU :=DRIFT ;
IF (E =Z) AND (DE =N) THEN SITU :=DRIFT ;
IF (E =N) AND (DE =P) THEN SITU :=POSITIVE;
IF (E =P) AND (DE =N) THEN SITU :=POSITIVE ;
IF (E =N) AND (DE =Z) THEN SITU :=PASSIVE ;
IF (E =P) AND (DE =Z) THEN SITU :=PASSIVE .

The dynamic behavior of the plant is nontrivial, as the start of the control of a set point change needs other PID parameters than the end of a set point change. For instance, to move the propelling arm from the horizontal position to a positive angle, a positive value for the actuating variable is needed to start the movement in the correct direction. But the balance of the propelling arm in a positive angle requests a negative value for the actuating variable. The classical, linear PID controller is not able to do this control task satisfactorily. The fuzzy adaptor improves the behavior of the PID controller.

The singletons for the PID parameters were defined as a priori known PID parameters. Thus the resulting controller at the beginning of optimization is not adaptive, but linear. This initial PID forces the plant into all desired set points in a bad quality (Figure 16.17).

The optimization of the singletons can be treated as the learning of optimal adaption rules. While the situation recognition level is designed optimally by the optimization of the membership functions, the second tuning level is designed by moving the singletons to their optimal positions or also the learning of the corresponding adaption rules. The global optimum was found by the evolution strategy. The optimum membership functions and the corresponding 3-dimensional characteristics are shown in Figures 16.18–21.

This optimally designed fuzzy adaptive PID controller achieves much better results than the linear PID controller (Figure 16.22).

The controller behavior is very fast at the beginning of the set point change. For the exact adjustment of the new set point, however, a slow integrating behavior is used. The optimal fuzzy adaption system, designed by the software tool **IFT** [14,15] (Ilmenauer Fuzzy Tool), can be also exported to various hardware platforms by the **IFT** too (Figure 16.23).

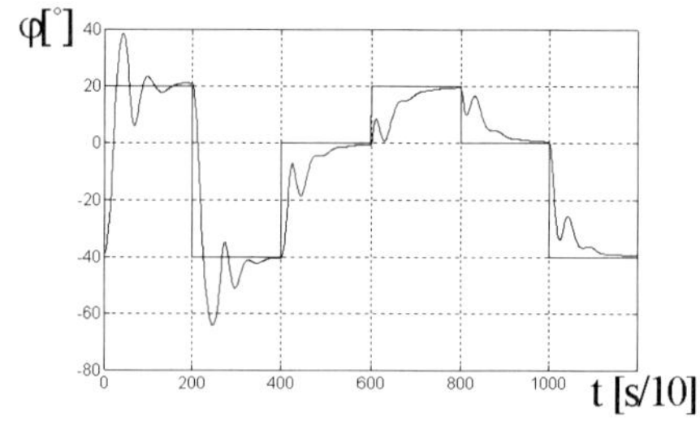

Figure 16.17 The initial controller at the beginning of optimization.

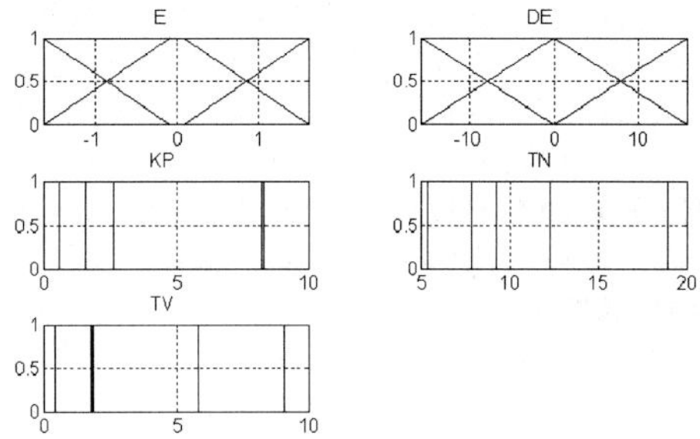

Figure 16.18 The optimum membership functions.

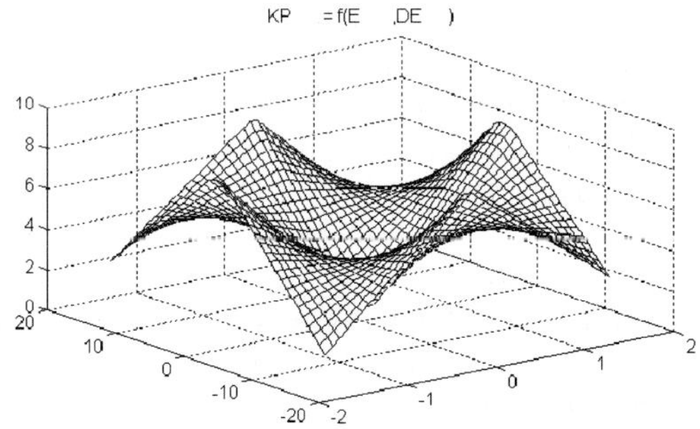

Figure 16.19 3D-characteristic for KP.

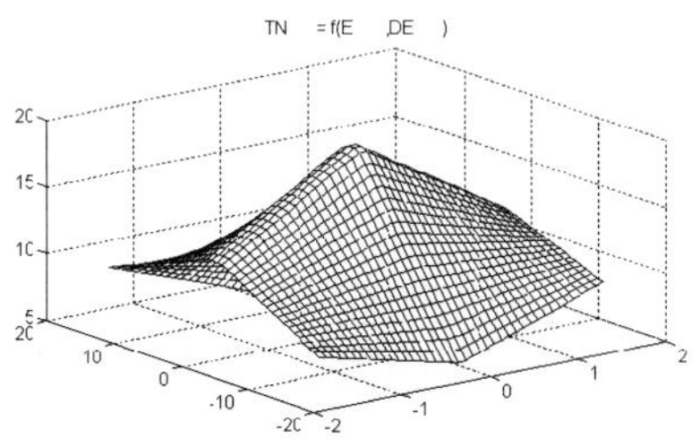

Figure 16.20 3-D characteristic for TN.

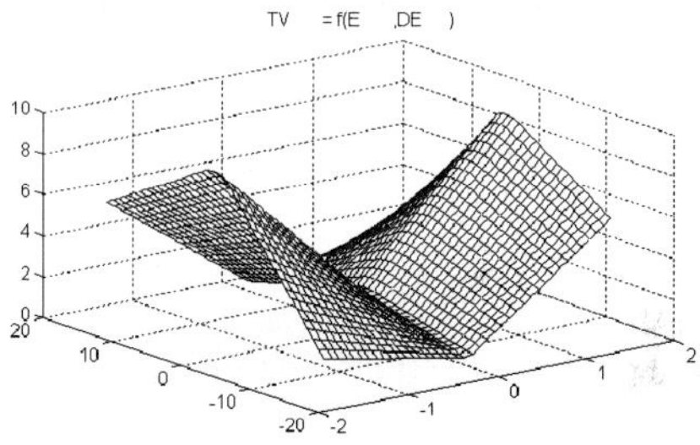

Figure 16.21 3-D characteristic for TV.

In this application case an instruction list for the PLC A120 by AEG Modicon was generated. This instruction list was implemented on the PLC, together with a common PID block. The real control results show not only the improved quality of the PID controller, but also the fact that optimal fuzzy adaptive PID controllers can be implemented on low-cost PLC's by the **IFT**.

The control loop was sampled by $T = 0.1$ s. But the computing time for both the fuzzy adaptor and the PID controller is about 35 ms. This is a very good result for a low-cost PLC.

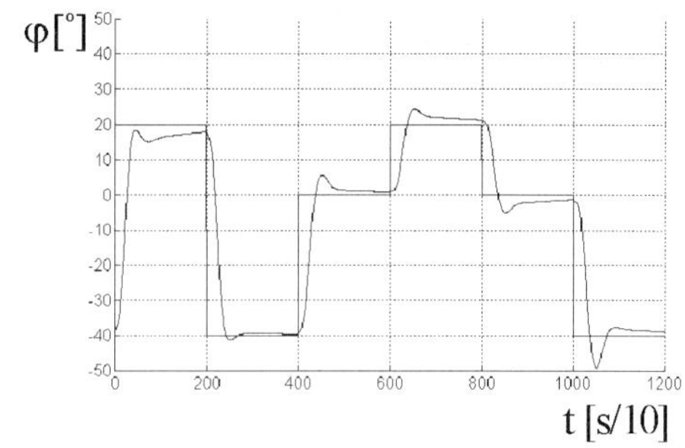

Figure 16.22 The behavior of the optimum adaptive controller.

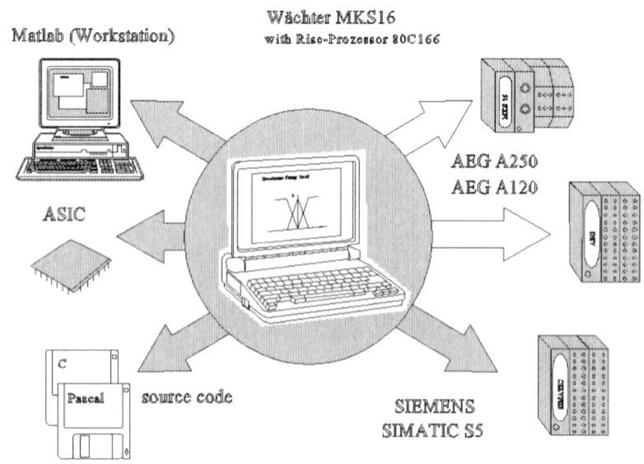

Figure 16.23 Export abilities of IFT.

16.5 REFERENCES

1. Zadeh, L. A. et. al. *Fuzzy Sets and Their Application to Cognitive and Decision Processes.* Academic Press, 1975.

2. Gupta, M. M. et. al. *Advances in Fuzzy Set Theory and Applications.* North-Holland, 1979.

3. Sugeno, M. *Industrial Applications of Fuzzy Control.* North-Holland, 1985.

4. Zimmermann, H.-J. *Fuzzy Set Theory and its Applications.* Klüwer-Nijhoff, 1985.

5. Larsen, M. P. *Industrial Applications of Fuzzy Logic Control.* Academic Press Inc., 1980.

6. King, P. J. and Mamdani, E. H. "The Application of Fuzzy Control Systems to Industrial Processes." *Automatica.* Vol. 13., 1977, pp. 235-242.

7. Koch, M. Ein Vergleich von Entwurfsverfahren für Klassifikatoren auf der Basis der Bayes- und Fuzzy-Theorie zur Prozeßsteuerung. at 41 (1993) 12, S. 461-466.

8. Koch, M. Klassifikatorkonzepte zur Steuerung dynamischer Systeme. Promotion A, Technische Universität Ilmenau, Ilmenau 1994.

9. Bocklisch, S. F. Prozeßanalyse mit unscharfen Verfahren. Verlag Technik Berlin, 1987.

10. Kuhn, T. and Wernstedt, J. "A method of optimal design of fuzzy controllers." *First European Congress on Fuzzy and Intelligent Technologies.* pp. 1323-1328, Aachen, Germany, September 1993.

11. Kuhn, T. and Wernstedt, J. SOFCON - Eine Strategie zum optimalen Entwurf von Fuzzy-Regelungen. at 42 (1994) 3, S. 91-99.

12. Papageorgiou, M. Optimierung: statische, dynamische, stochastische Verfahren für die Anwendung. München: Oldenburg 1991.

13. Schwefel, H. P. Nichtlineare Optimierung von Computer-Modellen mittels Evolutionsstrategie. Birkhäuser Basel und Stuttgart, 1977.

14. Kopacek, P.; Krenn, G.; Kuhn, T.; Wernstedt, J. Optimal Fuzzy Control Design for Robotics. Preprints zum 39. Wissenschaftlichen Kolloquium vom 27.-30.09.1994 der TU Ilmenau, Band III, Ilmenau 1994.

15. Kuhn, T., Marquardt, R., Wernstedt, J. Das Ilmenauer Fuzzy Tool-Benutzeranleitung, Technische Universität Ilmenau; system engineering GmbH ilmenau, Ilmenau 1993.

16. Kuhn, T. Die Ilmenauer Fuzzy Toolbox für Matlab/Simulink; system engineering GmbH ilmenau, Ilmenau 1994.

17

FUZZY TRACKING METHODS FOR MOBILE ROBOTS

A. Ollero
Universidad de Sevilla
SPAIN

A. García-Cerezo, J. L. Martínez and A. Mandow
Universidad de Málaga
SPAIN

This chapter deals with the application of fuzzy logic to autonomous navigation of mobile robots. Two different tracking problems are considered: tracking of previously computed explicit paths, and tracking of walls and other natural features in the robot environment. Fuzzy logic allows the management of heuristic rule-base knowledge, imprecise information from sensors, and the uncertainties in the knowledge about the environment. The chapter includes the application of the tracking methods to the mobile robots RAM-1 and AURORA.

17.1 INTRODUCTION

The number of applications of fuzzy logic to mobile robot and autonomous vehicle control has increased significantly in the last years. The best known arguments supporting fuzzy control are the ability to cope with imprecise information in heuristic rule based knowledge and sensor measurements, the interpolative nature of the fuzzy controllers, and the flexibility in the definition of nonlinear control laws.

Motion planning and control of autonomous vehicles involves several problems related with environment perception and representation, path planning, path tracking, velocity control and low level motion control. This chapter deals with the applications of fuzzy logic to mobile robot tracking of references, which include:

a) tracking of previously defined explicit paths, using dead reckoning and navigation sensors.

b) tracking of walls and other natural features in the environment using external sensors.

In (a) the tracking objective is to generate the control commands for the vehicle to follow the previously defined path by taking into account the actual position and the constraints imposed by the vehicle and its lower-level motion controllers. Fuzzy logic can be used both to supervise conventional path trackers, as proposed in [1], and for direct fuzzy control. This Chapter considers the second case. That is, the fuzzy controller directly generates the steering command from the sensor inputs [2].

Approach (b) is related to fuzzy reactive approaches for mobile robot navigation. In this case the controller inputs are mainly provided by the external sensors.

The chapter is organized as follows. The following section is devoted to fuzzy direct explicit path tracking. The section includes the description of the method and the application to the mobile robot RAM-1. The next section is related to reactive navigation. Fuzzy tracking is considered in the framework of a behavior-based architecture for mobile robot control. The application of this method to the AURORA mobile robot is considered in the next section. The two last sections of the chapter are the conclusions and references.

17.2. FUZZY EXPLICIT PATH TRACKING

The application of fuzzy control methods to solve the tracking of explicit paths is motivated by difficulties to apply in real time realistic control laws derived from complex kinematics and dynamic models. Explicit path tracking is based on position estimation produced by the mobile robot's dead reckoning system and inertial navigation system. This estimation is not very accurate but allows the vehicle controller to quickly close the control loop. The general scheme for fuzzy explicit path tracking is shown in Figure 17.1.

In this section the problem of steering the vehicle to track a previously defined path in spite of external perturbations is considered. This path can be defined off-line prior to navigation or computed on-line by the autonomous navigation system. Today several path

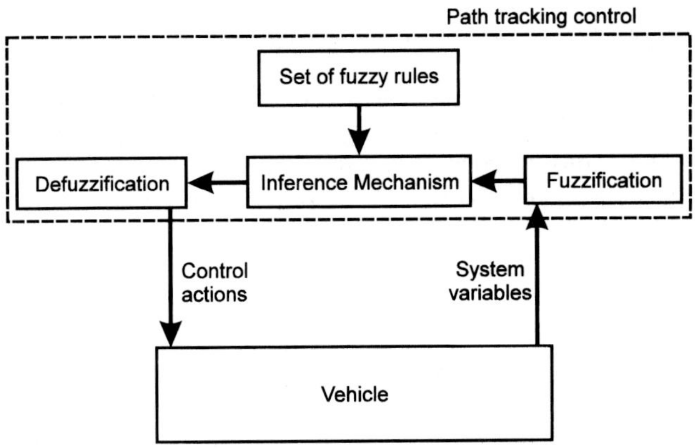

Figure 17.1 Fuzzy explicit path tracking.

generation methods can be applied in real time, providing a path with good kinematic properties to be followed by the real vehicle [3].

17.2.1 Description of the Method

The inputs to the fuzzy controller are variables defining the state of the vehicle with respect to the path, and the output is the steering command to be executed by the low-level motion controllers. Consider a goal point in the path at a lookahead distance away as shown by Figure 17.2(a) [4]. The goal point has a negative x coordinate (negative Δx) and the path tangent in this point (path heading) is the same as the vehicle's heading (zero heading in the vehicle's coordinate frame or $\Delta\phi = 0$). Then, the vehicle must be turned to the left by a certain amount as shown in Figure 17.2(a).

Consider now the situation illustrated by Figure 17.2(b). In this case Δx is zero, but $\Delta\phi$ is positive and the vehicle must be turned again to the left by a certain amount. It seems clear that fuzzy logic could be used to interpret this heuristic in order to generate the steering command γ_d from the values of Δx and $\Delta\phi$.

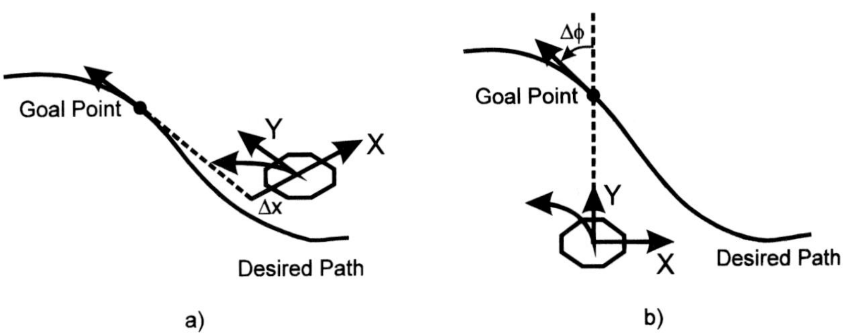

Figure 17.2 Motivation of the fuzzy path tracking strategy.

In addition to Δx and $\Delta\phi$, it is advantageous [5] to consider as antecedents the distance between the vehicle and the nearest point in the path Δs, the difference between the curvature at the goal point and the robot's curvature $\Delta\gamma$, and the vehicle's velocity v. The curvature has to be increased when $\Delta\gamma$ is positive, and decreased when the error in the curvature is negative. When velocity is high, the desired curvature should change in a smooth way. When Δs is shorter the curvature can change more abruptly. When conflicts appear in applying the above heuristic rules, the error in position Δx will be the most significant, while the error in curvature will be the least.

Since the vehicle cannot correct the errors that it has with the nearest point in the path to follow, the goal point is chosen some distance ahead D from the nearest point. This distance is measured over the previously recorded desired path in order to make the search easier (see Figure 17.3). The lookahead distance allows the vehicle to anticipate since it knows the path to follow in advance. A larger lookahead distance implies smoother control but it also means worse tracking. A smaller lookahead can reduce the tracking errors; however, command control increases and can even become unstable. So, it would be useful to consider the lookahead as a parameter that has to be adjusted depending on the vehicle´s situation with respect to the desired path. Changes in the lookahead distance must be accomplished smoothly in order to change the desired curvature in the same way from one control period to the next.

The influence of the antecedents on the lookahead distance is the following:

- Larger errors in position, orientation, or curvature imply that the lookahead D should be increased, while smaller ones should give a decrease in D.
- As the vehicle moves faster, D should be increased. The opposite occurs when the vehicle decreases its velocity.
- The lookahead D has to be increased when Δs is bigger, and decreased when Δs is smaller.

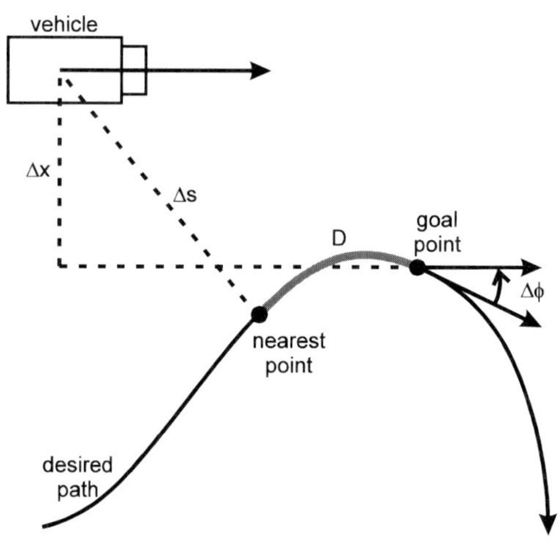

Figure 17.3 Variables of the fuzzy controller.

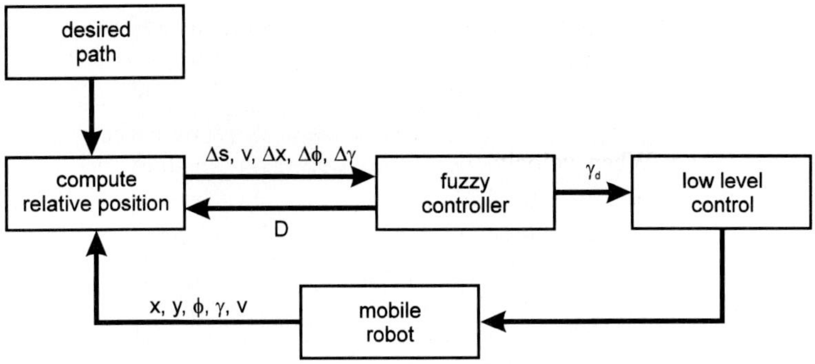

Figure 17.4 Fuzzy direct path tracking.

In case of conflict in applying the above rules, the worst case that increases D would be considered the dominant.

Thus, a set rules such as:

R_i: If Δx is NEGATIVE_ LARGE and $\Delta\phi$ is POSITIVE, $\Delta\gamma$ is ZERO,
v is MEDIUM and Δs is SMALL,
then γ_d is POSITIVE_ LARGE and D is MEDIUM.

are used, where SMALL, POSITIVE, NEGATIVE_LARGE, MEDIUM and NEAR_ZERO are labels associated with the corresponding variables and R_i is the i-th rule of the set of fuzzy rules of the controller. These labels are represented by fuzzy subsets. The set of fuzzy rules represents a fuzzy relation between the antecedents and consequents.

From the control law point of view, the fuzzy direct controller is defined by a nonlinear function:

$$(\gamma_d, D) = \Phi(\Delta x, \Delta\phi, \Delta\gamma, \Delta s, v)$$

Then, the controller inputs are the vehicle's lateral errors with respect to the goal point in the path to follow, which are: errors in position Δx, orientation $\Delta\phi$, and curvature $\Delta\gamma$, the vehicle's distance to the nearest point in the path Δs, and the mobile robot's speed v. The outputs are the lookahead distance D and the steering command γ_d. Figure 17.4 shows the block diagram of the fuzzy direct path tracking method.

17.2.2 Application To RAM-1

RAM-1 is an autonomous mobile robot designed as a testbed for the automation of surveillance, manipulation and small part transportation [6]. The RAM-1 vehicle (see Figure 17.5) has four wheels located in the vertices of a rhombus with a diagonal in the longitudinal axis. The two parallel wheels are driven by DC motors. The front and rear

Figure 17.5 The autonomous mobile robot RAM-1.

wheels are steered by a DC motor with a kinematic rigid link. The locomotion system can provide a zero turning radius. For path tracking the front and rear wheels are steered and the parallel wheels are used with differential drive. The top speed that is 1.6 m/sec only can be reached when the vehicle moves along a straight line, and automatically decreases as the curvature of the vehicle increases. The steering time constant for the RAM-1 was measured to be 0.1 seconds. RAM-1 has several sensors including a scanner laser, 2 video cameras, a sonar ring, and navigation sensors. The RAM-1 control architecture is shown in Figure 17.6.

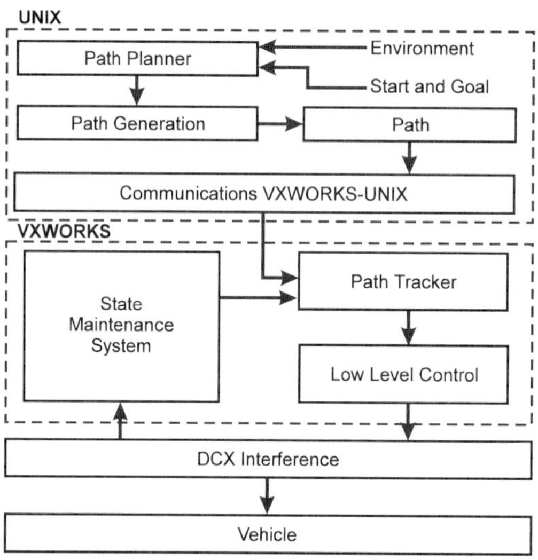

Figure 17.6 RAM-1 control architecture.

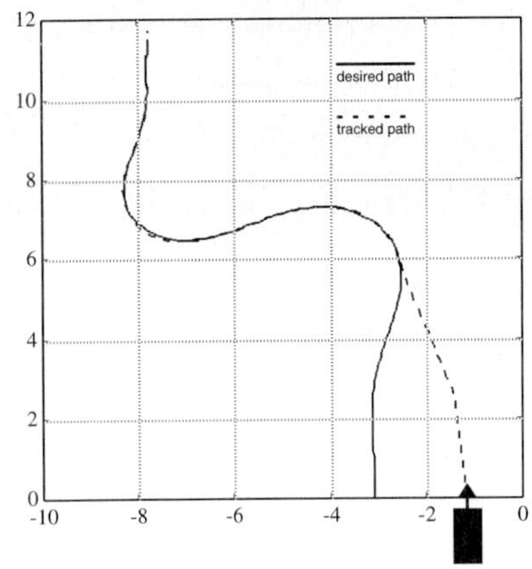

a) Path

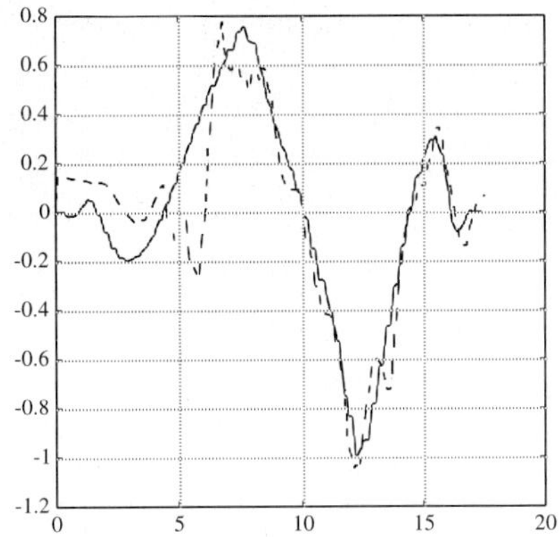

b) Curvature

Figure 17.7 Fuzzy direct path tracking.

The steps that have been taken to build the RAM-1 fuzzy controller are the following:

- Representation in a rule based form of the heuristic knowledge described above.
- The use of a RAM-1 simulation program in real time that allows testing and improving the fuzzy controller very easily, quickly, and without danger.
- Application of the fuzzy controller to the real vehicle and, eventually, tuning some parameters.

The fuzzy controller is embedded in the real-time controller of the vehicle, which runs on conventional hardware, and is programmed in the C language. The real-time controller can compensate for perturbations, such as terrain irregularities, while tracking a path.

Direct fuzzy control provides smooth transitions on the variation of curvature and of the lookahead distance between one control period and the next (20 ms for RAM-1). Path tracking is robust, even when the mobile robot begins far from or badly orientated to the desired path. Experimental results with the RAM-1 show an excellent performance and a robust behavior with large speed variations.

Several experiments have been done confirming the above comments. Figure 17.7 shows the results in the tracking of a reference path at a speed of 1.4 m/s when RAM-1 starts 1.8 meters away ($\Delta x = -1.8$) with a zero error in orientation and curvature. The efficiency of the method can be observed when tracking a difficult path with sharp curves.

18 3. FUZZY TRACKING FOR BEHAVIOR-BASED ARCHITECTURES

Mobile robots are increasingly required to navigate and perform purposeful autonomous tasks in more complex domains, where the environment is uncertain and dynamic. These requirements demand reactive capacity in their navigation systems in order to attain speed and reliability, which implies a strong dependency on sensed information about the robot's surroundings. Therefore, these applications require full use of the dynamic capabilities of sensor systems, which result from the exchange of information between different sensors, each taking limited or partial observations of the environment [7].

Much of the work on reactive navigation has been inspired by the layered control system of the subsumption architecture [8], which tightly couples sensing and action. The emergent behavior of the robot is the result of the cooperation of independent reactive modules, each one specialized in a particular basic behavior. This approach of competing and cooperative agents has also been applied in a fuzzy frame [9]. However, since reactivity alone does not suffice to meet the needs of real world tasks, hybrid solutions have been devised that blend low level reactive behaviors with a high level plan, used to supervise overall mission execution by activating and deactivating behaviors either sequentially [10] or by means of fuzzy transitions [11].

But the obtainment of actor data (command reference values) from sensor data (observation variables) for autonomous mobile robot navigation is a complex task. One interesting approach aiming to reduce complexity and to add flexibility is to deal with a fuzzy system that encapsulates sets of rules in modules and performs inference in parts, with intermediate fuzzy variables [12,13]. For instance, the basic navigation level in the mobile robot control architecture could be divided into two groups of fuzzy rules. The first one would provide a fuzzy interpretation of the sensed data appropriate for each particular navigation concern, while the aim of the second group, a set of navigation behaviors, would be to generate a reference value for the navigation of the robot according to the fuzzy sensor interpretation of the environment. Figure 17.8 sketches the main components of this basic navigation level, in which each behavior makes use of a particular fuzzy interpretation of the sensors needed to accomplish its goal.

The focus of the remaining part of this chapter is on how basic reactive behaviors for navigation can be implemented by means of fuzzy rule bases, considering issues of both measurement (fuzzy range sensor fusion) and action (fuzzy tracking of sensed features). By way of illustration, a real application with the AURORA mobile robot and some experiments are presented.

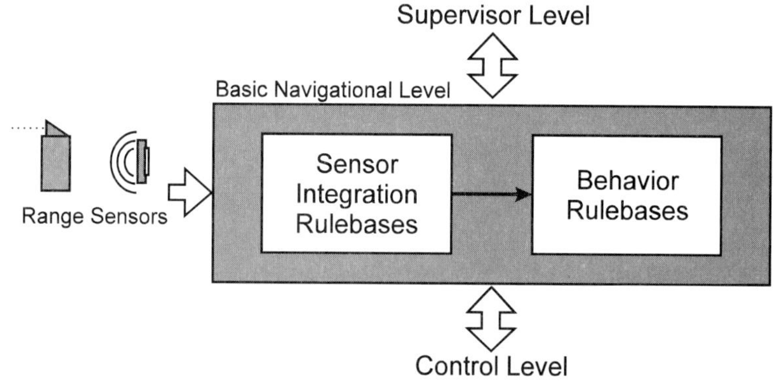

Figure 17.8 The basic navigation level.

17.3.1. Fuzzy Sensors for Navigation Behaviors

Different range sensors usually applied in mobile robot navigation, like laser scanners, sonars and infrared, offer distinct characteristics in reliability, operating limits, accuracy, bearing angle, operating frequency, and last but not least, cost. Robotic applications usually require the combination of sensory data from multiple sources so that reliable and robust interpretations of the environment are available for navigation and operation. The integration of disparate sensory information in order to obtain a coherent description of the world around the robot is rather complex. Fuzzy logic provides a promising way for dealing with conflicts, complementary information, and other issues involved in sensor coordination. Furthermore, a fuzzy description of the sensed environment can be integrated in a fuzzy navigation system.

The nature of sensor interpretation also depends on the use given to it. By adopting a reactive approach, the aim of the system is not to build a map of the environment from an integration of all sensed information [14], but to obtain some immediate indication about the presence of objects around the robot. The meaning given to that indication is task-dependent and environment-dependent, since some of the characteristics of task and environment are used implicitly for navigation. When dealing with the tracking of environmental features such as walls, corridors and corners, an interesting solution is to use just global information, obtained from the integration of different sensors, about particular regions around the robot, e.g., its front and both sides. It must be noted that for task-oriented navigation in open environments it should be necessary to have long-range information as well, so that long-range targets can be detected.

The problem of multiple-sensor processing is fit for a fuzzy approach, since it implies dealing with uncertainty in three different and combined ways:

- Uncertainty in the distance to surrounding objects, which can be modeled by terms such as "far" or "close." It arises from the inaccuracy of sensors, that are very sensitive to such external factors as the quality of the reflecting surfaces or environmental conditions such as light and temperature.
- Uncertainty about the reliability of sensed data from which the distance can be inferred. Sonars, for instance, provide multiple erroneous readings and do not always detect the presence of objects. Despite stochastic models, it is not always possible to feasibly estimate the reliability of an individual sensor. However, an estimation can be inferred from a combined interpretation by the degree of agreement between different redundant sensors.
- Mobile applications involve specific sources of uncertainty: the distance to surrounding objects varies during the measurement process due to the robot's motion, and when data are available for use they are not completely up-to-date, because of the time required for acquisition and processing.

From the navigation point of view, several efficient reactive systems have been implemented which make use of range sensor information of an homogeneous type (ultrasonic rings, laser range finders, etc.). Nevertheless, an appropriate combination of several sources of sensor data may offer more information with increased reliability, but still an homogeneous interpretation is preferable to be used immediately by the reactive navigation system. In other words, the design of a particular navigation behavior (e.g. wall following) is quite simple if it makes use of just one input (i.e., the distance to the wall) provided by a single sensor, whereas by using inputs from more sensors the behavior should be more reliable, but at the expense of increased complexity.

As a result of this, the introduction of a model based on virtual fuzzy sensors offers a reliable solution for the integration of disparate sensor information with an autonomous navigation system. A virtual fuzzy sensor provides a fuzzy range estimation that results from fusing information produced by a set of heterogeneous real sensors (that can be complementary in range, redundant, of different quality, etc.) covering the region assigned to it. From the navigation point of view there would be no difference between using the data from a sonar ring or from a virtual ring of fuzzy sensors, except for the improved reliability of the latter solution.

Moreover, the modularity of this approach allows new behaviors to be added to cope with new tasks or environmental requirements, which in turn may require the introduction of new virtual fuzzy sensors in the rule base to meet their particular perception requirements. This is very interesting from the designer's point of view, since simple independent behaviors and virtual sensors consisting of a reduced number of fuzzy rules can be easily developed and tested before being merged into a comprehensible navigation control system.

17.3.2. Application to the AURORA Mobile Robot

This section presents the implementation of fuzzy reactive navigation behaviors for AURORA, a mobile robot which is equipped with a sensing system based on ultrasonic sensors for autonomous navigation. Ultrasonic sensors can be adjusted to work at different range distances and can thus be combined to sense both the immediate surroundings (a few centimeters) and mid-range distances (a few meters) from the robot.

AURORA (see Figure 17.9) is an autonomous wheeled mobile robot for greenhouse operations designed and built at the Málaga University to perform operations in greenhouses [15]. AURORA must be capable of operating in a variety of different greenhouses without imposing any alterations on them at all. The locomotion system is a modification of the RAM-1 dual configuration [6]. The mechatronic system consists of an octagonal mobile platform 80 cm in width and 140 cm in length that accommodates a spraying device, the power system, standard electronic and computer enclosures, and a variety of sensors for intelligent operation and navigation.

For autonomous navigation, AURORA carries a heterogeneous configuration of ultrasonic sensors that was chosen considering the characteristics of constrained environments, where the robot has to navigate through narrow spaces among a high obstacle density [16]. Instead of a typical ring of homogeneous sonars, its sensor configuration is a combination of three different types of sonars placed at different heights on the front half of the robot, covering the ranges shown in Figure 17.10. The number and types of sonars that have been used are the following: four Short-Range Digital (SRD) sensors, two Mid-Range Digital (MRD) sensors, and four Mid-Range Analog (MRA) sensors. It must be noted how analog range sensors are combined with digital ones that do not provide a range measurement but a truth value, depending on the

Figure 17.9 The AURORA mobile robot.

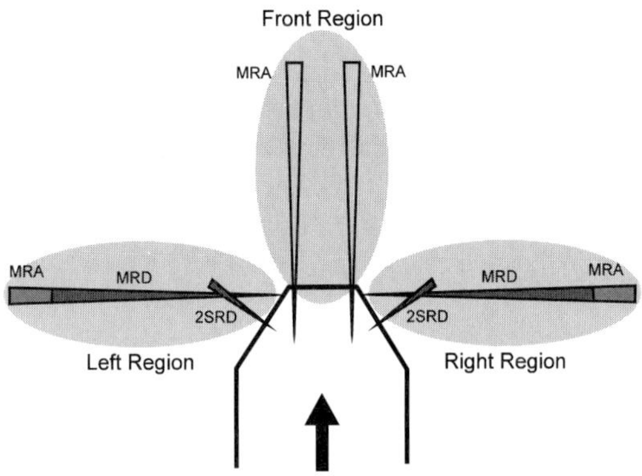

Figure 17.10 Operating range limits of ultrasonic sensors.

presence or not of an object within its working range. Technical details about these types of ultrasonic sensors are presented in Table 17.1.

This is a case where sensor fusion is necessary in order to produce an immediate sensorial interpretation of the robot's surroundings. The fact that all sensors are ultrasonic somehow simplifies the problem of fuzzy sensor fusion, but their heterogeneity allows that the results presented here could be extended, without much effort, for the integration of different kinds of range sensing devices.

The concept of behavior-based fuzzy navigation system is illustrated by Figure 17.11 that shows the sets of basic navigation behaviors and virtual fuzzy sensors implemented for AURORA. The following sections describe the design and implementation of some of these modules, together with some illustrative experimental results.

Sensor integration. In the presented application, the sensor configuration allows for the definition of three fuzzy sensors, covering three regions around the front half of the robot,

Table 17.1 Technical specifications of ultrasonic sensors

	SRD	MRD	MRA
Sensing range	6 to 30 cm	20 to 100 cm	20 to 130 cm
Switching frequency	8 Hz	4 Hz	4 Hz
Ultrasonic frequency	400 kHz	200 kHz	200 kHz
Repeat accuracy	± 0.45 mm	± 1.5 mm	± 2 mm
Resolution	–	–	5 mm
Angle	5°	5°	5°

as can be seen in Figure 17.10. The front area is covered by two similar sensors, separated by a distance so that they provide complementary information. The side areas are covered by a mixture of disparate sensors, so they offer an interesting case for sensor fusion, and will be considered below in more detail.

Each fuzzy system dedicated to sensor interpretation must obtain a conclusion about the existence or proximity of objects around the robot from the information provided by a set of heterogeneous ultrasonic sensors. A major cause for uncertainty in the distance to surrounding objects arises from the fact that some of the sensors used in this robot are digital, i.e., they merely provide two-valued information about the presence of an object within a range, and the possibility of having an object is the same along the whole working range of the sensor. Nevertheless, this range can be easily represented in rule consequents as a crisp interval along the sensor range (as the sets SHO and LON in Figure 17.12. Through sensor fusion, that uncertain information, insufficient to produce an accurate description of the environment, can be used to add reliability to that from analog sensors.

Heuristic knowledge about each kind of sensor can be introduced in the fuzzy sensor

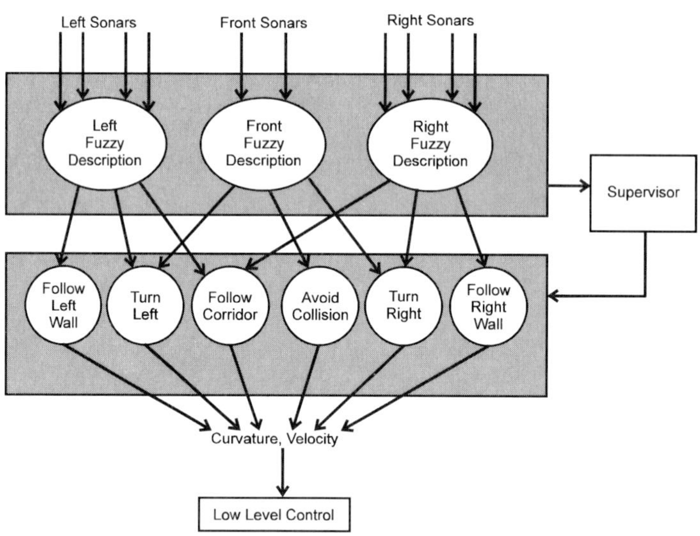

Figure 17.11 Elements of fuzzy navigation with sensor fusion.

rule base so that a feasible result is inferred. In the particular case of ultrasonic sensors, there are two types of erroneous readings: First, analog sensors may provide an erroneous range, and second, a sensor, either analog or digital, may be mistaken about the presence or not of an object. Besides, a sonar detecting something is more reliable that an other that does not. With the exception of echoes and multiple reflections, the reason an ultrasonic sensor detects the presence of something is because something is really there; on the other hand, sensors do not always detect the presence of an object, depending on the quality or the angle of the reflecting surface. For that reason, heuristic approach to deal with sonar reliability, when some kind of redundancy is available, is to give more credibility to those sensors that detect something, while relying less on those that fail to detect it. One way to represent this knowledge is to assign each rule a weight corresponding to its relative reliability with respect to other rules, according to the following criteria:

- Rules for digital sensors have a lower value, depending on the range covered.
- Rules for close ranges have the highest values, and they decrease according to distance. The closer a range detected, the more reliable it will be. Reflected beams make a longer fly and result in erroneous distant readings.

The fuzzy rule base designed for the side fuzzy sensors (which can be used for both left and right sides, since their configuration is symmetrical) consists of 12 rules with decreasing weights along the range of the analog sensor, and two additional rules for the digital sensors. Figure 17.12 shows an example of some of these rules with their membership functions and how they are activated by a particular set of input data (MRA = 0.27, SRD1 = 1, SRD2 = 0, MRD = 1). The output defuzzified value (0.479 in the

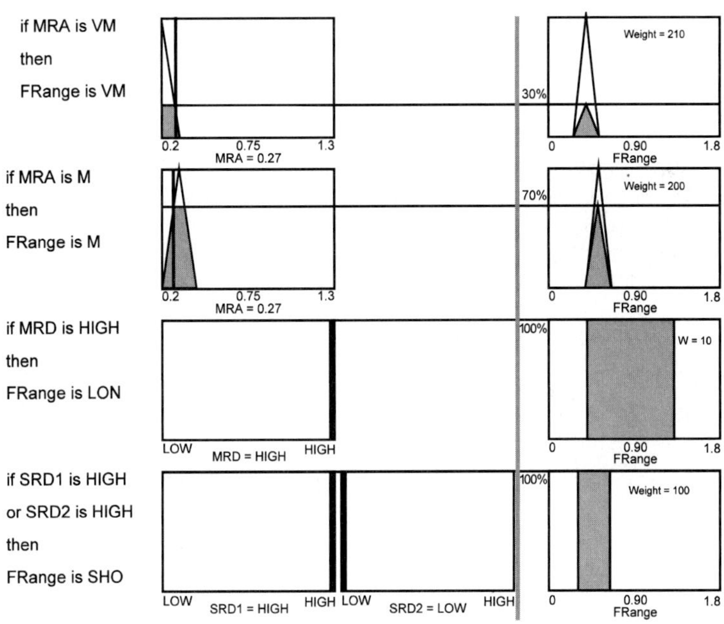

Figure 17.12 Fuzzy outputs of each active rule of the side fuzzy sensor.

presented example) is the result of applying the max-dot/height method, whose value is given in the following equation:

$$v^* = \frac{\sum_{i=1}^{n} \alpha_i H_i C_i W_i}{\sum_{i=1}^{n} \alpha_i H_i W_i}$$

where v^* is the defuzzified output value, α_i is the degree of membership computed for the premise of rule i, W_i is the weight assigned to rule i, H_i is the height of the membership function assigned to the consequent v in rule i, and C_i is the center of gravity of the membership function assigned to v in rule i. The intervals for the sets SHO and LON are obtained from the sensing ranges for SRD and MRD presented in Table 17.1.

Figure 17.13 shows the result for the side fuzzy sensor while the robot diverged from a wall as compared with the original readings from all side sensors (analog and digital). If any of the digital sensors fails to detect anything, the incidence on the result is very small. On the other hand, if the failing sensor is the analog one (i.e. the less uncertain), the result is not accurate, but still offers a rough approximation to reality.

Figure 17.13(a) shows how the side analog sensor fails to detect a lateral obstacle in the last section of readings. The fuzzy sensor [Figure 17.13(b)] provides integrated data, referred to the center of the robot, from all side sensors, using information from the digital ones to compensate the erroneous readings of the analog.

Tracking natural features. The use of the virtual fuzzy sensors presented above allows the design of much simpler navigation behaviors, such as wall following, corridor following or turning.

Such is the case of the set of rules shown in Table 17.2, which are used for wall tracking. Two inputs are defined; the first one is the distance that must be kept as reference while following the wall. For the sake of simplicity in task specification, this is a crisp variable with only three possible values: Very Close, Close, and Middle. The

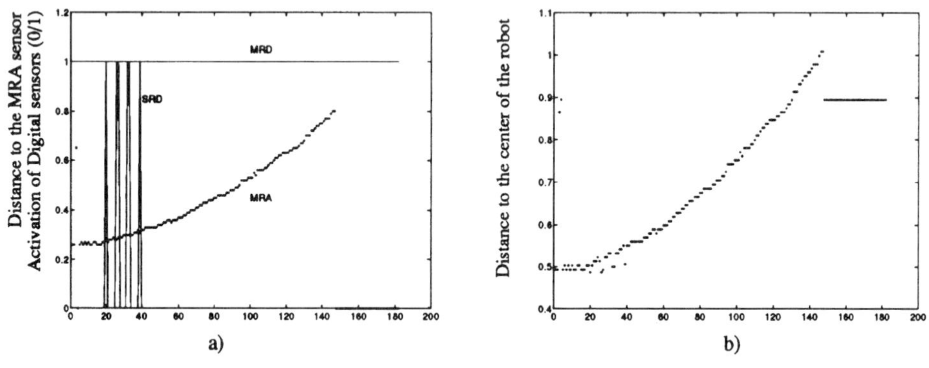

Figure 17.13 Experimental results (a) with the side sensor and (b) the fuzzy sensor.

Table 17.2 Fuzzy Inference Rules for Wall Following

Reference	Measured Range				
	TOOCLOSE	VCLOSE	CLOSE	MIDDLE	FAR
VCLOSE	PM	Z	N	NM	NM
CLOSE	PM	P	Z	N	NM
MIDDLE	PM	PM	P	Z	NM

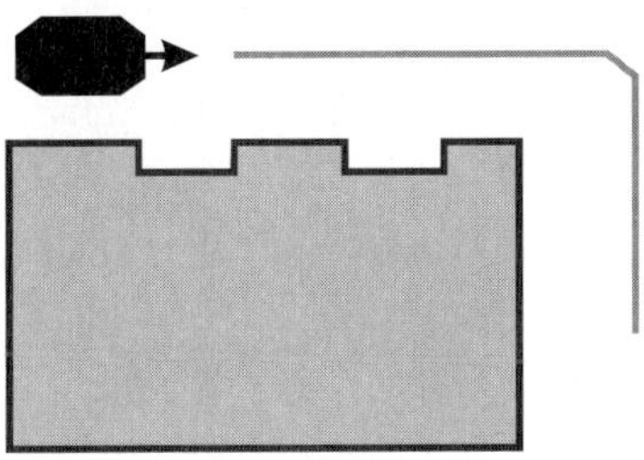

Figure 17.14 Physical layout of the experiment.

second input is that of the virtual fuzzy sensor providing the distance to the wall, which has been defined by means of five triangular fuzzy sets. The output is a reference value for the curvature of the robot, which is defined by five triangular sets in the universe of discourse [–0.8, 0.8]. The defuzzification method used for these rules has been the center of gravity of the unioned scaled output. The corridor following behavior is similar, but it receives inputs from two fuzzy sensors, representing distances to walls at both sides of the robot.

Figure 17.14 shows the physical layout of an experiment in which AURORA accomplished the task "Follow close right wall, turn right, and follow close right wall." Figure 17.15 shows the measures provided by the "right fuzzy sensor" and the vehicle's curvature during the experiment. The result produced by the fuzzy right sensor is used to generate the curvature commands by the navigation behaviors involved in the task. The –2.5/m peak in curvature corresponds to the turn, which takes place when the fuzzy sensor detects free space at the right.

After several tests with this setup, AURORA repeatedly detected the same environmental landmarks specified in the task (corridor, turning point and wall), accomplishing the mission independently of the robot's starting position or the exact path followed.

17.4 CONCLUSIONS

Fuzzy logic can be used for both path tracking of paths previously computed by path planners and tracking of walls and other natural features encountered in the robot environment. In the first case a fuzzy controller implements heuristic rules to generate the steering commands to take or maintain the path from the knowledge of the vehicle's position and orientation provided by dead reckoning and navigation sensors. The fuzzy

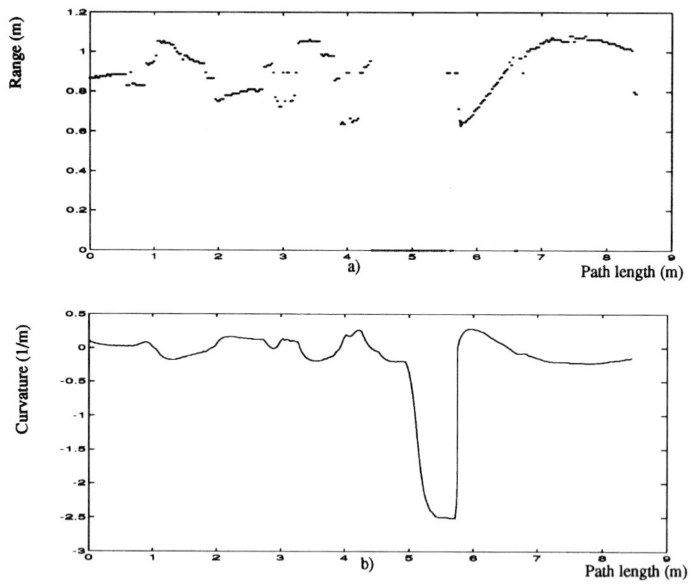

Figure 17.15 Experimental results. a) Measures provided by the "right fuzzy sensor." b) Vehicle's curvature.

controller also updates the lookahead distance on the path to track, which is the critical parameter in path tracking. Thus, the resulting fuzzy controller is robust and can obtain good performance for both compensating the effect of perturbations (e.g., due to terrain interaction) and tracking sharp paths. The implementation of this strategy in the RAM-1 mobile robot has validated these conclusions.

On the other hand, fuzzy rules can be efficiently used to implement basic reactive behaviors based on external sensor information. Particularly, this chapter shows how the so-called "virtual fuzzy sensors" offer a reliable solution for the integration of disparate sensor information in an autonomous navigation system. This virtual sensor provides a fuzzy range estimation that results from fusing information from a number of heterogeneous real sensors. The efficiency and robustness of this approach have been demonstrated in wall tracking experiments with the AURORA mobile robot using three types of ultrasonic sensors. Some results are also included in the Chapter.

The same approach will be implemented in the near future for other different navigation behaviors such as following corridors and avoid collisions. Furthermore, fuzzy logic will be also used to coordinate the different behaviors in the reactive architecture.

17.5 REFERENCES

1. Ollero A., Garcia-Cerezo, A. and Martínez, J. L.. "Fuzzy Supervisory Path Tracking of Autonomous Vehicles." *Control Engineering Practice*. Vol. 2, No. 2, 1994, pp. 313-319.

2. Martínez J. L., Ollero, A. and García-Cerezo, A. "Fuzzy Strategies for Path Tracking of Autonomous Vehicles." *Proc. First European Congress on Fuzzy and Intelligent Technologies*, Vol. 1, pp. 24-30, Aachen, Germany, 1993.

3. Muñoz V., Ollero, A. Simón, A. and Prado, M.. "Mobile Robot Trajectory Planning with Kinematic and Dynamic Constraints." *Proc. IEEE International Conference on Robotics and Automation*, Vol. 4, pp. 2802-2807, 1994.

4. Amidi, O. "Integrated Mobile Robot Control". Carnegie-Mellon Univ. Robotics Institute. Technical Report CMU-RI-TR-90-17, 1990.

5. Garcia-Cerezo, A., Martinez, J. L. and Ollero, A. "Fuzzy direct path tracking of explicit paths in mobile robots. Application to RAM-1." *Proceedings of the EURISCON'94 Conference*. Vol. 1, pp. 253-260. Málaga, Spain, 1994.

6. Ollero, A., Simón, A. García, F. and Torres, V. "Integrated Mechanical Design of a New Mobile Robot." *Proc. of the First IFAC International Symposium on Intelligent Components and Instruments for Control Applications*, pp. 461-466, Pergamon Press, 1993.

7. Durrant-Whyte, H. F. *Integration Coordination and Control of Multi-sensor Robot Systems*. Vol. SECS 36, International Series in Engineering and Computer Science Kluwer Academic Publishers, 1988.

8. Brooks, R. A. "A Robust Layered Control System For a Mobile Robot." *IEEE Journal on Robotics and Automation,* Vol. RA-2, No. 1, March, 1986.

9. Goodridge, S. G. and Luo, R. C. "Fuzzy Behavior Fusion for Reactive Control of an Autonomous Mobile Robot: MARGE." *IEEE International Conference Robotics and Automation*, San Diego, California, 1994.

10. Ollero, A., Mandow, A. Muñoz, V. F. and Gómez de Gabriel, J. "Control Architecture for Mobile Robot Operation and Navigation." *Robotics & Computer Integrated Manufacturing*. Vol. 12. Elsevier Science, Pergamon, 1995.

11. Ruspini, E.H., Saffioti, A. and Konolige, K. "Progress in Research on Autonomous Vehicle Motion Planning." *Industrial Applications of Fuzzy Logic and Intelligent Systems* (Yen, J., Langari, R. and Zadeh, L. A., eds.). IEEE Press, 1995.

12. von Altrock, C., Krause, B. and Zimmerman, H. J. "Advanced Fuzzy Logic Control of a Model Car in Extreme Situations." *Fuzzy Sets and Systems.* Vol. 48, North-Holland, 1992, pp. 41-52.

13. Mauris, G., Benoit, Foullow, E. L. and Josserand, J. F. "Fuzzy Range Sensor Fusion for the Navigation of Mobile Robots," *Proc. EUFIT'94*, Aachen, Germany, 1994.

14. Poloni, M., Ulivi, G. and Vendittelli, M. "Fuzzy Logic and Autonomous Vehicles: Experiments in Ultrasonic Vision." *Fuzzy Sets and Systems*. Elsevier, Vol. 69, 1995, pp. 15-27.

15. Ollero, A., Martínez, J. Simon, L. Gómez de Gabriel, A. J. and Muñoz, V. F. Mandow, A., García-Cerezo, A. García, F., Martínez, M. A. "The Autonomous Robot for Spraying: AURORA." *4th Workshop on Robotics in Agriculture and the Food Industry IARP'95*, Toulouse, France, 1995.

16. Mandow, A., Martínez, J. L. Muñoz, V. F. Gómez de Gabriel, J. Ollero, A. and Reina, A. J. "Reactive Navigation in Constrained Environments with the AURORA Mobile Robot". *7th International Conference on Advanced Robotics, ICAR 95*, Sant Feliu de Guixols, Spain, September 1995.

18

FUZZY CONTROL FOR AUTOMOTIVE ENGINE—IDLE SPEED CONTROL

S. Boverie, C. Raymond and JM. Le Quellec
SIEMENS Automotive S. A.
FRANCE

A. Titli*
INSA/LAAS CNRS
FRANCE

Idle speed control for automobile engines is always considered a difficult problem. The hardening of the rules against pollutant emissions leads to a reduction of the engine idle speed. Unfortunately the stability of the engine at low speed is difficult to maintain due to mechanical frictions, misfiring, great sensitivity to load disturbances, etc. On the other hand the reaching phases to the idle speed regulation set point are difficult to control, and random load disturbances such as power steering, air conditioner, lights, etc., could modify the behavior of the engine and could involve some undesirable effects like noise, vibration, jerking, etc. The idle speed control (including reaching phases and regulation) must be efficient both when the vehicle is stationary and when it is moving.

---

* Currently on leave at NASA Center for Autonomous Control Engineering, The University of New Mexico, Albuquerque, NM, USA.

The dynamic responses of stationary and moving vehicle are quite different which lead in fact to consider two different processes. In addition, the engine process is difficult to model due to numerous engine variables, high nonlinearity of the process, unobservability of some variables, etc.

Moreover, it is a time-varying process with respect to environmental conditions (like air temperature, water cooling temperature, etc.), aging and it shows a large variation between engines of the same class. Finally real-time constraints and cost constraints are severe. In order to achieve idle speed regulation goal, conventional control techniques like supervised PID are usually carried out. These structures are generally complex and their tuning, for each engine class, is tedious and time-consuming. Also, they often show some lack of robustness with respect to load disturbances, for example.

In order to tackle these problems, fuzzy control seems to be a suitable alternative. On the one hand, this technique could use the knowledge from the tuning engineers like a "qualitative modeling" in order to reduce the tuning time, to improve the tuning methodology and to take into account subjective criteria like comfort. On the other hand it is now well known that the nonlinear control structure provided by the fuzzy controller is robust with respect to internal and external disturbances. But, unfortunately, conventional fuzzy controllers cannot take into account all the engine constraints. So, in this chapter an original control structure based on the analogy between sliding mode and fuzzy controllers is proposed.

- The first part of the chapter presents the idle speed control principles.
- The second part is dedicated to the design of the fuzzy idle speed controller,
- The third deals with implementation aspects.
- Some results and conclusions are presented in the fifth and sixth parts

18.1 INTRODUCTION

For nearly a century the automobile has been the principal means of transport in the industrial world. During the past twenty years, the implementation of strict regulations related to emission controls, fuel consumption, and security, as well as the search for improved driveability and comfort have inevitably led to the need for more and more precise control of vehicle operation. Therefore, today, many vehicle functions are currently performed through strategies implemented in various types of electronic control units (ABS ECU, Injection ECU). For a few years, the stiffening of the requirements in terms of safety and comfort has led to the development of more sophisticated systems based on controlled actuators (injectors, semi-active dampers) driven By Electronic Control Units (ECUs). These refinements have been performed in parallel with the search and the development of sophisticated control laws which are implemented inside these ECUs. The conventional control methods used in these ECUs are often unsuccessful due to a lack of robustness, the need for mathematical models of complex processes or phenomena and, also, the need of elaborate "tuning" procedures. Indeed, the control of these different subfunctions is a very difficult problem because of:

- the high nonlinearities of the system,
- the interactions between the different subsystems,
- the multivariable aspects,

- the integration of qualitative constraints like comfort,
- the necessity to find compromise between antagonist goals.

In addition, on-board automotive systems have to deal with severe real-time constraints. Besides, they are drastically constrained with respect to financial costs. For example, the sophisticated solutions which have been developed for aeronautic field can not be used in this case. So, compromise solutions must be found between high performance and cost.

The techniques of fuzzy control, based on expert knowledge and possessing intrinsic qualities like robustness and ease of implementation (many dedicated toolboxes are available), seem to be an interesting alternative [1,2]. At SIEMENS Automotive SA, a number of studies have already been performed related to the aspects of engine control as well as driving security and comfort.

In this chapter we are going to present one of those applications, i.e., idle speed control.

The requirements which must be met by this subsystem are multiple:

- low consumption,
- minimization of pollutant emissions,
- good comfort.

18.2 IDLE SPEED CONTROL PRINCIPLES

18.2.1 Introduction

Idle speed function includes, in fact, several engine running phases (Figure 18.1):

- the regulation around the engine idle speed set point,
- the reaching phases toward the speed set point (when accelerator pedal is released and the engine speed is below an offset N_0),
- the starting phases (including engine cold temperature phases).

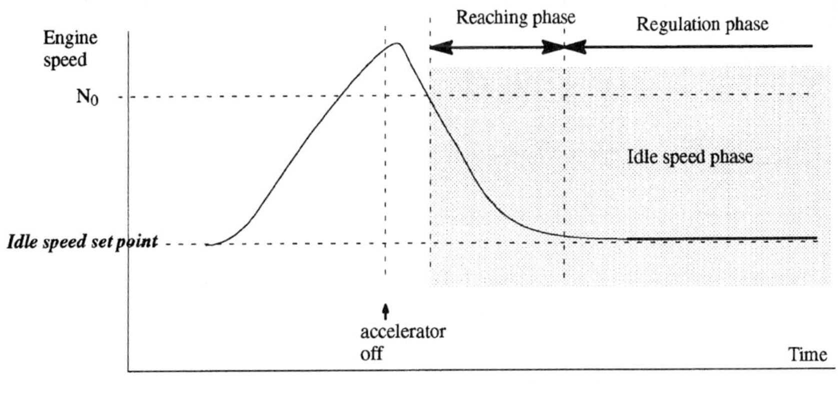

Figure 18.1 Definition of the idle speed phase.

In addition, idle speed control (including reaching phases and regulation) must be efficient both when the vehicle is stationary and when it is moving. The achievement of a good idle speed control implies to take into account several requirements which may sometimes be antagonist; on the one hand the optimization of the driveability and the comfort and, on the other hand, the minimization of the consumption and the pollutant emissions, for example:

- the stiffening of the rules against pollutant emissions has led to a reduction of the engine idle speed set point. Unfortunately the stability of the engine at low speed is difficult to maintain due to mechanical frictions, misfiring, great sensitivity to load disturbances, etc. and these phenomena can produce undesirable effects for the driver.
- the reaching phases to the idle speed regulation set point are difficult to control, Random load disturbances such as power steering, air conditioner, lights, etc., could modify the behavior of the engine, and could also involve some undesirable effects like noise, vibration, jerking, etc.

In addition, engine is a time varying process with respect to environmental conditions like air temperature, water cooling temperature, etc. and to aging. It also presents a large scattering between engines of the same class.

Finally, the engine process is difficult to model due to numerous engine variables, high nonlinearity of the process, unobservability of some variables, etc.

In conclusion one can say that idle speed control for automobile engines is considered as a difficult problem.

18.2.2 Idle Speed Requirements

18.2.2.1 Driveability and Comfort

Driveability and comfort requirements lead to suppressing or to minimizing the engine and vehicle behaviors which can bother the user (engine vibration, hiccup, undershoot, overshoot) and to reduce the sensitivity to load disturbance. The driver perception of these different phenomena is mainly subjective and difficult to evaluate. Nevertheless, a qualitative "idealistic" behavior can be defined by the specialists when reaching the idle speed regulation set point and during the regulation phase (see Figure 18.2). This reference takes into account minimum reaching time considerations, and defines acceptable variation of the engine speed around the reference set point.

18.2.2.2 Consumption and Pollutant Emissions

The reduction of the vehicle pollutant emissions is one of the greatest challenge of the end of this century. The stiffening of low emissions rules (LEV, ULEV, ZEV) are forcing car manufacturers to continuously improve the engine performance. Currently, idle speed phases represent an important part of engine running (30% of the running time for an urban emission test cycle, see Figure 18.3). So, one of the major objectives of car

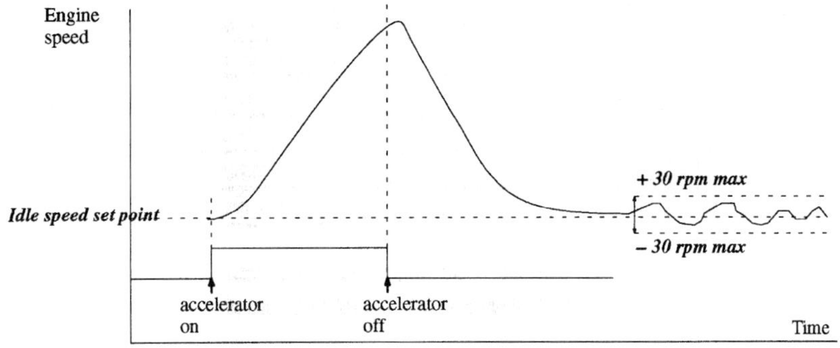

Figure 18.2 Idle speed behavior requirements.

manufacturers is to minimize emissions during idle phases. This can be achieved by reducing the engine idle speed set point which will lead to a fuel consumption reduction and therefore to a reduction of direct emissions (before catalyst) and residual emissions (after catalyst).

A reduction of engine idle speed can be achieved, on the one hand, by an improvement to the mechanical characteristics of the engine and, on the other hand, by the design of more sophisticated and robust control strategies able to handle random disturbances and stability problems.

18.2.3 System Analysis

Engine process at idle can be analyzed in terms of input/outputs (Figure 18.4).

The main control variables are:

- the injection time, t_i, which manages the fuel quantity entering into the cylinders,
- the ignition advance angle which determines the ignition time of air/fuel mixture,
- the opening of the idle speed valve which determines the air mass entering into the cylinders.

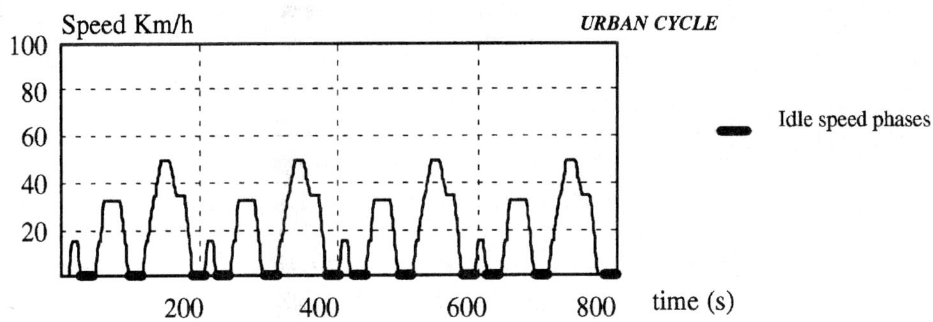

Figure 18.3 European emission cycle.

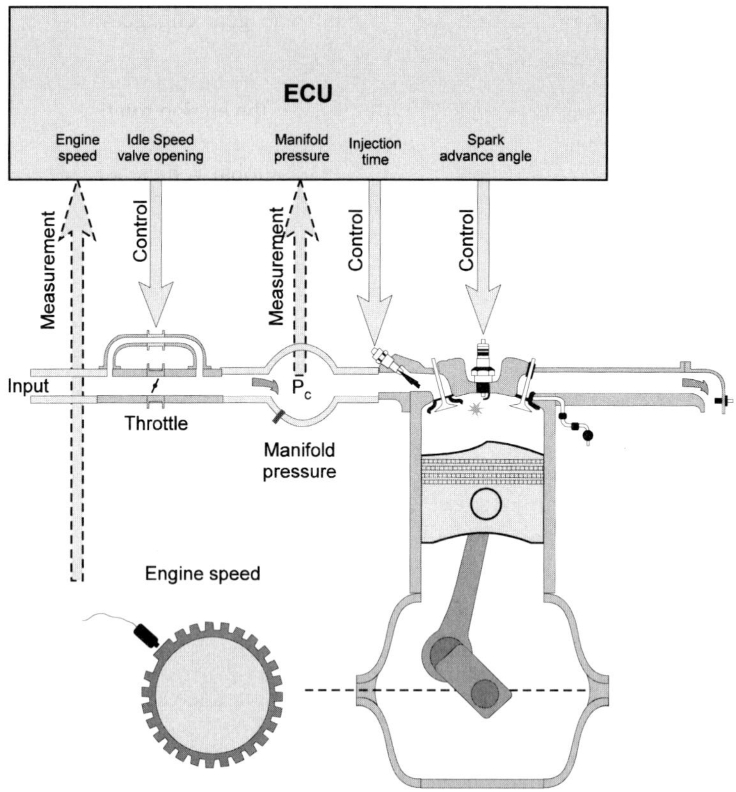

Figure 18.4 Process description.

The control variables can be classified according to their authority and response time:

	Authority	Response time
Ignition advance angle	MEDIUM (restricted around operating point)	Fast
Idle speed valve	High	Medium
Injection time	High	Large

Measurable outputs are:

- the manifold pressure P_c which is representative of the air mass at the inlet of the engine (other sensors like a mass air flow meter can also be used to value the inlet air quantity),
- the engine rotational speed, N.

The characteristics of the system to be controlled are in fact linked to the different vehicle running phases.

If the vehicle is stationary (engine running but no gear engaged) the process is only constituted of the engine (engine rotational inertia). The process is very sensitive to load disturbances (power steering, air conditioner) and its response time is short. Small changes of the control variables have a big impact on the engine rotational speed.

If the vehicle is moving (with a gear engaged), it is necessary to take into account the global vehicle dynamics (engine rotational inertia + global vehicle's inertia). The process response time is slow and the load disturbances are filtered by the vehicle inertia.

Finally, the last case deals with transient phases during gear changes.

18.2.4 State of the Art

Current idle speed control strategies are generally based on engine speed closed loop structures. Control variables are mainly air mass (idle speed valve opening) and ignition advance angle.

Conventional control techniques, like supervised PID, are generally carried out. These structures are often complex and their tuning, for each engine class, is tedious and time-consuming. Also, they often show some lack of robustness with respect to load disturbances.

18.3 DEFINITION OF THE IDLE SPEED FUZZY CONTROLLER

To manage engine speed at idle, we have chosen to control simultaneously the amount of air at the inlet of the engine, through a bypass valve and the spark ignition advance. The fuzzy controller only controls the air input. A conventional strategy is preserved to control the spark ignition. The main input of the fuzzy controller is the engine speed (see Figure 18.5).

The global structure of the fuzzy controller is given in Figure 18.6 [3,4]. The fuzzy core provides a variation of control variable Δu which is integrated [3,4]. RCO_{nom} is an offset which is added to the control value in order to limit the influence of the integrator.

The idle speed controller is activated when the three following conditions occur:

- accelerator pedal is released,
- the engine speed is below a fixed threshold N_0,
- the derivative of the error is positive.

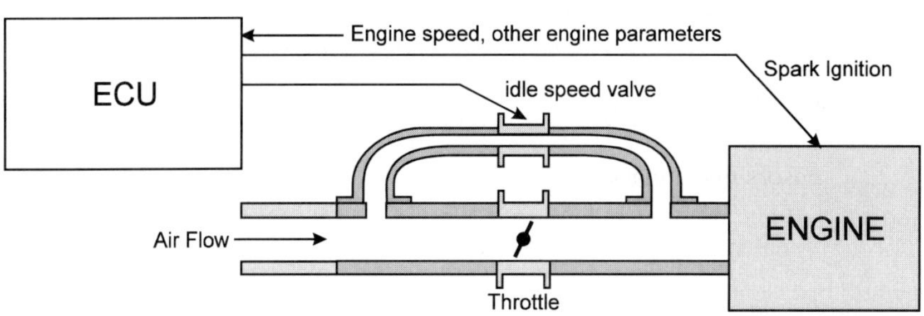

Figure 18.5 Controlled process.

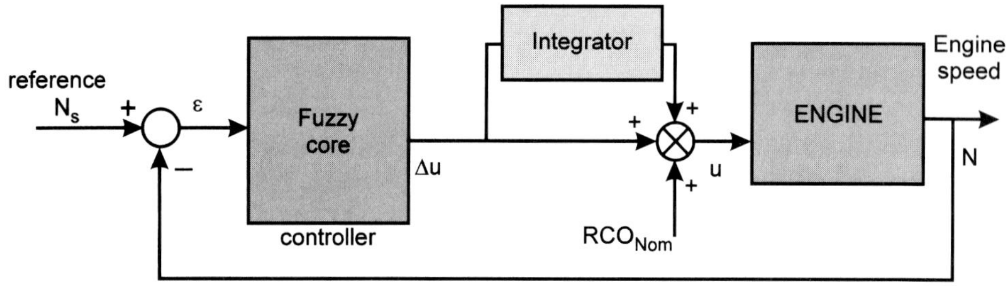

Figure 18.6 Idle speed fuzzy controller structure.

In the following section we describe in more details the structure of the fuzzy core [5,6].

A first controller is proposed in Section 18.3.2 but it only works when the vehicle is stationary. In Section 18.3.3 this controller is extended to take into account the overall running conditions.

18.3.1 General Structure Of A Fuzzy Controller

Current fuzzy controllers, as Mc Vicar Whelan ones, can be seen as nonlinear PD (PI) controllers with saturated areas or, perhaps, an analogy can be made with robust nonlinear techniques as sliding mode control. Figures 18.7a to d show a comparison of the different control surfaces generated by these different control methods.[7].

The structure of a conventional fuzzy PD (PI) controller uses as inputs the error ε and the derivative of the error ε. Mc Vicar Whelan has suggested a symmetrical control structure which can be analyzed in a qualitative phase plane [Figure 18.7(c)]. The fuzzy control value is equal to zero on the diagonal line, positive (negative) above (below) this line: the value of the control increases progressively in the distance from this line. It is possible to represent the equivalent control surface in a three dimensional space $(\Delta u, \varepsilon, \dot{\varepsilon})$. This surface is nonlinear depending on the choice of the modal points of ε and $\dot{\varepsilon}$. The intersection between this surface and the plane gives us the zero control curve $S(\varepsilon, \dot{\varepsilon}) = 0 \,/\, \Delta u = 0$. The control action can be determined as a nonlinear function: $\Delta u = F(S(\varepsilon, \dot{\varepsilon}))$.

This zero control curve defines the nominal behavior of the system [3]. The control value Δu applied to the process tries to bring the state vector $(\varepsilon, \dot{\varepsilon})$ on this curve.

18.3.2 Definition of the Fuzzy Core for Stationary Vehicle

When the vehicle is stationary, the process to be controlled is only the engine.

The engine idle speed control laws must take into account different engine running phases:

- The regulation phase at the engine idle speed set point,
- The reaching phase to the engine idle speed set point.

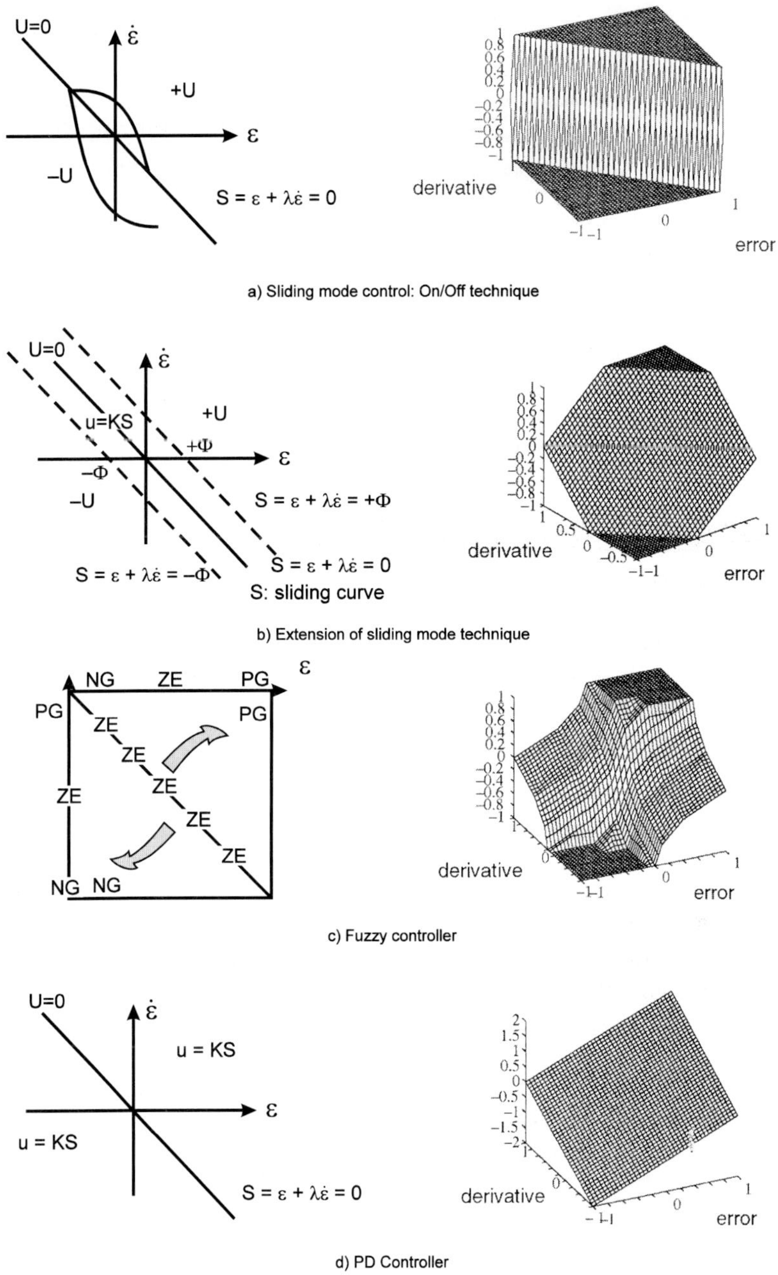

Figure 18.7 Analogy between sliding mode, PD, and fuzzy controllers.

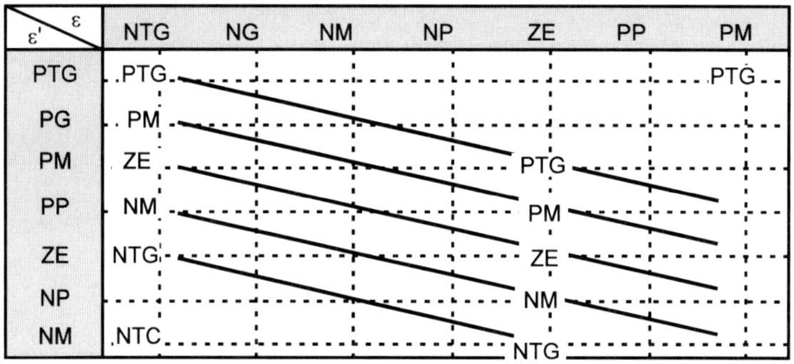

Figure 18.8 Nondiagonal control table for nonzero initial conditions.

The regulation phase is not a strong problem and a conventional fuzzy controller structure, like a Mc Vicar Whelan one, produces good results very quickly, even when loads are applied. It is not the same during reaching phases. When the idle speed control law is activated, the applied control action must be non-zero and must depend on the running conditions of the engine before releasing the accelerator pedal (for the same engine speed error ε the variation of the error Δε is not the same depending on the initial conditions). In addition, reachable engine speeds are not equally distributed around idle speed set point. So, a simple Mc Vicar Whelan structure is not able to take into account these constraints and it is necessary to design a nondiagonal table (see Figure 18.8).

The diagonal of zeros represents the "ideal" state vector trajectory, defined with respect to comfort and driveability constraints. Control actions around this sliding curve try to bring the state vector on it or in its vicinity in order to respect these constraints (Figure 18.9).

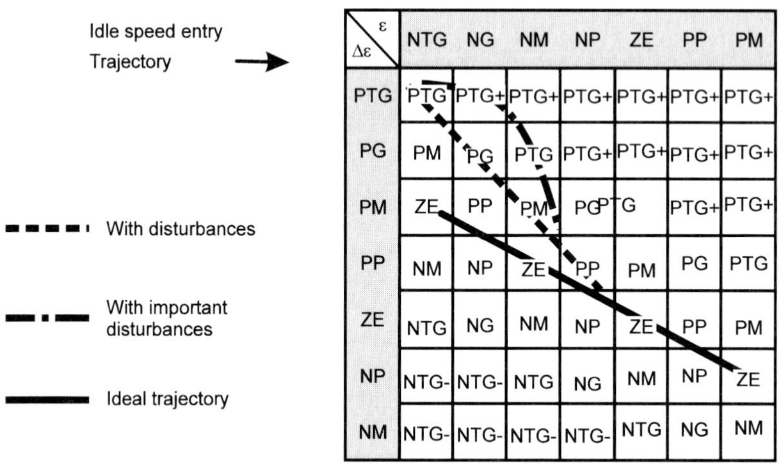

Δε \ ε	NTG	NG	NM	NP	ZE	PP	PM
PTG	PTG	PTG+	PTG+	PTG+	PTG+	PTG+	PTG+
PG	PM	PG	PTG	PTG+	PTG+	PTG+	PTG+
PM	ZE	PP	PM	PG	PTG	PTG+	PTG+
PP	NM	NP	ZE	PP	PM	PG	PTG
ZE	NTG	NG	NM	NP	ZE	PP	PM
NP	NTG-	NTG-	NTG	NG	NM	NP	ZE
NM	NTG-	NTG-	NTG-	NTG-	NTG	NG	NM

Figure 18.9 State vector behavior for different initial conditions.

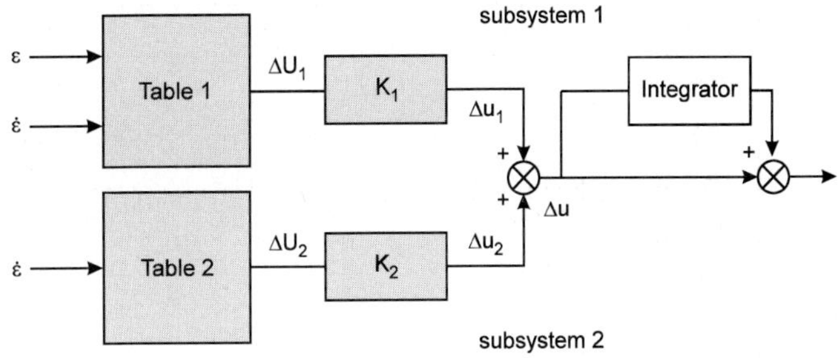

Figure 18.10 Fuzzy controller for a large dynamic process.

18.3.3 Extension of a Standard Fuzzy Controller to the Overall Runing Conditions

The idle speed controller must be applied successfully both when the car is stationary and when it is moving. In the first case, the process to be considered is only the engine dynamic but in the second case it is the engine plus the vehicle dynamics. This can be considered as two different processes. Unfortunately, the previous fuzzy core is not able to tackle both problems. A more sophisticated control structure has been designed. This structure allows taking into account the previous problem of nonzero initial conditions and the problem of process dynamic changes.

If we consider a large dynamic process, this involves some constraints in the choice of the "zero control curve." In order to take into account the different dynamics, it is necessary to rotate the control surface around the point $(\varepsilon = \dot{\varepsilon} = 0)$. To do this, it is possible to consider the following structure(see Figure 18.10).

By modifying the values of the gains K_i, this structure allows a global rotation of the control surface around the point $(\varepsilon = \dot{\varepsilon} = 0)$. In some cases, it should be suitable to choose K_1 and K_2 as $K_1 + K_2 = \text{cte}$. This allows avoiding any modification of the maximum value of the control variable.

18.3.4 Final Structure Of The Fuzzy Idle Speed Controller

The final structure of the fuzzy idle speed controller is in fact equivalent to that of a nonlinear supervised PID controller (Figure 18.11). This structure can be divided in several parts:

- preprocessing of the inputs,
- fuzzy core (described in the previous sections),
- supervision block which adapts the output gains of the fuzzy core to the process modification (stationary, moving, transient phases),
- the output PI block.

An example of the effect of the K_i gain change on the control surface is given Figure 18.12 and Figure 18.13.

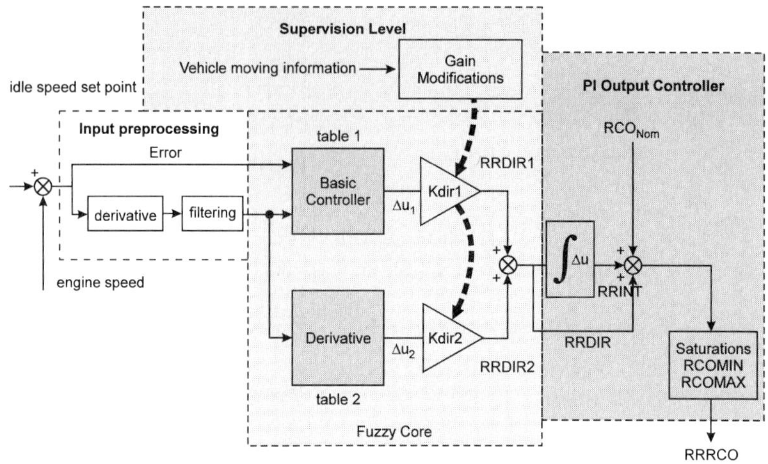

Figure 18.11 Global fuzzy idle speed controller structure.

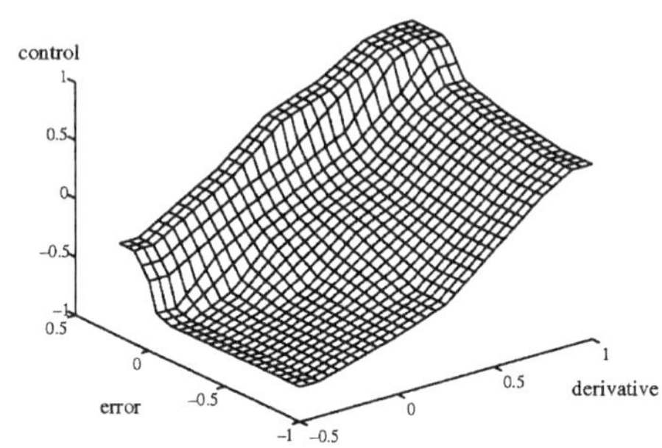

Figure 18.12 Control surface example for stationary vehicle.

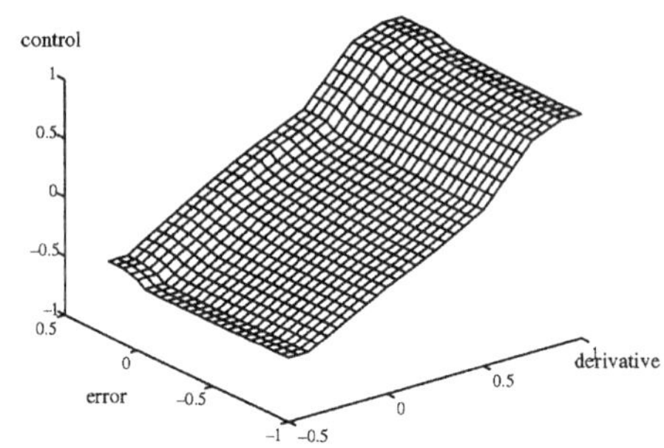

Figure 18.13 Control surface example for moving vehicle.

This control structure, as defined, is based on the use of conventional fuzzy inference methods. Nevertheless, its implementation on real-time constrained ECUs is not without problems. In addition to this work, we were interested in simplifying this algorithm in order to make its implementation easier.

18.4 IMPLEMENTATION OF THE FUZZY IDLE SPEED CONTROLLER

Implementation of control strategies in a real-time automotive constrained environment is not without problems (time calculation, memory space). So, it could be suitable to use a simplified control structure which can respect these constraints. In the next section, a simplified control structure which exhibits important advantages for real-time implementation will be proposed and compared to conventional one [6].

A fuzzy control system can be considered as a nonlinear control structure. In fact, the nonlinearities can be decomposed in two levels. The first nonlinearity is the result of the association of several elementary cells. This nonlinearity is the stronger one and can be controlled during the tuning phase. On the other hand, the local nonlinearities (between each modal point) can be explained by the use of the conventional fuzzy operators (min-max, product-max). Many researchers have tested these different methods on process control problems, and the results are generally equivalent [8,9].

18.4.1 Hypothesis and Definitions

The comparison is done with a conventional 2-input (ε, $\Delta\varepsilon$) – 1-output Mc Vicar Whelan's structure (Figure 18.14), but it can be extended to less conventional tables.

18.4.1.1 Controller Definition

The membership functions of the two input variables are defined Figure 18.15.
The different fuzzy sets are written:

$$B_1,\ldots,B_i,B_{i+1},\ldots,B_m \text{ for } \varepsilon \text{ and } A_1,\ldots,A_j,A_{j+1},\ldots,A_m \text{ for } \Delta\varepsilon$$

For a couple of input variables, the degrees of membership are:

α_i and α_{i+1} are the membership degrees of ε_0 for the sets B_i and B_{i+1}

$\Delta\varepsilon$ \ ε	NB	NM	NS	ZE	PS	PM	PB
PB	ZE	PS	PM	PB	PB	PB	PB
PM	NS	ZE	PS	PM	PS	PB	PB
PS	NM	NS	ZE	PS	PM	PB	PB
ZE	NB	NM	NS	ZE	PE	PM	PB
NS	NB	NB	NM	NS	ZE	PS	PM
NM	NB	NB	NB	NM	NS	ZE	PS
NB	NB	NB	NB	NB	NM	NS	ZE

Figure 18.14 Mc Vicar Whelan's table.

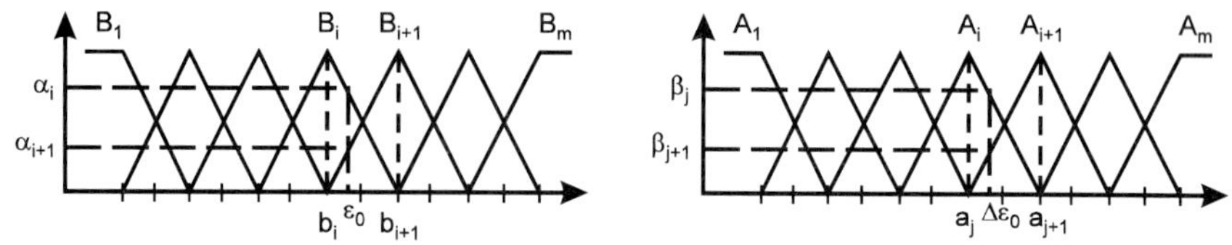

Figure 18.15 Membership functions of the inputs.

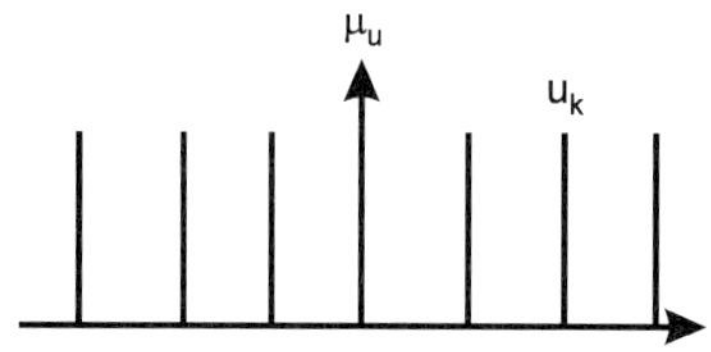

Figure 18.16 Output membership functions.

β_j and β_{j+1} are the membership degrees of $\Delta\varepsilon_0$ for the sets A_j and A_{j+1}

The output values are chosen as crisp values [10,11] (see Figure 18.16), so the rules can be expressed as:

If ε is B_i and $\Delta\varepsilon$ is A_j then U is u_k

The defuzzyfication method used is the center of gravity method.

$$U = \frac{\sum_{k=0}^{k=n} \mu_{u_k} u_k}{\sum_{k=0}^{k=n} \mu_{u_k}}$$

18.4.1.2 Definition of an Elementary Cell

In order to compare the different methods, it is convenient to define an elementary cell, and to write on this cell the value of the different control actions. An elementary cell is defined by four nodes (see Figure 18.17).

On an elementary cell, four rules are activated:

If ε is B_i and $\Delta\varepsilon$ is A_j then U is Z_{ij}
If ε is B_{i+1} and $\Delta\varepsilon$ is A_j then U is Z_{i+1j}
If ε is B_i and $\Delta\varepsilon$ is A_{j+1} then U is Z_{ij+1}
If ε is B_{i+1} and $\Delta\varepsilon$ is A_{j+1} then U is Z_{i+1j+1}

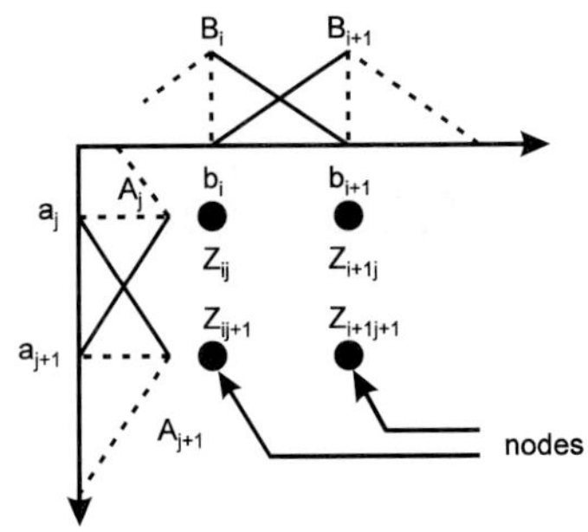

Figure 18.17 Structure of an elementary mesh.

In Mc Vicar Whelan tables, two cases of study can be distinguished. The first one concerns elementary cells belonging to the center of the table (Figure 18.14, white table part), where the control actions are symmetrical, the second one, to cells belonging to the dark table part where the control actions are nonsymmetrical.

18.4.2 Control Equations For The Different Operators

In the symmetrical part of the table ($Z_{ij} = Z_{i+1j+1}$) the expression of the control action on an elementary cell for the min-max method is given by the Equation (18-1) and by Equation (18-3) for the product-max method in a nonsymmetrical case ($Z_{ij} = Z_{i+1j+1} = Z_{ij+1}$, for example), the expression of the control action on an elementary cell for the min-max method is given by the Equation (18-2) and by Equation (18-4) for the product-max method .

For the symmetrical case we have:

$$U_{MinMax} = \frac{\max\left[\min(a_i,\beta_j);\min(a_{i+1},\beta_{j+1})\right]\cdot Z_{ij}}{\max\left[\min(a_i,\beta_j);\min(a_{i+1},\beta_{j+1})\right] + \min(a_{i+1},\beta_j) + \min(a_i,\beta_{j+1})} + \frac{\min(a_{i+1},\beta_j)\cdot Z_{i+1j} + \min(a_i,\beta_{j+1})\cdot Z_{ij+1}}{\max\left[\min(a_i,\beta_j);\min(a_{i+1},\beta_{j+1})\right] + \min(a_{i+1},\beta_j) + \min(a_i,\beta_{j+1})} \tag{18-1}$$

For the non symmetrical case it comes:

$$U_{MinMax} = \frac{\max\left[\min(a_i,\beta_j);\min(a_{i+1},\beta_{j+1});\min(a_i,\beta_{j+1})\right]\cdot Z_{ij}}{\max\left[\min(a_i,\beta_j);\min(a_{i+1},\beta_{j+1});\min(a_i,\beta_{j+1})\right] + \min(a_{i+1},\beta_j)} + \frac{\min(a_{i+1},\beta_j)\cdot Z_{i+1j}}{\max\left[\min(a_i,\beta_j);\min(a_{i+1},\beta_{j+1});\min(a_i,\beta_{j+1})\right] + \min(a_{i+1},\beta_j)} \tag{18-2}$$

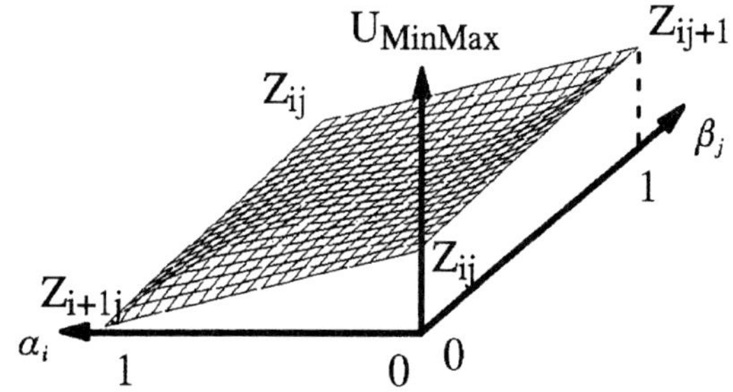

Figure 18.18 Shape of the control action on an elementary cell with (min-max) method [2]. (symmetrical case).

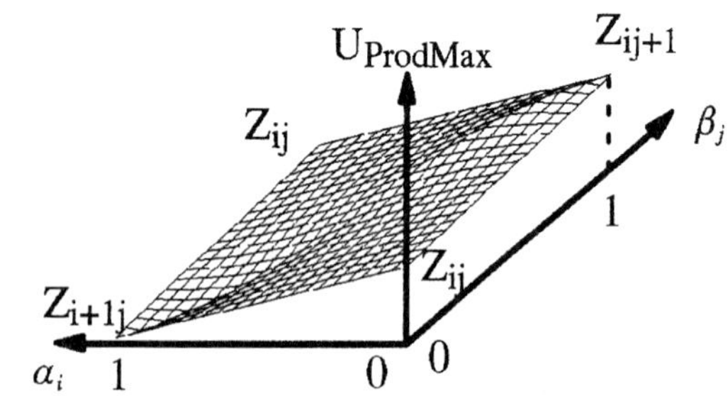

Figure 18.19 Shape of the control action on an elementary cell with (product-max) method [5]. (symmetrical case).

For the symmetrical case we have:

$$U_{\text{ProdMax}} = \frac{\max(a_i \cdot \beta_j ; a_{i+1} \cdot \beta_{j+1}) \cdot Z_{ij} + a_{i+1} \cdot \beta_j \cdot Z_{i+1j} + a_i \cdot \beta_{j+1} \cdot Z_{ij+1}}{\max(a_i \cdot \beta_j ; a_{i+1} \cdot \beta_{j+1}) + a_{i+1} \cdot \beta_j + a_i \cdot \beta_{j+1}} \quad (18\text{-}3)$$

For the nonsymmetrical case it is:

$$U_{\text{ProdMax}} = \frac{\max(a_i \cdot \beta_j ; a_{i+1} \cdot \beta_{j+1} ; a_i \cdot \beta_{j+1}) \cdot Z_{ij} + a_{i+1} \cdot \beta_j \cdot Z_{i+1j}}{\max(a_i \cdot \beta_j ; a_{i+1} \cdot \beta_{j+1} ; + a_i \cdot \beta_{j+1}) + a_i \cdot \beta_{j+1}} \quad (18\text{-}4)$$

18.4.3 Definition of the Simplified Structure

The simplified structure is based on the use of a double linear interpolation calculation (Figure 18.20).

Instead of calculating the control value from the analysis of the input membership degrees, a double linear interpolation is done between the modal points (b_i, b_{i+1}, a_j, a_{j+1}).

For an input couple (ε, $\Delta\varepsilon$), the output value for the symmetrical case can be written as:

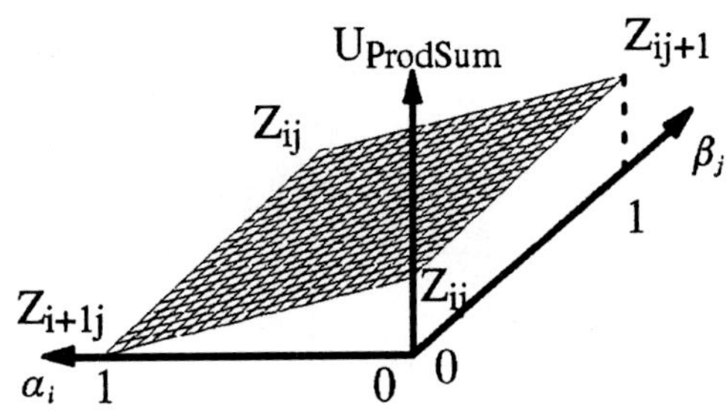

Figure 18.20 Shape of the control action on an elementary cell with double interpolation method [6]. (symmetrical case).

$$U = \frac{(b_{i+1} - \varepsilon)\cdot(a_{j+1} - \Delta\varepsilon)\cdot Z_{ij} + (\varepsilon - b_i)\cdot(a_{j+1} - \Delta\varepsilon)\cdot Z_{i+1j}}{(b_{i+1} - b_i)\cdot(a_{j+1} - a_j)}$$
$$+\frac{(b_{i+1} - \varepsilon)\cdot(\Delta\varepsilon - a_j)\cdot Z_{ij+1} + (\varepsilon - b_i)\cdot(\Delta\varepsilon - a_j)\cdot Z_{ij}}{(b_{i+1} - b_i)\cdot(a_{j+1} - a_j)}$$

With the membership degrees (α_i, α_{i+1}, β_j, β_{j+1}) of the triangular functions as defined in Section 18.4.1.1, U can be written as follows:

$$U = (\alpha_i\cdot\beta_j + \alpha_{i+1}\cdot\beta_{j+1})\cdot Z_{ij} + \alpha_{i+1}\cdot\beta_j\cdot Z_{i+1j} + \alpha_i\cdot\beta_{j+1}\cdot Z_{ij+1}$$

In the nonsymmetrical case:

$$U_{ProdSum} = U = (\alpha_i\cdot\beta_j + \alpha_{i+1}\cdot\beta_{j+1} + \alpha_i\cdot\beta_{j+1})\cdot Z_{ij} + \alpha_{i+1}\cdot\beta_j\cdot Z_{i+1j}$$

18.4.4 Comparison of the Different Reasoning Methods

When we compare the control values provided by the simplified structure based on the use of linear interpolation and those provided using the conventional fuzzy operators, we see that the difference between the control laws is not very important. The error is a function of an error factor which depends on the operators and of the width of the cell (Figure 18.21 to Figure 18.24). In fact, we could not justify the influence of these local nonlinearities on the global quality of the closed loop control. Some distortions will appear between saturated and non-saturated areas, nevertheless the errors are relatively small and have no influence on the closed loop control behavior. This work shows, that it is possible to approximate the global structure of the fuzzy controller with a piecewise function. This approximation is very interesting under strong real-time constraints, so it can be useful to implement this simplified structure which exhibits computation time advantages for practical application.

Error comparison in the symmetrical case

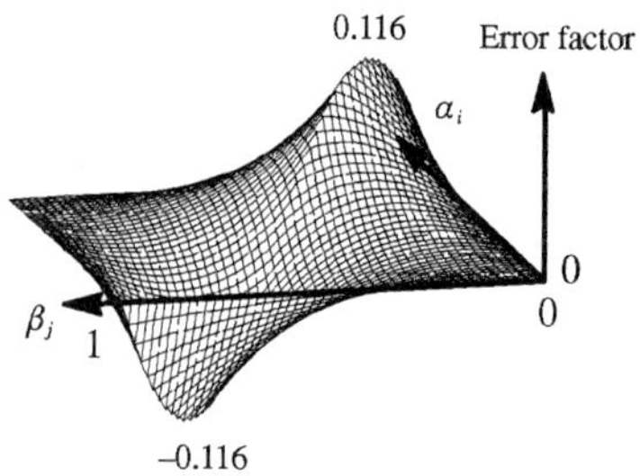

Figure 18.21 Error factor between (product-max) method and interpolation.

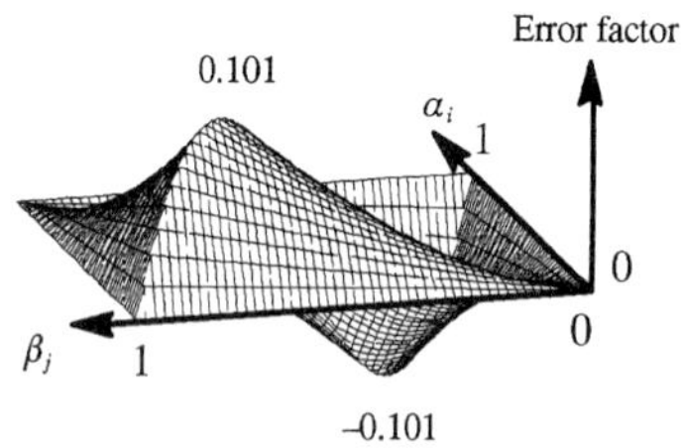

Figure 18.22 Error factor between (min-max) method and interpolation.

Error comparison in the nonsymmetrical case

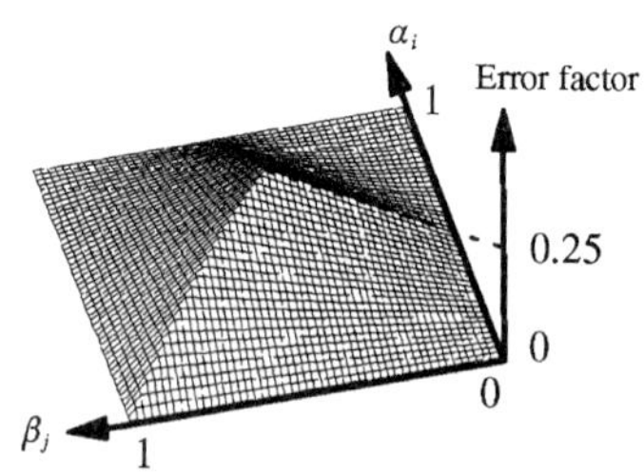

Figure 18.23 Error factor between (product-max) method and interpolation.

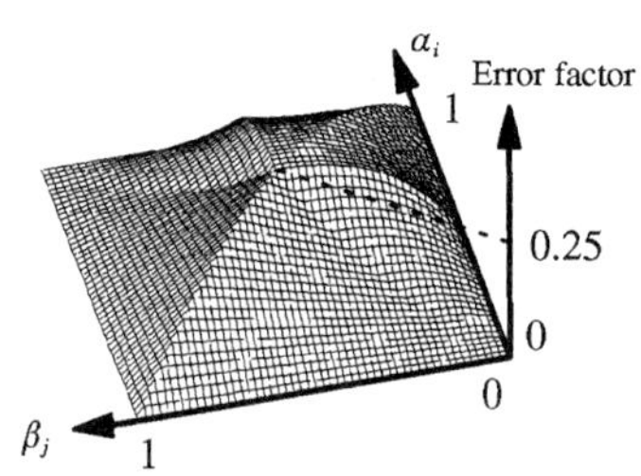

Figure 18.24 Error factor between (min-max) method and interpolation.

18.5 RESULTS

This control structure has been applied to the idle speed controller. The nondiagonal table can be directly provided considering K_1 and K_2 different from 0. A first set of values (K_{10}, K_{20}) has been tuned for a stationary vehicle. These values are adapted by a supervisory level in order to take into account the moving phases and weak accelerations with respect to the relation $K_1 + K_2 = cte$. Figure 18.12 shows the control surface for the nominal gains K_{10} and K_{20}. Figure 18.13 shows the control surface for a moving vehicle.

The control laws have been implemented on a real-time ECU and successfully tested on a car (Renault 19 16 valves).

Figure 18.25 and Figure 18.26 show the behavior of the engine speed in the regulation phase. When a load is applied (power steering), the engine comes back very quickly to the regulation set point and the transient speed reduction is very low (80 rpm).

Figure 18.27 and Figure 18.28 show the reaching phases with and without load (power steering + cooling ventilator). In the two cases the quality of the engine speed behavior is equivalent and very good.

Figure 18.29 shows a reaching phase when the car is moving. We notice the good adaptation of the control strategies which allow a very good behavior of the engine speed with respect to the idle speed requirements defined in 18.2.2.

18.6 CONCLUSION

This work has been dedicated toward the design of an idle speed controller for thermic engines. Idle speed control is considered by the specialist as a difficult problem, with respect to engine stability, engine consumption, and emission. In addition, idle speed control must be efficient both when the vehicle is moving or when it is stationary, which

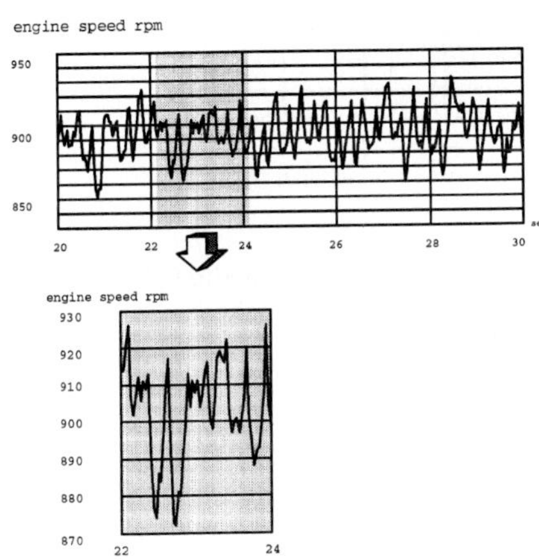

Figure 18.25 Idle speed regulation without load.

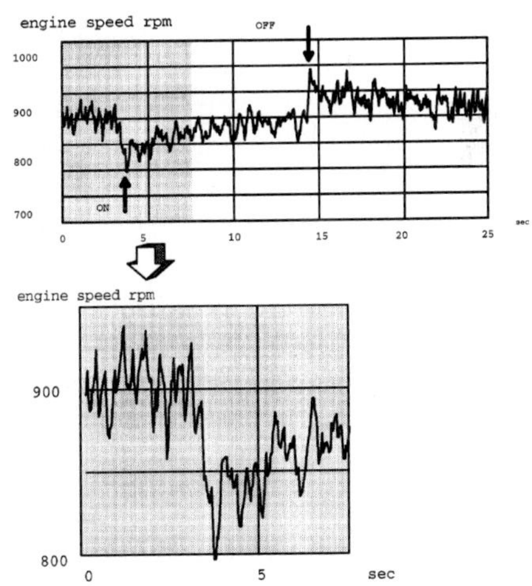

Figure 18.26 Idle speed regulation with load (power steering).

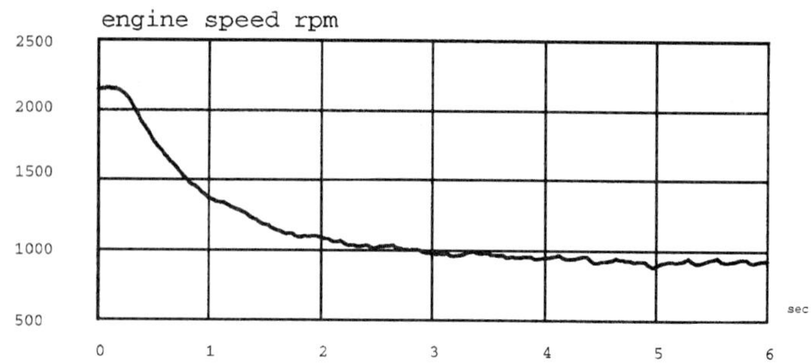

Figure 18.27 Reaching idle speed set point without load.

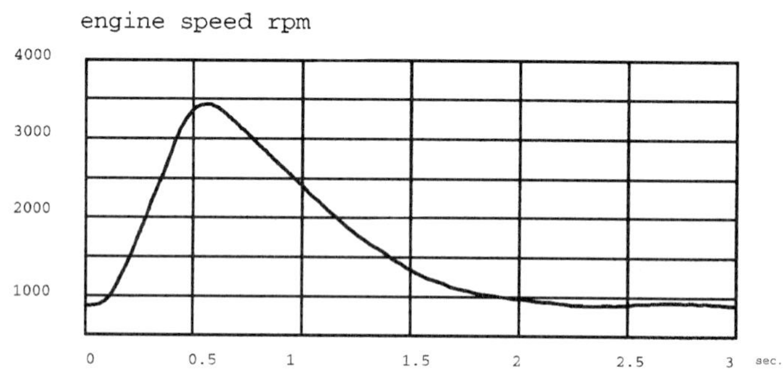

Figure 18.28 Reaching idle speed set point with load (power steering + ventilator).

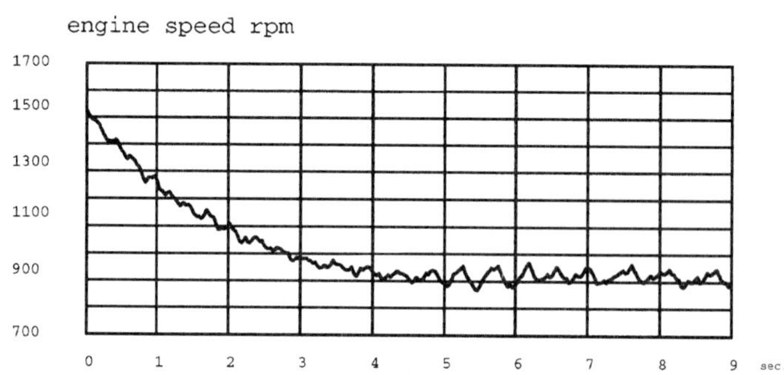

Figure 18.29 Reaching idle speed set point with load and with running vehicle.

means two very different processes to be controlled. In this study, we have developed an original controller based on an analogy between a nonlinear control technique like sliding mode and the fuzzy control technique. This structure allows taking into account large dynamic variations of the processes and adapting the controller structure to these different running conditions. This controller has been implemented in a real-time ECU; taking into account the real-time constraints and the cost constraint; we have also proposed a simplified structure of this controller.

Finally, the control, algorithms have been successfully tested and validated on a car.

18.7 REFERENCES

1. Zadeh, L. A. "Fuzzy Algorithms." *Information and Control*, Vol. 12, pp. 1968, 94-112.

2. Mamdani, E. H. "Application of Fuzzy Algorithm for Control of a Simple Dynamic Plant." *Proc. IEEE.* Vol. 121, 1974, pp. 1585-1588.

3. Cerf, P., Boverie, S. and Vernières, F. "Fuzzy Sliding Mode Control—Application to Large Time Varying System." *SICICA*, Budapest, Hungary, June, 1994.

4. Boverie, S., Cerf, P. and Le Quellec, JM. "Fuzzy Sliding Mode Control—Application to Idle Speed Control" *FUZZ' IEEE*, Vol. 2, pp. 974-978 Orlando, FL, 1994.

5. Boverie, S., Demaya, B., Le Quellec, JM. and Titli, A., "Performance Evaluation of Fuzzy Control Through an International Benchmark." *IFSA 93'*, Seoul Korea, July 93.

6. Raymond, C., Boverie, S. and Le Quellec, JM. "Practical Realization of Fuzzy Controller—Comparison with Conventional Methods." *Proceedings of the EUFIT Symposium*, Aachen, Germany, Sept. 1993.

7. Palm, R. "Sliding Mode Fuzzy Control." *FUZZ-IEEE '92*, San Diego, CA, March, 1992.

8. Mizumoto, M. "Realization of PID Controls by Fuzzy Control Methods." *FUZZ' IEEE*, pp. 709-715, San Diego, CA, 1992.

9. Mizumoto, M. "Fuzzy Control Under Various Fuzzy Reasoning Methods." *Information Sciences*, Vol. 45, 1988, pp. 129-151.

10. Foulloy L. and Galichet S., "Représentation des Contrôleurs Flous," *Les Applications des Ensembles Flous, Journíes Nationales les Nîmes*. 1992, pp. 129-136.

11. Takagi, T. and Sugeno, M. "Fuzzy Identification of Systems and its Applications to Modeling and Control." *IEEE Trans and Systems, Man and Cybernetics*. Vol. SMC 15, No. 1, pp. 116-132.

19

FUZZY CONTROL OF THE SHUTTLE TRAINING AIRCRAFT

S. R. Swanson
NASA—Johnson Space Center
USA

To train the pilot astronauts to land the Shuttle, an airborne Shuttle simulator is needed. NASA accomplishes this task with the Shuttle Training Aircraft (STA), a modified Gulfstream II business jet. The simulation is performed with the use of a model-following control system, which consists of a math model of the Shuttle and an aircraft control system to make the STA "follow" the Shuttle model. As one would expect this simulation is performed in realtime by on-board computers. The aircraft control system is now performed with fuzzy control.

19.1 INTRODUCTION

The Shuttle in its landing phase is a unique aircraft. It is a large glider with a fly-by-wire control system utilizing attitude rate inputs from the astronaut. The first time an astronaut tries to land the shuttle, it is going to land, there are no go-arounds. Therefore the astronauts must be very well trained, a task too great for a ground simulator alone. This is why NASA uses an airborne shuttle simulator called the Shuttle Training Aircraft (STA) to train the astronauts to land the Shuttle. The STA, a modified Gulfstream II, can simulate the shuttle landing profile from an altitude of 35,000 feet to shuttle main gear touchdown. Just prior to starting the simulation approach the main gear of the STA are lowered, to add more drag to the STA, and the engine thrust reversers are deployed to give the capability of controllable drag. During the simulation the STA, following the same profile as the Shuttle, descends upon an 18° or 20° glidepath toward a point 7500 feet in front of the runway. At 2000 feet Above Ground Level (AGL), a preflare is initiated to put the STA on a 1.5° glidepath. Starting at approximately 50 feet AGL, the final flare is started and the simulation ends at the point when the Shuttle's main gear would touch down. At this time the STA is approximately 22 feet above the ground. The entire simulation approach lasts approximately three minutes. Once the simulation has ended, the thrust reversers are stowed, the gear is raised, and the STA climbs up for another approach. The STA usually performs ten simulation approaches during each training mission. A picture of the STA is shown in Figure 19.1.

To perform the simulation, the STA processes the astronaut's inputs to create a new state of a mathematical model of the shuttle. Then the model-following control system strives to make the STA's parameters match the shuttle model's parameters. The parameters matched by the model-following control system are;

1. pitch axis (pitch and pitch rate),
2. roll axis (roll and roll rate),

Figure 19.1 Shuttle training aircraft.

3. yaw axis (yaw rate),
4. body x axis (true airspeed),
5. flight path angle (gamma), and
6. normal acceleration of the cockpit.

When the STA was first developed in 1975, conventional linear control systems were used. Although these systems performed well, there were areas that could be improved. The use of fuzzy control was investigated with the conclusion that implementing it in the STA would improve the control system performance. The use of fuzzy control allows for a design based on the physical characteristics of the plant, or STA, as opposed to the previous design based on an approximate mathematical model of the plant. This, plus the inherent structure of fuzzy control, allows for an easier implementation of a complex nonlinear control system. The nonlinear characteristic of fuzzy control systems is the biggest advantage over the old linear control system.

The STA has four flight control surfaces: the elevator, aileron, flaperon, and flap surfaces. The elevator is on the horizontal stabilizer, or tail, and it controls pitch along the body axis. The other three surfaces are located on the trailing edge of the wing. The aileron is the most outboard surface, the flaperons are in the middle, and the flaps are the inboard most surface. The aileron and flaperon together control roll along the body axis. The flaperon and flap work together as flaps but contrary to ordinary flaps, which only move down, they travel from 20° down to 30° up. Together they control the flight path angle of the STA. The flaps by themselves are fast acting, moving at 50° per second, and are used to control the normal acceleration and small deviations in the flight path angle of the STA. Along with these flight control surfaces, the STA utilizes an autothrottle system to help control the thrust produced by the engines. However, since the STA's thrust reversers are deployed, the autothrottle system actually controls the drag of the STA. All of these together allow the STA to simulate the proper lift over drag ratio and handling characteristics of the Shuttle.

There are six model-following control systems in the STA. They are;

1. Elevator control,
2. Aileron control,
3. Rudder control,
4. Direct Lift Control (DLC),
5. Direct lift control trim, and
6. Autothrottle control.

The elevator control system performs the pitch axis model-following, the aileron control system performs the roll axis model-following, and the rudder control system performs the yaw rate model-following. The DLC and DLC trim control systems combine to perform gamma and normal acceleration model-following. The autothrottle control system performs true airspeed model-following.

Each control system actually controls a servo, which is an AC electro-mechanical motor. The servo manipulates a hydraulic actuator which moves the aircraft flight control surface. The output command of the control systems are an absolute servo position.

Consequently, the control system calculates a delta command and then adds the current servo position to it to obtain an absolute command. In the simulation mode, the servo and surface positions are directly related; therefore the control systems, for design purposes, are directly controlling the flight surfaces.

19.2 INFERENCE ENGINE FOR THE STA FUZZY LOGIC CONTROL SYSTEM

Since each control system uses at least one fuzzy inference engine and the cycle time of the STA software is 50 milliseconds and the control system is budgeted approximately 3 milliseconds, the speed of the inference engine is a very major design requirement. That in turn leads to simplicity. For that reason, membership functions are triangular, a discrete input can be in at most two membership functions, and the summation of the degrees of memberships over the universe of discourse is one. Plus, to keep the number of rules low, the number of inputs into an inference engine have been limited to three. If more than three inputs are needed, the inputs are logically divided into two sets and used as inputs into two different inference engines. The outputs are then added together to get the overall command. For conjunction the hard min operator is used and for disjunction the max operator is used. The defuzzification scheme is a weighted mean-of-max given below:

$$u_0 = \frac{\sum_{i=\{FS\}} \left(y_i * u_i(y_i)\right)}{\sum_{i=\{FS\}} u_i(y_i)} \tag{19-1}$$

where $\{FS\}$ is the set of fuzzy subsets in the universe of discourse, y_i is the value at which the ith fuzzy subset is at a maximum, and $u_i(y_i)$ is the weight or degree of membership of the corresponding maximum.

Each universe of discourse can have any number of membership functions. Each membership function *FS* is defined by three values: its maximum point, the left zero point, and the right zero point. The left and right zero points correspond to the maximums of the membership functions on each side of *FS*. Therefore the number of parameters needed to describe the membership functions equals the number of membership functions.

For example, assume a universe of discourse has five fuzzy subsets: large negative (LN), small negative (SN), zero (ZO), small positive (SP), and large positive (LP). Then five values are needed to define the membership functions of the fuzzy subsets (Figure 19.2). If a value moves it not only moves its membership function's maximum but also the zero points of the membership functions on either side of it.

19.3 ELEVATOR CONTROL SYSTEM

The pitch axis of the STA is the most difficult to control because it is much less stable in pitch than any other axis. For that reason the elevator control system has the most inputs and took the longest to develop. The elevator control system is broken into two parts: the pitch command and the pitch predict command. The pitch command uses the

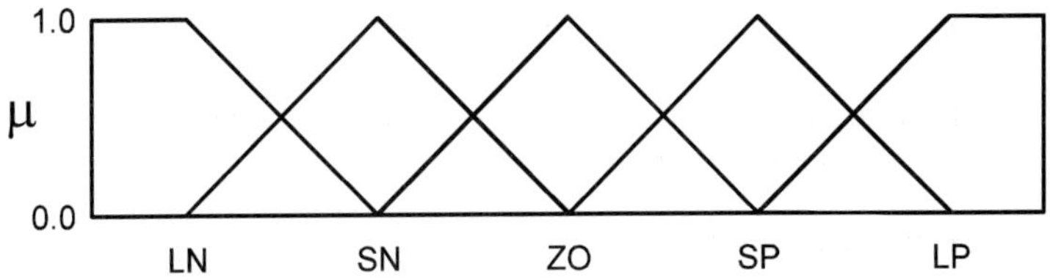

Figure 19.2 Fuzzy membership function example.

differences or deltas (Δ) between the Shuttle model's parameters and the STA's parameters. The parameters used are pitch attitude (theta), body pitch rate (*q*), and body pitch acceleration. The delta pitch attitude is offset by a constant amount of 4.65° because the STA cannot safely attain the angles of attack the Shuttle can fly. The output of the pitch command is multiplied by a gain which is linearly dependent upon dynamic pressure (*qbar*). This allows for larger elevator commands at lower dynamic pressures where more elevator surface movement is needed. Due to the STA not having a pitch angular accelerometer and because the pitch rate gyros are quite noisy, the STA pitch acceleration is obtained from differentiating a filtered pitch rate derived from the pitch rate gyros, differentiated pitch attitude, and differentiated true heading. Since the pitch acceleration and pitch rate terms are noisy, the gains on the delta pitch acceleration and delta pitch rate terms must be low. Due to these low gains, the elevator control system with the pitch command by itself has a large lag. This is why a pitch predict command is needed.

The pitch predict command is obtained from Shuttle model parameters only. They are Shuttle body pitch rate (Shuttle *q*), Shuttle body pitch acceleration, and the change in Shuttle elevator command. The change in the delta Shuttle elevator command is one of the first indications of a coming change in pitch rate of the Shuttle model. Thus the delta elevator command term is used to initiate a change in STA pitch rate, but only lasts for a short time. The Shuttle pitch acceleration term then grows, helping to drive the STA's pitch rate to that of the Shuttle model's. If the Shuttle model's pitch rate is sustained, the Shuttle pitch rate term aids in holding the STA pitch rate. The inputs, Shuttle *q* and Shuttle pitch acceleration, are both multiplied by a dynamic pressure gain, or *qbar* adjusted. The Δ Shuttle RHC pitch command input is not *qbar* adjusted because it changes with *qbar*. The elevator control system is given in Figure 19.3.

19.4 AILERON CONTROL SYSTEM

The roll axis of the STA is more stable than the pitch axis and only uses four inputs. However, the aileron control system is similar to the elevator control system and is also broken into two parts:the roll command and the roll predict command. As with the pitch command, the roll command uses the differences or deltas (Δ) between the Shuttle model's parameters and the STA's parameters. The parameters of each used are roll attitude (phi) and body roll rate (*p*). The body roll acceleration delta term has moved to the roll predict command to even the number of inputs in the roll and roll predict fuzzy schemes. The acceleration term was moved since it is more of a predict term than the roll

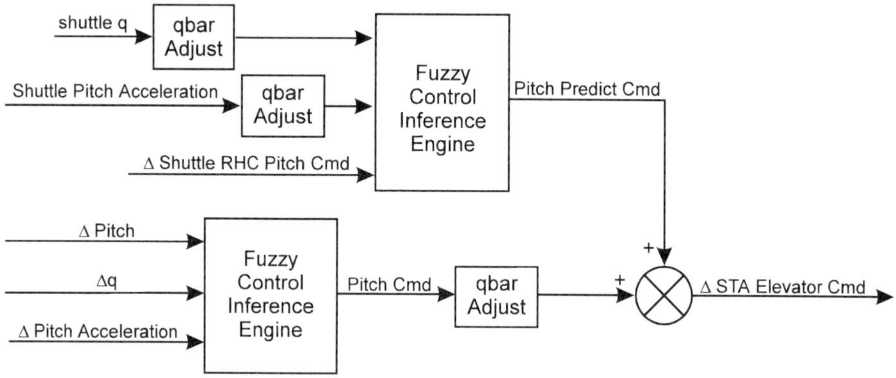

Figure 19.3 STA elevator control system.

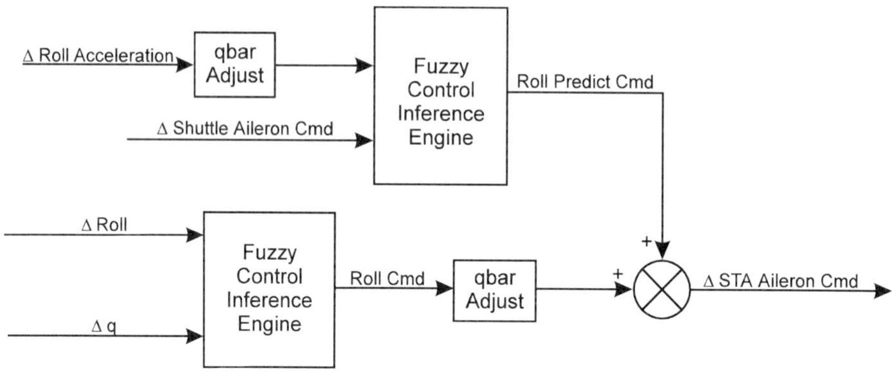

Figure 19.4 STA aileron control system.

attitude or roll rate terms. As with the pitch, the STA roll acceleration input is obtained from differentiating a filtered roll rate derived from the roll rate gyros, differentiated roll attitude, and differentiated true heading. And the roll command is also multiplied by a *qbar* gain.

The roll predict command is obtained from the change in the Shuttle aileron command and the difference in STA roll acceleration and Shuttle roll acceleration. As with the Shuttle delta elevator command, the delta Shuttle aileron command is one of the first indications of a change in roll rate and is dependent upon *qbar* and hence is not *qbar* adjusted. The aileron fuzzy control system is given in Figure 19.4.

19.5 RUDDER CONTROL SYSTEM

The STA rudder control system tries to match the STA's yaw rate to the Shuttle model's yaw rate. There are three inputs into this system: the difference in the STA's and Shuttle model's yaw rate, the STA's roll rate, and the STA roll command. Since the roll

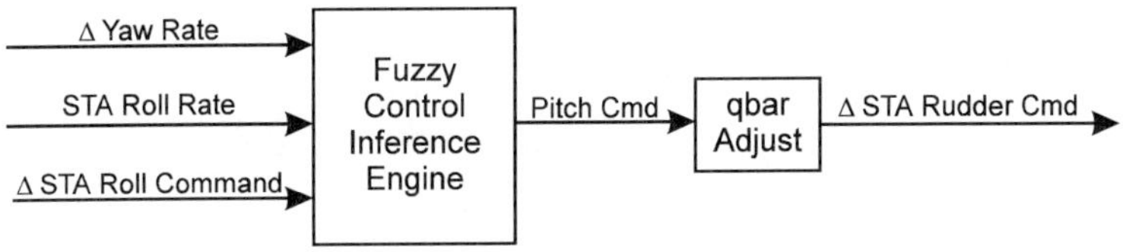

Figure 19.5 STA rudder control system.

and yaw rates of an aircraft are coupled due to turn coordination and the STA roll rate and the Shuttle model's roll rate are very similar, the STA roll rate and roll command are good prediction terms for the rudder control system. The output of the fuzzy control scheme is also *qbar* adjusted. The rudder control system is given in Figure 19.5.

19.6 DIRECT LIFT CONTROL SERVO CONTROL SYSTEM

The direct lift control system has a fast-acting actuator that moves the flaps at 50° per second. The control system is therefore designed more for quick responses to *Nz* and gamma errors. The direct lift control trim system is designed for the low frequency errors. There are three inputs at a time into the DLC control system. They are the change in Shuttle pilot station normal acceleration (pilot *Nz*), the difference between STA pilot station *Nz* and the Shuttle model's pilot station *Nz*, and either the difference in flight path angle (gamma) or the difference in vertical velocity. Gamma and vertical velocity are very closely related. There are related by Equation (19.2)

$$\sin(\text{gamma}) = \frac{\text{hdot}}{\text{true air speed}}. \tag{19-2}$$

So if true airspeeds are equal, theoretically there is not much difference in modeling gamma or vertical velocity. However, at altitudes higher than 100 feet AGL, the flight profile is determined by gamma or flight path angle and consequently Δ gamma is used. Below 50 ft. AGL, in the final flare region, gamma becomes very close to zero and the simulation pilot is concentrating on vertical velocity and therefore Δ vertical velocity is used. Between 50 and 100 ft. AGL, there is a linear transition from Δ gamma to Δ vertical velocity. Since both gamma and vertical velocity are in the vertical plane, these terms must be multiplied by the cosine of the STA roll attitude to translate the inputs into the z body axis.

The primary input for the DLC control system is Δ pilot *Nz* since the principal job of the DLC surfaces is to control the normal acceleration of the STA. The change in the Shuttle pilot *Nz* term is used as a prediction term to anticipate future Δ pilot *Nz* values. The output of the fuzzy scheme is also *qbar* adjusted. The DLC control system is given in Figure 19.6.

19.7 DIRECT LIFT CONTROL TRIM SERVO CONTROL SYSTEM

The DLC trim servo moves the flaps and flaperons together at an approximate rate of 5° per second. The DLC trim control system is very similar to the DLC control system. However since the DLC trim control system has a different function, there are a few differences. They are the pilot station normal acceleration terms are replaced by center of gravity normal acceleration terms and the DLC trim control system is designed to act as a trim surface for the DLC system. More specifically as a trim surface, it is designed to off-load the DLC servo to keep the average command approximately zero. The main input for the DLC trim control system is Δ gamma or Δ vertical velocity since these errors have a much lower frequency (e.g., they do not change very quickly) than the *Nz* terms, and are the inputs that need to be off loaded from the DLC control system. The change in the Shuttle *Nz* term and Δ *Nz* term are used as prediction terms to anticipate future Δ gamma or Δ vertical velocity values. The output of the fuzzy scheme is *qbar* adjusted. The DLC Trim control system is given in Figure 19.7.

19.8 AUTOTHROTTLE CONTROL SYSTEM

The drag of the STA needs to be variable since the Shuttle has a speedbrake which affects the drag of the Shuttle and the STA's *x* body acceleration is dependent upon weight and specific aircraft. This is the job of the autothrottle control system. The autothrottle control system determines the throttle setting of the engines with the left and right throttle settings always being equal with the goal of matching the STA's true airspeed to that of the Shuttle model's. There are two inputs into the fuzzy inference engine of the autothrottle control system. They are the difference between the STA's true airspeed and Shuttle model's true airspeed and the difference in the STA's body *x* axis acceleration and the Shuttle model's body *x* axis acceleration. The *x* body axis acceleration terms are adjusted for gravitational acceleration when the pitch attitude is

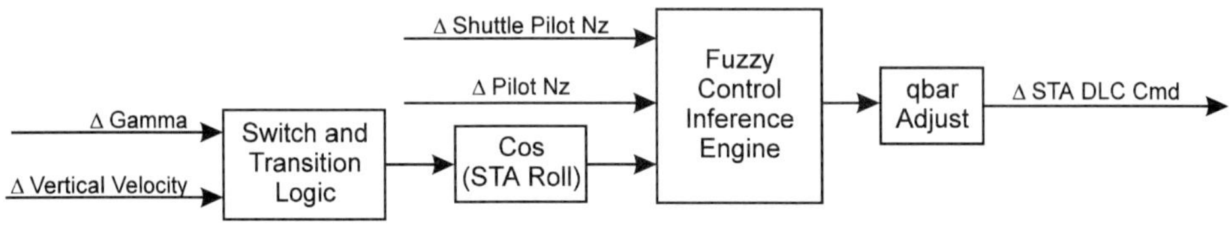

Figure 19.6 STA DLC control system.

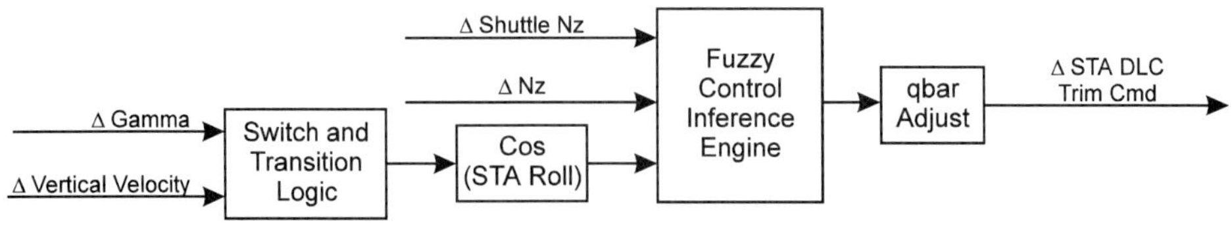

Figure 19.7 STA DLC trim control system.

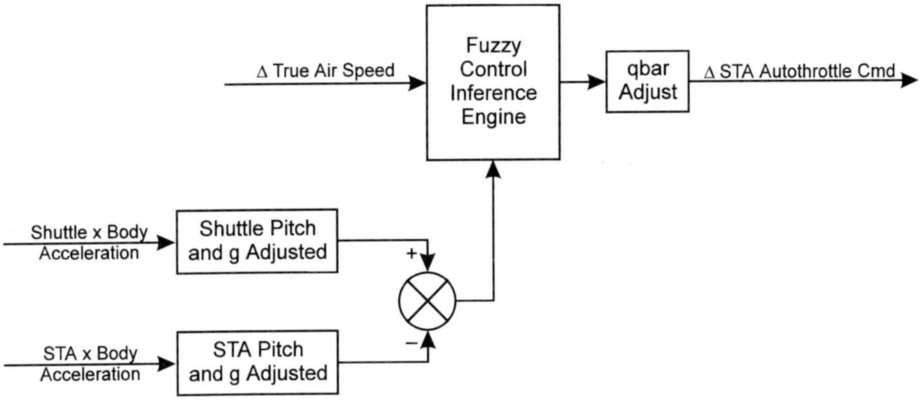

Figure 19.8 STA Autothrottle Control System.

nonzero. Since the pitch attitudes are different due to the bias term and pitch attitude errors, the adjustments are performed separately and then the difference between the two terms are calculated. This Δx body axis acceleration is used as a prediction term in the autothrottle control system. The autothrottle control system is given in Figure 19.8.

19.9 SUMMARY

The STA is a very complex and nonlinear plant which is very hard to accurately describe mathematically and therefore very difficult to design and implement a control system using linear control system methods. However, when designing the fuzzy control systems, the physical characteristics of each system were analyzed and not a math model of the STA. A computer model of the STA was used to tune the membership functions of the controllers. In the end. the fuzzy control system's overall performance is better, it is more than the original linear control system. The fuzzy control has improved the simulation fidelity of the STA and consequently astronaut training.

19.10 REFERENCES

1. Langari, R. and Berenji, H. "Fuzzy Logic in Control Engineering." Draft.

2. Lee, C. C. "Fuzzy Logic in Control Systems: Fuzzy Logic Controller—Part I." *IEEE Trans. on Systems, Man, and Cybernetics*, Vol. 20, No. 2, March/April 1990.

3. Lee, C. C. "Fuzzy Logic in Control Systems: Fuzzy Logic Controller—Part II." *IEEE Trans. on Systems, Man, and Cybernetics*. Vol. 20, No. 2, March/April 1990.

4. Mandami, E. H. "Twenty Years of Fuzzy Control: Experiences Gained and Lessons Learned." *Proceedings of the IEEE Conference on Fuzzy Logic*, pp. 339-344, 1993.

5. Zimmermann, H.-J. *Fuzzy Set Theory and Its Applications*, 2nd ed. KAP, 1991.

20

EXPERT SYSTEMS AND FUZZY MOVES

A. Kandel and Y.-Q. Zhang
University of South Florida
USA

Based on the Theory of Moves (TOM) and classical game theory, the Theory of Fuzzy Moves (TFM) has been developed to make a better decision in a fuzzy game under an uncertain and fuzzy environment. A methodology of transforming an original game to a global game is proposed so as to describe all players' globally strategic goals as objectively as possible. The hybrid decision-making systems for fuzzy moves are also presented. Finally, some examples of making fuzzy moves are discussed.

20.1 INTRODUCTION

Brams [2-4] developed what he calls the Theory of Moves (TOM), which adds a dynamic dimension to classical game theory. TOM focuses on interdependent strategic situations in which the outcome depends on the choices that all players make. It radically alters the rules of play, enabling players to look ahead—sometimes several steps—before making a move.

The traditional methods assume that both players are *egotists*, in the sense that they only seek the maximization of their own gains. If each player tries to reach an outcome that is most advantageous to only *his own* side [14], they are said to have *local goals* and the game is called a *local game*. Both game theory and TOM can be used to solve a local game. If the contenders have different goals, the game's payoffs must be reevaluated and the outcomes may be quite different from those of a local game. For this reason, we propose a *global* game with global goals, including two subgoals that are (1) an outcome which is as advantageous to *one's own side* as possible based on *his* payoffs, and (2), an outcome which is as disadvantageous to *his opponent* as possible based on *the opponent's* payoffs. Therefore, the original local game will be transformed into other global games according to different global goals. In this sense, the outcome of a game depends not only on the original payoffs but also on the global strategic goals of each of the players. In a global game, a player can evaluate each of the subgoals in a fuzzy manner, considering, for instance, that the first subgoal very important and the second is less important. There is a lot of fuzzy data and uncertain information in real-world games, so fuzzy set theory is a useful approach to the problem [6,9,17]. Another question arises, which is how to develop a theory to deal with both precise data and certain information, as well as fuzzy data and uncertain information in real situations [7,11,12,18]. Therefore, it is necessary to find a method to select relatively rational and globally optimal moves to get a player's own payoff as high as possible as well as make its opponent's payoff as low as possible based on fuzzy data in the dynamic decision environment. TOM is used to solve only the local game such as the traditional Prisoner's Dilemma (PD). Therefore, we are developing the Theory Of Fuzzy Moves (TFM) to solve the global game with fuzzy data and uncertain information.

20.2 THEORY OF FUZZY MOVES

The six new rules of the Theory of Moves give players the chance to look ahead simultaneously, make decisions globally, and select a relatively rational move or a final optimal move [2-4].

Because of the complexity and uncertainty of the real world, we need to develop a theory which is able to completely analyze the *absolute* order payoffs and the *relative* fuzzy payoffs of the players, and finally achieve the global goal of moves.

20.2.1 Motivation

We analyze a simple example.

Example 20.1: The original game called Chicken is illustrated in Figure 20.1(a). The global goals of the two players are:

- The row player's global payoff = 0.6 × [its own payoff] – 0.2 × [its opponent's payoff] (i.e., maximizing its own payoff is more important than minimizing its opponent's payoff.)
- The column player's global payoff = 0.4 × [its own payoff] – 0.6 × [its opponent's payoff] (i.e., maximizing its own payoff is less important than minimizing its opponent's payoff.)

(70, 70)	(50, 100)	Chicken Transformed to →	(42, –14)	(20, 10)
(100, 50)	(20, 20)		(70, –40)	(12, –4)

a. Original Game **b. Global Game**

Figure 20.1 The global game transformed from the original game.

(a_{11}, b_{11})	(a_{12}, b_{12})
(a_{21}, b_{21})	(a_{22}, b_{22})

Figure 20.2 An original game.

Then the original game is transformed into the global game shown in Figure 20.1(b). Both players have different global goals which can change the original payoffs and further affect the final outcome of the game.

In order to make the correct decision in a game, we must consider the following key problems which are:

1. How to objectively construct the payoff matrix for the game?

2. What are the global goals that the players will choose?

For the above problems, we first discuss the general structure of the 2×2 game (or, for brevity, simply *game*), and then develop the methodology of fuzzy moves.

20.2.2 Global Games

20.2.2.1 Original Games

In the matrix, a_{ij} is player A's payoffs, and b_{ij} is player B's payoffs, $a_{ij}, b_{ij} \in (-\infty, +\infty)$ for $i, j \in \{1, 2\}$. (Note: Player A corresponds to the rows, and player B corresponds to the columns.)

According to the range of the values in the matrix of Figure 20.2, different kinds of games are hierarchically listed in Table 20.1.

20.2.2.2 Global Goals

In Section 20.2.1, we discussed a simple model of the global goal that is a typical linearly weighted function. Now we define the general concepts as follows.

Definition 2.1: The local goal of each player in a game is to reach an outcome that is as advantageous to his own side as possible only based on his own payoffs. A game with only local goals is called a local game.

Table 20.1 Hierarchy of the Original Games

Original Game	Payoffs (a_{ij}, b_{ij})
n-valued Game	$a_{ij}, b_{ij} \in \{p_1, p_2, ..., p_n\}$ $(n \geq 2)$
Normal Game	$a_{ij}, b_{ij} \in [0, 1]$
Fuzzy Game	$a_{ij}, b_{ij} \in$ {Fuzzy Linguistic Terms}
Real-Valued Game	$a_{ij}, b_{ij} \in (-\infty, +\infty)$

Definition 2.2: In the global game, a player's decision is based on the importance, or weight, it assigns to each of the subgoals, which together constitute the global goal:

1. *maximizing* his own payoff with the corresponding weight, and

2. *minimizing* his opponent's *payoff* with the corresponding weight.

For player A, the weights regarding subgoals 1 and 2 are α and α', respectively. The weight α' is defined as the *negation* of α, that is,

$$\alpha' = 1 \sim \alpha, \quad \alpha, \alpha' \in [0,1]$$

The same criterion applies to player B,

$$\beta' = 1 \sim \beta, \quad \beta, \beta' \in [0,1]$$

where ~ is the negation operation.

A game with the global goal is called a *global* game.

The typical values of α, α', β, β' are shown in Table 20.2.

The two types of the weights (i.e., α, α', β, β') from Table 20.2 are:

(a) *Normal Weights*

- $\alpha, \alpha', \beta, \beta' \in [0, 1]$, with the typical negation functions $\alpha' = 1 - \alpha$ and $\beta' = 1 - \beta'$.

In this case, α and β can completely describe the global goals. Obviously, when $\alpha = \beta = 1$, the global game is reduced to the local game. The local game is an extreme case of the global game.

Table 20.2 Hierarchy of the Global Goals

Global Goal	Weights of Global Goal
Normal Global Goal	$\alpha, \beta \in [0, 1]$ and $\alpha' = 1 - \alpha$, $\beta' = 1 - \beta$.
Fuzzy Global Goal	$\alpha, \alpha', \beta, \beta' \in$ {fuzzy linguistic terms}

(b) *Fuzzy Weights*

- $\alpha, \alpha', \beta, \beta'$ are fuzzy linguistic variables whose values are fuzzy sets on [0, 1].

For example, $\alpha \in$ {*smallest, very small, medium, very large, largest*}. Intuitively, if α = *very small*, then α', the *negation* of α (i.e., 1 ~ α), will be *very large*. The fuzzy weights can be used to build the fuzzy rule bases.

For convenience, we will use only α and β to represent the weights of the global goals since α' (β') can mathematically or logically be derived from the former based on the definition of negation.

20.2.2.3 Representation of Global Games

(a) *Original Games.* A game can be compactly represented by:

$$G_{2\times2} = \left[(a_{ij}, b_{ij})\right]_{2\times2}$$

where a_{ij} designate player A's payoffs, and b_{ij} stand for player B's payoffs. Generally, a_{ij}, $b_{ij} \in (-\infty, +\infty)$ for $i, j \in \{1, 2\}$. For clarity, the payoff matrix is shown in Figure 20.3.

(b) *Global Games.* In this case, we have:

$$\overline{G}_{2\times2}^{\alpha,\beta} = \left[(\overline{a}_{ij}, \overline{b}_{ij})\right]_{2\times2}^{\alpha,\beta}$$

where $\overline{a}_{ij}$ are player A's global payoffs, and $\overline{a}_{ij} = F(a_{ij}, b_{ij}, \alpha)$ for $i, j \in \{1, 2\}$. Here, $F(a_{ij}, b_{ij}, \alpha)$ is called the transformation function of player A's globally strategic goal; b_{ij} are Player B's global payoffs, and $b_{ij} = F(a_{ij}, b_{ij}, \alpha)$ for $i, j \in \{1, 2\}$. Similarly, $G(a_{ij}, b_{ij}, \alpha)$ is called the transformation function of player B's globally strategic goal.

The game transformation is a procedure for mapping an original game into the global game, such that

$$G_{2\times2} = \left[(a_{ij}, b_{ij})\right]_{2\times2} \xrightarrow{\text{Game Transformation}} \overline{G}_{2\times2}^{\alpha,\beta} = \left[(\overline{a}_{ij}, \overline{b}_{ij})\right]_{2\times2}^{\alpha,\beta}$$

The concise representation of the global game is illustrated in Figure 20.4.

	Strategy +	Strategy –
Strategy +	(a_{11}, b_{11})	(a_{12}, b_{12})
Strategy –	(a_{21}, b_{21})	(a_{22}, b_{22})

Figure 20.3 The original game.

α	β	Strategy +	Strategy −
Strategy +		$(\overline{a}_{11}, \overline{b}_{11})$	$(\overline{a}_{12}, \overline{b}_{12})$
Strategy −		$(\overline{a}_{21}, \overline{b}_{21})$	$(\overline{a}_{22}, \overline{b}_{22})$

Figure 20.4 The Global Game.

(c) *Hierarchy of Global Games.* According to the payoffs and the globally strategic goals, we briefly discuss the global games from the simplest to the most complex structure. Table 20.3 indicates the hierarchical structures of the global games.

20.2.3 Play Rules of TFM Games

Generally, in order to obtain the most favorable results in a game, a player should determine both his own globally strategic goal and his opponent's goal. TOM only emphasizes an extreme kind of fuzzy moves games where both players choose a same *locally* strategic goal (i.e., each of them pursues an optimal outcome which is as advantageous *as possible* to only *his own side*, even if his opponent achieves the best one). But TFM analyzes what globally strategic goals the players choose, then indicates to a player the fuzzy moves to be chosen in a specific situation.

1. Fuzzy Real-Valued Game

$$G_{2\times 2} = \left[(a_{ij}, b_{ij})\right]_{2\times 2} \xrightarrow{\text{Fuzzy Real-Valued Game Transformation}} \overline{G}^{\alpha,\beta}_{2\times 2} = \left[(\overline{a}_{ij}, \overline{b}_{ij})\right]^{\alpha,\beta}_{2\times 2}$$

$$a_{ij},\ b_{ij},\ \overline{a}_{ij},\ \overline{b}_{ij} \in (-\infty, +\infty),\ \alpha,\ \beta \in [0,1]$$

2. Fuzzy Linguistic-Valued Game

$$G_{2\times 2} = \left[(A_{ij}, B_{ij})\right]_{2\times 2} \xrightarrow{\text{Fuzzy Linguistic-Valued Game Transformation}} \overline{G}^{\alpha,\beta}_{2\times 2} = \left[(\overline{A}_{ij}, \overline{B}_{ij})\right]^{\alpha,\beta}_{2\times 2}$$

where A_{ij}, B_{ij}, $\overline{A}_{ij}$, $\overline{B}_{ij}$ $\in \{F_i \mid F_i$ are fuzzy sets for $i = 1, 2, ..., n\}$, α, β $\in \{G_i \mid G_i$ are

Table 20.3 The Global Games

Global Games	Payoffs (a_{ij}, b_{ij})	Weights (α, β)
Normal-valued Game	$\{p_1, p_2, ..., p_n\}$	[0, 1]
Fuzzy-valued Game	$\{p_1, p_2, ..., p_n\}$	{fuzzy linguistic terms}
Normal Game	[0, 1]	[0, 1]
Fuzzy Normal Game	[0, 1]	{fuzzy linguistic terms}
Normal Fuzzy Game	{fuzzy linguistic terms}	[0, 1]
Fuzzy Game	{fuzzy linguistic terms}	{fuzzy linguistic terms}
Normal Real-valued Game	(−∞, +∞)	[0, 1]
Fuzzy Real-valued Game	(−∞, +∞)	{fuzzy linguistic terms}

fuzzy sets for $i = 1, 2, ..., m\}$.

The fuzzy game transformation function given by the player may be linear or nonlinear.

TOM has 6 basic rules [2-4]. Similarly, TFM has 9 basic fuzzy moves rules:

Rule 0: Determine what globally strategic goals the players choose.

Rule 1: A fuzzy moves game starts at an initial state which is a row-and-column intersection of a fuzzy payoff and an order payoff from the respective matrices.

Rule 2: Either player can switch to a new strategy, thereby generating a new outcome; the first player to move is called player 1.

Rule 3: The other player, player 2, can then move.

Rule 4: The players respond alternately until neither switches strategies. The resulting outcome is the final state.

Rule 5: A player will not make a move unless it leads to the globally optimal goal, based on the analysis of the fuzzy payoffs and on his anticipation of the final state.

Rule 6: A player considers the rational calculations of the other players before moving, taking into account their possible moves and countermoves, and so on.

Rule 7: The fuzzy data in a fuzzy payoff vector matrix may be dynamically changed as the new fuzzy situations appear.

Rule 8: A player can make several consecutive moves if the others don't make any moves.

The fuzzy moves rules deal with some practical situations. For example, a player may move first according to Rule 5; but if this player can do better by letting the other player move first, and if it is rational for that player to do so, then the first player will await his opponent's move, according to Rule 6 [2]. The first player can change his move which was just made when the others don't implement any moves, i.e., the players have the same chances to dynamically make new moves, based on Rule 8. Temporary equilibria, or stable outcomes, can be broken when real conditions change. In this case, all players have competitive opportunities to actively choose new strategies; one player may make several moves while another does nothing. So, every player will make moves in realtime, according Rule 7 and Rule 8.

20.2.4 Fuzzy Moves Knowledge Base on Original Game

If appropriate game transformation functions cannot be found or if we want to play the global game directly based on the original game, the fuzzy moves knowledge base will come from the original game.

First, let us consider the game:

$$\mathrm{G}_{n\times m} = \left[(a_{ij}, b_{ij})\right]_{n\times m}$$

Definition 2.3:

(1) Player A's gain of payoffs between two states:

$$P_a(ij,kl) = a_{kl} - a_{ij}, \text{ from } (a_{ij}, b_{ij}) \text{ to } (a_{kl}, b_{ij});\ P_a(ij,kl) \in [-1,1]$$

(2) Player B's gain of payoffs between two states:

$$P_b(ij,kl) = b_{kl} - b_{ij}, \text{ from } (a_{ij}, b_{ij}) \text{ to } (a_{kl}, b_{kl}).\ P_b(ij,kl) \in [-1,1]$$

Definition 2.4:

(1) Player A's gain of fuzzy payoffs between two states:

$$P_a(ij,kl) = a_{kl} - a_{ij}, \text{ from } (a_{ij}, b_{ij}) \text{ to } (a_{kl}, b_{kl}).\ P_a(ij,kl) \in \{a_1, a_2, \ldots, a_{n-1}, a_n\}$$

here, a_i ($i = 1,2,\ldots,n$) are fuzzy linguistic terms.

(2) Player B's gain of fuzzy payoffs between two states:

$$P_b(ij,kl) = b_{kl} - b_{ij}, \text{ from } (a_{ij}, b_{ij}) \text{ to } (a_{kl}, b_{kl}).\ P_a(ij,kl) \in \{b_1, b_2, \ldots, b_{n-1}, b_n\}$$

here, b_i ($i = 1,2,\ldots,n$) are fuzzy linguistic terms.

The membership functions for $P_a(ij,kl)$ and $P_b(ij,kl)$ must be determined by the player. For clarity, the typical triangle-type membership functions are given for the 5 fuzzy sets shown in Figure 20.5.

According to the fuzzy rules shown in Tables 20.4 and 20.5, we are able to make an approximate reasoning (e.g. the compositional rule of inference) for each possible move from (a_{ij}, b_{ij}) to (a_{kl}, b_{kl}) and then find if the fuzzy move is Best, Good, Middle, Bad, or Worst, and finally decide whether we will move or not.

The 25 sample fuzzy rules for A and for B are described in Tables 20.4 and 20.5,

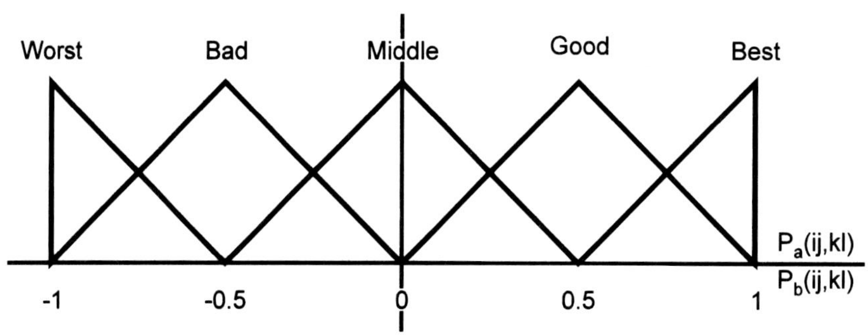

Figure 20.5 The membership functions for $P_a(ij,kl)$ and $P_b(ij,kl)$.

respectively. (1 = Worst, 2 = Bad, 3 = Middle, 4 = Good, 5 = Best)

- Player A's Fuzzy Moves Rules ($\alpha = 0.75$) in Table 20.4
  Rule RA11: IF P_a = Worst and P_b = Worst THEN A_{fm} = Bad
  Rule RA12: IF P_a = Worst and P_b = Bad THEN A_{fm} = Worst
  ......
  Rule RA55: IF P_a = Best and P_b = Best THEN A_{fm} = Good

- Player B's Fuzzy Moves Rules ($\beta = 0.25$) in Table 20.5
  Rule RB11: IF P_a = Worst and P_b = Worst THEN B_{fm} = Good
  Rule RB12: IF P_a = Worst and P_b = Bad THEN B_{fm} = Good
  ......
  Rule RB55: IF P_a = Best and P_b = Best THEN B_{fm} = Bad

20.2.5 Fuzzy Moves Knowledge Base for the Global Game

If the game transformation functions are nonlinear, we can use neural networks to learn nonlinear functions. Here, we use the approximate reasoning method to perform this task.

Suppose X and Y are the fuzzy linguistic variables representing the payoffs of players A and B, respectively. $X \in \{x_1, x_2, ..., x_n\}$, $Y \in \{y_1, y_2, ..., y_m)$, $x_i(i = 1,2,...,n)$ and y_j $(j = 1,2,...,m)$ are fuzzy sets (or fuzzy terms). Let C be the fuzzy linguistic variable representing the conclusion, and $C \in \{c_1, c_2, ..., c_l\}$, $c_k(k = 1,2,...,l)$ are fuzzy sets.

Then the fuzzy rule based on the global goal of a player can be represented by

IF (X is x_i) *and* (Y is y_i) THEN (C is c_k)

Given the payoffs (x, y), we can finally get the global payoffs (x', y') through the

Table 20.4 The 25 Sample Fuzzy Moves Rules for Player A ($\alpha = 0.75$)

	1	2	3	4	5
1	2	1	1	1	1
2	3	2	2	1	1
3	4	4	3	2	2
4	5	5	4	3	3
5	5	5	5	4	4

Table 20.5 The 25 Sample Fuzzy Moves Rules for Player B ($\beta = 0.25$)

	1	2	3	4	5
1	4	4	5	5	5
2	3	4	4	5	5
3	2	3	3	4	5
4	1	2	2	2	4
5	1	1	2	2	2

fuzzification, fuzzy inference, and defuzzification. The fuzzy reasoning procedure is similar to that in Section 20.2.4.

20.3 HYBRID DECISION-MAKING SYSTEMS FOR FUZZY MOVES

A player must construct not only his own fuzzy rule base but also his opponent's fuzzy rule base in order to make fuzzy moves, and constructing double knowledge bases is more complex than constructing a traditional single knowledge base. For this reason, we developed several hybrid systems to make better fuzzy moves by incorporating two or more techniques.

20.3.1 Hybrid Precise Fuzzy Decision-making System

Based on the fuzzy inference approach and the precise inference approach mentioned above, the hybrid decision-making system for fuzzy moves is shown in Figure 20.6 [15].

The hybrid decision-making system for fuzzy moves consists of 3 parts:

1. *Fuzzy Information Preprocessing*: A player collects fuzzy data in the fuzzy moves game, determines what globally strategic goals he and his counterplayer will choose (i.e. selects rational and suitable values of α and β) and then transfers the global fuzzy data into the corresponding fuzzy payoff/order payoff matrix.

2. *Fuzzy Moves Knowledge Base*: The fuzzy inference system generates the fuzzy outcomes for fuzzy moves, and the precise inference system generates the precise outcomes for fuzzy moves.

3. *Decision Making for Globally Good Fuzzy Moves*: The player should choose the globally good fuzzy moves based on the fuzzy outcomes and the precise ones. If the outcome of the fuzzy inference is the same as that of the precise inference, the outcome is relatively rational and globally good; If the outcome of the fuzzy inference is different from that of the precise inference, the final outcome will be analyzed by the experts that would select relatively rational and globally good fuzzy

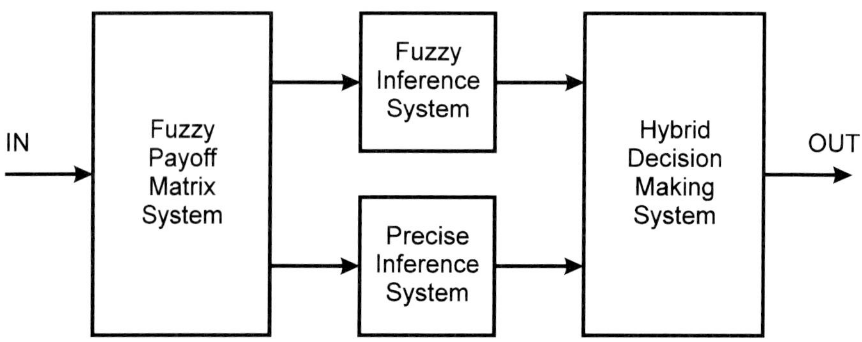

Figure 20.6 Hybrid decision-making system for fuzzy moves.

moves. This methodology may increase the reliability and correctness of the hybrid decision-making system.

20.3.2 Hybrid Fuzzy Neural Decision-making System

Through analysis, we have found that (1) for *direct* rules, the fuzzy system is fault-tolerant and stable; (2) for *vague* rules, the fuzzy system is sensitive for the relevant players' fuzzy payoffs and globally strategic goals since there are sensitive points or curves. In general, not only the artificial system but also the human brain may make a wrong decision in the sensitively fuzzy decision-making regions. Through transforming the original fuzzy moves' rules to direct and vague rules, the hybrid direct-vague fuzzy moves system with the precise rules can run more effectively and quickly than the fuzzy moves system using the conventional fuzzification, fuzzy inference, and defuzzification since only very simple mathematical operations are needed by the hybrid direct-vague fuzzy moves system [7]. The hybrid fuzzy-neural decision-making system is shown in Figure 20.7.

Generally, we give a concise algorithm for Player A which can perform the hybrid direct-neural inference. The algorithm for Player B is similar to that for Player A. The simplified algorithm is given below (see details in [7]),

Algorithm of Hybrid Direct-Vague Inference for Player A

Step 1: Direct Inference Using Direct Rule

If $\alpha = 0$ or $\alpha = 1$, directly use the corresponding direct rule.

Otherwise,

IF $C = 1$ THEN MOVE, GOTO STEP 3;
ELSE IF $C = 0$ THEN DON'T MOVE, GOTO STEP 3;
ELSE GOTO STEP 2.

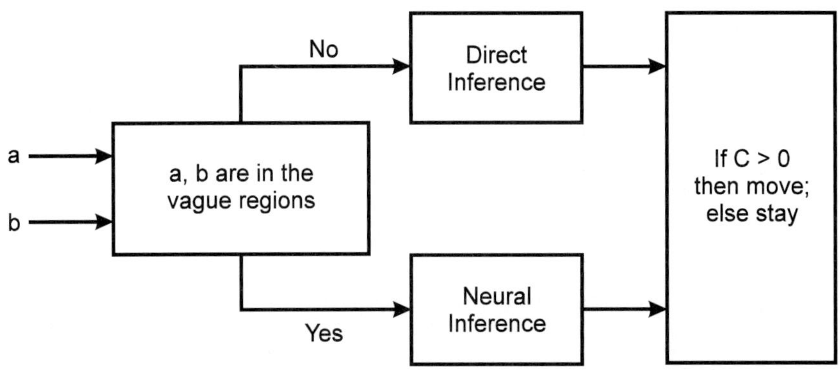

Figure 20.7 The fuzzy neural decision system for fuzzy moves.

Step 2: Neural Inference Using Vague Rule

For the data in one of the sensitive regions, the trained neural network can quickly make the neural inference to get the value of c_i ($i \in \{1, 2, 3, 6, 7, 8\}$).

If $c_i \leq 0.5$, then $C = 0$; if $c_i > 0.5$, then $C = 1$.

Step 3: Hybrid Decision

If $C = 0$, then don't move; if $C = 1$, then move.

20.3.3 Compensatory Neuro-Fuzzy Decision-Making System

A neuro-fuzzy system is capable of performing high-level fuzzy reasoning by using trained fuzzy neural networks which are constructed by learning from given sample data. Such a neuro-fuzzy integration brings high-level fuzzy IF-Then rules into neural networks, and incorporates low-level numerical learning mechanisms into fuzzy logic systems. Generally, the neuro-fuzzy system is much more effective than either neural networks or fuzzy logic systems since it can combine advantages of both of them. For this reason, we have designed a compensatory neuro-fuzzy system for fuzzy moves [16]. This system can adjust the parameters of membership functions based on the given sample data, and can learn the fuzzy If-Then rules after training. Finally, the trained neuro-fuzzy system has the ability to make better fuzzy moves since it is trained based on the high-level fuzzy If-Then rules and the low-level learning algorithms.

20.4 EXAMPLES OF FUZZY MOVES

Now we discuss two typical fuzzy moves games.

20.4.1 Fuzzy Prisoner's Delimma

The original PD is shown in Table 20.6. Suppose $\alpha = 0.75$ and $\beta = 0.25$.

We use the method shown in Section 20.2.5 to play the global fuzzy game. The membership functions for the payoffs of Players A and B are defined in Figure 20.8.

Now we use fuzzy logic to perform the nonlinear game transformation. First we show how to transform the state (0.4, 0.4) in the original PD in Table 20.6 to the relevant state of the global PD.

For Player A, we have

Table 20.6 The Original PD.

1	1	Confess	Not Confess
Confess		(0.4, 0.4)	(1, 0)
Not Confess		(0, 1)	(0.6, 0.6)

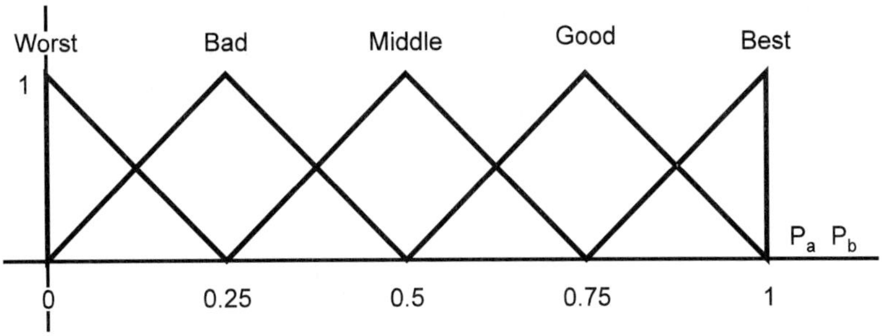

Figure 20.8 The membership functions of the payoffs (P_a, P_b)

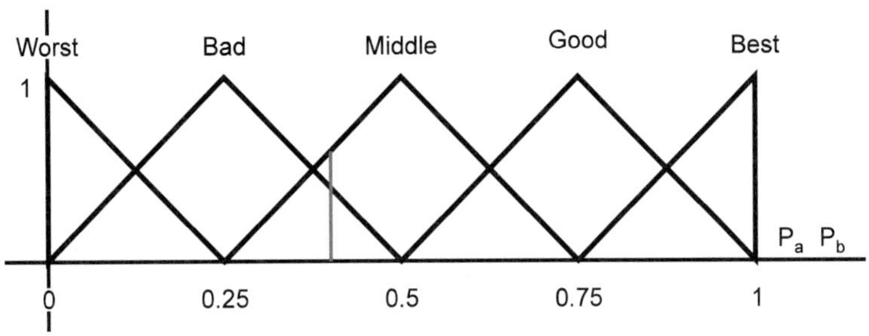

Figure 20.9 The fuzzification results of P_a = 0.4.

1. *Fuzzification* The fuzzification results are shown in Figure 20.9.

2. *Fuzzy Inference* The 4 relevant fuzzy rules in Table 20.4 are fired such that

Rule A22:	IF P_a = Bad	and P_b = Bad	THEN A_{fm} = Bad
Rule A23:	IF P_a = Bad	and P_b = Middle	THEN A_{fm} = Bad
Rule A32:	IF P_a = Middle	and P_b = Bad	THEN A_{fm} = Good
Rule A33:	IF P_a = Middle	and P_b = Middle	THEN A_{fm} = Middle

The inference results are shown in Figure 20.10.

3. *Defuzzification* The center of sums [5] is

$$C_a = \frac{0.25 \times 0.16 + 0.25 \times 0.16 + 0.5 \times 0.21 + 0.75 \times 0.16}{0.16 + 0.16 + 0.21 + 0.16} = 0.44$$

Similarly, we can get the center of sum for Player B C_b = 0.62.

So, the state (0.44, 0.62) is the global state of the original state (0.4, 0.4) through the fuzzy mapping.

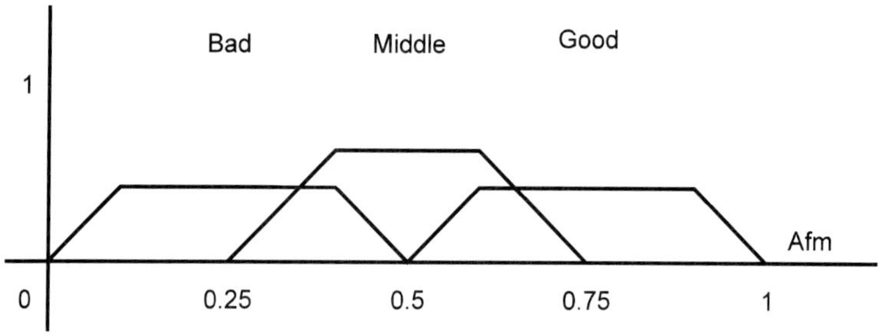

Figure 20.10 The inference results.

0.75	0.25

	Confess	Not Confess
Confess	(0.44, 0.62)	(1, 0)
Not Confess	(0, 1)	(0.5, 0.44)

Figure 20.11 The global PD by fuzzy mapping

Original PD

(0.4, 0.4)	(1, 0)
(0, 1)	(0.6, 0.6)

α,β $\rightarrow$

Total Conflict Game

(0.44, 0.62)	(1, 0)
(0, 1)	(0.5, 0.44)

Figure 20.12 The global PD transformed from the original PD.

Then we can get the other 3 global states from the relevant original ones by using the same fuzzy mapping method.

Finally, we have the global PD in Figure 20.11.

At this point, the original PD has been transformed to the global PD (or called total conflict game) shown in Figure 20.12.

According to TOM or game theory, the outcome of the global PD is (0.44, 0.62).

20.5 FUZZY GAME WITH FUZZY LINGUISTIC VALUES

The fuzzy linguistic terms F_i $(I = 1,2,...,7)$ in the payoff interval of [0, 1] are defined in Figure 20.13.

The original fuzzy game is shown in Figure 20.14. Suppose α = *very large* and β = *very small.*

We use the method shown in Section 20.4 to make fuzzy moves. In order to find the NMEs, we need to check all 4 cases which are:

Case 1: Initial state is (F1, F7);
Case 2: Initial state is (F3, F3);
Case 3: Initial state is (F7, F1);
Case 4: Initial state is (F6, F6).

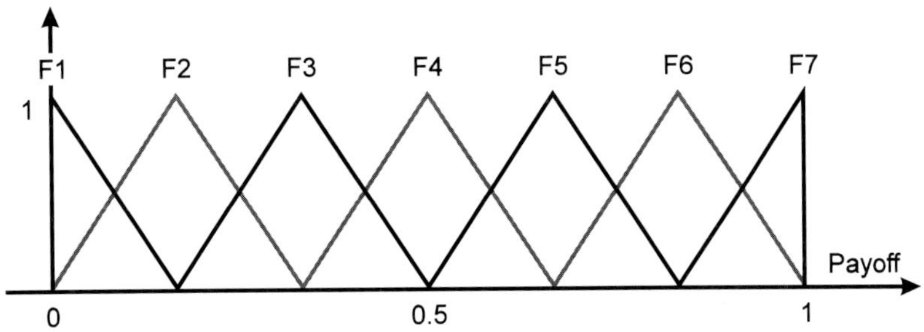

Figure 20.13 The Fuzzy Linguistic Terms (*F*1, *F*2,..., *F*7)

very large | *very small*	Confess	Not Confess
Confess	(F3, F3)	(F7, F1)
Not Confess	(F1, F7)	(F6, F6)

Figure 20.14 An Original Fuzzy Game.

At first we examine case 1; then similarly we can also analyze the other cases.

For case 1, a simple fuzzy inference method is used to make fuzzy moves by analyzing(1) the clockwise progression and (2) the counterclockwise progression.

The clockwise progression

Step 1:

	State 1	State 2	State 3	State 4	State 1
	A	B	A	B	
A Starts	(F1, F7)	(F3, F3)	(F7, F1)	(F6, F6)	(F1, F7)
Survivor				?	

(1) *Fuzzification*: In order to calculate the difference between two fuzzy linguistic values such as (very good-very bad = ?), we define a mapping rule table for player B in Table 20.7, where the new fuzzy terms a1, *a*2, ... , *a*9 are defined in Figure 20.15.

If player A moves from (F6, F2) to(F1, F7), according to the mapping rule table for the difference between two fuzzy linguistic values, we have $P_a(21, 11)$ = F1 – F6 = a1, $P_b(21, 11)$ = F7 – F6 = *a*6. The fuzzification results are shown in Figure 20.16.

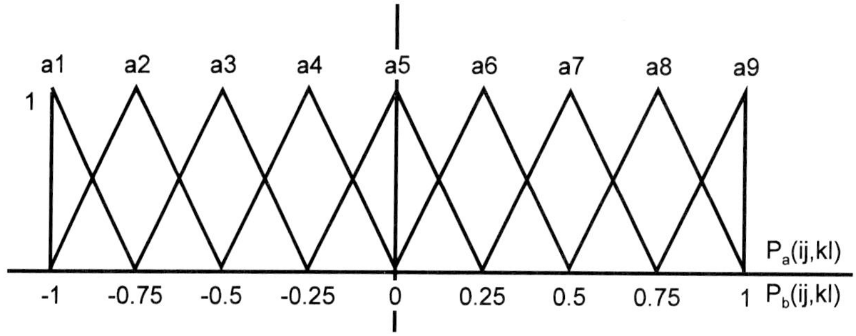

Figure 20.15 Fuzzy terms a_i (i = 1,2...,9).

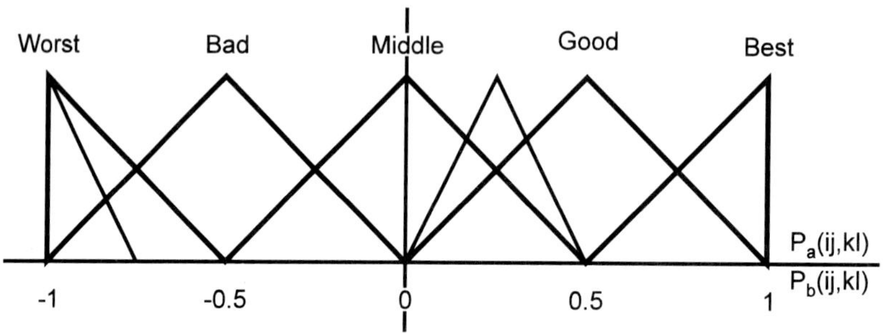

Figure 20.16 The fuzzification results of $P_a(21, 11) = a1$, $P_b(21, 11) = a6$.

(2) *Fuzzy Inference*: The fuzzy rules for A and for B are described in Tables 20.8 and 20.9, respectively.

Four relevant fuzzy rules from Table 20.9 are fired such that:

Rule B23	IF P_a = Bad	and P_b = Middle	THEN B_{fm} = Good
Rule B24:	IF P_a = Bad	and P_b = Good	THEN B_{fm} = Best
Rule B13:	IF P_a = Worst	and P_b = Middle	THEN B_{fm} = Best
Rule B14:	IF P_a = Worst	and P_b = Good	THEN B_{fm} = Best

The inference results are shown in Figure 20.17.

(3) *Defuzzification*: The center of sums is:

$$c = \frac{0.5 \times 0.28 + 0.82 \times 0.22 + 0.5 \times 0.28 + 0.82 \times 0.22}{0.28 + 0.22 + 0.28 + 0.22} = 0.64$$

As $c = 0.64 > 0$, the better survivor is (F1, F7). So we have:

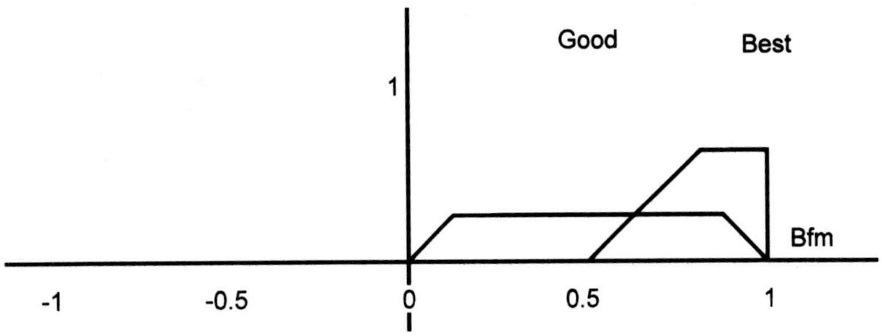

(a) Conclusions for rules B13 and B23

(b) Conclusions for rules B14 and B24

Figure 20.17 The inference results

Table 20.8 The 25 Sample Fuzzy Moves Rules for Player A (α = *very large*)

	1	2	3	4	5
1	2	1	1	1	1
2	3	3	1	1	1
3	4	4	3	2	2
4	5	5	4	3	3
5	5	5	5	4	4

Table 20.9 The 25 Sample Fuzzy Moves Rules for Player B (β = *very small*)

	1	2	3	4	5
1	4	4	5	5	5
2	3	4	4	4	5
3	2	3	3	4	5
4	1	2	2	2	4
5	1	1	2	2	2

	State 1	State 2	State 3	State 4	State 1
	A	B	A	B	
A Starts	(F1, F7)	(F3, F3)	(F7, F1)	(F6, F6)	(F1, F7)
Survivor				(F1, F7)	

Step 2:

	State 1	State 2	State 3	State 4	State 1
	A	B	A	B	
A Starts	(F1, F7)	(F3, F3)	(F7, F1)	(F6, F6)	(F1, F7)
Survivor			?	(F1, F7)	

Similarly, the better survivor is (F7, F1). So we have:

	State 1	State 2	State 3	State 4	State 1
	A	B	A	B	
A Starts	(F1, F7)	(F3, F3)	(F7, F1)	(F6, F6)	(F1, F7)
Survivor			(F7, F1)	(F1, F7)	

Step 3:

	State 1	State 2	State 3	State 4	State 1
	A	B	A	B	
A Starts	(F1, F7)	(F3, F3)	(F7, F1)	(F6, F6)	(F1, F7)
Survivor		?	(F7, F1)	(F1, F7)	

The survivor at state 2 is (F4, F5). So we have:

	State 1	State 2	State 3	State 4	State 1
	A	B	A	B	
A Starts	(F1, F7)	(F3, F3)	(F7, F1)	(F6, F6)	(F1, F7)
Survivor		(F3, F3)	(F7, F1)	(F1, F7)	

Step 4:

	State 1	State 2	State 3	State 4	State 1
	A	B	A	B	
A Starts	(F1, F7)	(F3, F3)	(F7, F1)	(F6, F6)	(F1, F7)
Survivor	?	(F3, F3)	(F7, F1)	(F1, F7)	

The survivor at state 2 is (F4, F5). Finally we have:

	State 1	State 2	State 3	State 4	State 1
	A	B	A	B	
A Starts	(F1, F7)	(F3, F3)	(F7, F1)	(F6, F6)	(F1, F7)
Survivor	(F3, F3)	(F3, F3)	(F7, F1)	(F1, F7)	

Therefore, (F3, F3) is a fuzzy NME by the clockwise progression.

The counterclockwise progression. For conciseness, according to the same fuzzy inference method, the final result is given as follows:

	State 1	State 2	State 3	State 4	State 1
	B	A	B	A	
B Starts	(F1, F7)	(F6, F6)	(F7, F1)	(F3, F3)	(F1, F7)
Survivor	(F1, F7)	(F6, F6)	(F3, F3)	(F3, F3)	

Therefore, (F1, F7) is a fuzzy NME by the counterclockwise progression.

For case 1 [i.e., when the initial state is (F1, F7)], the outcome is (F3, F3) which is just the Confess-Confess outcome.

Similarly, the outcome is (F3, F3) for cases 2 and 3, and (F6, F6) for case 4.

According to TOM, the outcome is (F6, F6) for cases 1, 3 and 4, and (F3, F3) for case 2.

20.6 CONCLUSIONS

The Theory of Fuzzy Moves is a general extension of the theory of moves by extending the local goal of TOM to the global one of TFM through a game transformation. Before we make fuzzy moves, we first analyze which globally strategic goals the players choose (e.g., $\alpha = 0.9$ and $\beta = 0.3$, α = very small, and β = very large, etc.). Secondly, make fuzzy inferences based on the fuzzy rules given by the players, and finally make fuzzy moves so as to achieve the best outcome that may be not only as advantageous as possible to its own side but also as disadvantageous as possible to its opponent, based on both the original payoffs and the globally strategic goals the players choose.

Generally, neuro-fuzzy systems can be used to make better fuzzy moves by learning nonlinear fuzzy rules [7,16]. In the future, hybrid systems incorporating fuzzy logic, neural networks, and genetic algorithms will effectively be applied to make good decisions under uncertain situations.

20.7 ACKNOWLEDGMENT

The authors would like to thank Dr. Friedman, Dr. Borges and Miller for their very constructive comments.

20.8 REFERENCES

1. Axelrod, R. "The Evolution of Cooperation." Penguin Books, 1984, pp. 3-54.

2. Brams, Steven J. "Theory of Moves", American Scientist, Vol.81, Nov., 1993.

3. Brams, Steven J. "Theory of Moves", Cambridge, Cambridge University Press, 1994.

4. Brams, Steven J. and Mattli, Walter. "Theory of Moves: Overview and Examples." *Conflict Management and Peace Science*. Vol. 12, No. 2, 1993, pp. 1-39.

5. Driankov D., Hellendoorn, H. and Reinfrank, M. *An Introduction to Fuzzy Control.* Springer-Verlag, 1993.

6. Kacprzyk, J. and Fedrizzi, M. "Multiperson Decision Making Models Using Fuzzy Sets and Possibility Theory." Kluwer Academic Publishers, 1990.

7. Kandel, A., Zhang, Y.-Q. and Miller, T. "Fuzzy Neural Decision System for Fuzzy Moves." *3rd World Congress on Expert Systems*, 1996.

8. Binmore, Ken. *Game Theory and the Social Contract Volume I.* MIT Press, 1994.

9. Kickert, Walter J. M. *Fuzzy Theories on Decision-Making*. 1978.

10. Poundstone, W. *Prisoner's Dilemma*. Doubleday, 1992.

11. Raiffa, L. *Games and Decisions*. 1958.

12. Rasmnsen, Eric. *Games and Information.* Blackwell, 1994.

13. Schellenberg, J.-A. *Primitive Games.* Westview Press, 1990.

14. Wang, Jianhua. *The Theory of Games*, 1988.

15. Zhang, Y.-Q., Kandel, A. and Friedman, M. "Hybrid Decision-Making Systems for Fuzzy Moves." *Proc. of FUZZ-IEEE/IFES'95*, Vol. II, pp. 621-626, 1995.

16. Zhang Y.-Q. and Kandel A., "Compensatory FDNF-FCNF Neurofuzzy Systems for Fuzzy Games." Submitted.

17. Zimmermann, H.-J. *Fuzzy Sets, Decision Making, and Expert Systems.* Kluwer Academic Publishers, 1986.

18. Zimmermann, H.-J. *Fuzzy Set Theory and Its Applications.* Kluwer Academic Publishers, 1991.

INDEX

D

E

F

G

H

Please send more information on the following :

- ❏ *Fuzzy Logic Software*
- ❏ *Neuro-Fuzzy*
- ❏ *Logic Software*
- ❏ *Fuzzy Logic Educational, Training Opportunities and Tools*
- ❏ *Soft Computing Meetings and Journals*

Name: ____
Institution: ____
Address: ____
____
Phone: ______________________ FAX: ______________________
e-mail: ______________________

CAD Laboratory - UNMP. O. Box 14126
Albuquerque, NM 87191-4126 USA
Phone: + 505 298 5817 FAX: + 505 291 0013
email: wac@houdini.unm.edu

Please send more information on the following :

- ❏ *Fuzzy Logic Software*
- ❏ *Neuro-Fuzzy Logic Software*
- ❏ *Fuzzy Logic Educational, Training Opportunities and Tools*
- ❏ *Soft Computing Meetings and Journals*

Name: ____
Institution: ____
Address: ____
____
Phone: ______________________ FAX: ______________________
e-mail: ______________________

CAD Laboratory - UNM
P. O. Box 14126
Albuquerque, NM 87191-4126 USA
Phone: + 505 298 5817 FAX: + 505 291 0013
email: wac@houdini.unm.edu

Place
Stamp
Here

CAD Laboratory - UNM
P. O. Box 14126
Albuquerque, NM 87191-4126
USA

Place
Stamp
Here

CAD Laboratory - UNM
P. O. Box 14126
Albuquerque, NM 87191-4126
USA